Lecture Notes in Computer Science

Lecture Notes in Artificial Intelligence 15048
Founding Editor

Jörg Siekmann

Series Editors

Randy Goebel, *University of Alberta, Edmonton, Canada*
Wolfgang Wahlster, *DFKI, Berlin, Germany*
Zhi-Hua Zhou, *Nanjing University, Nanjing, China*

The series Lecture Notes in Artificial Intelligence (LNAI) was established in 1988 as a topical subseries of LNCS devoted to artificial intelligence.

The series publishes state-of-the-art research results at a high level. As with the LNCS mother series, the mission of the series is to serve the international R & D community by providing an invaluable service, mainly focused on the publication of conference and workshop proceedings and postproceedings.

Elmar Nöth · Aleš Horák · Petr Sojka
Editors

Text, Speech, and Dialogue

27th International Conference, TSD 2024
Brno, Czech Republic, September 9–13, 2024
Proceedings, Part I

 Springer

Editors
Elmar Nöth
Friedrich-Alexander-Universität
Erlangen, Germany

Aleš Horák
Masaryk University
Brno, Czech Republic

Petr Sojka
Masaryk University
Brno, Czech Republic

ISSN 0302-9743   ISSN 1611-3349 (electronic)
Lecture Notes in Artificial Intelligence
ISBN 978-3-031-70562-5   ISBN 978-3-031-70563-2 (eBook)
https://doi.org/10.1007/978-3-031-70563-2

LNCS Sublibrary: SL7 – Artificial Intelligence

© The Editor(s) (if applicable) and The Author(s), under exclusive license
to Springer Nature Switzerland AG 2024

This work is subject to copyright. All rights are solely and exclusively licensed by the Publisher, whether the whole or part of the material is concerned, specifically the rights of translation, reprinting, reuse of illustrations, recitation, broadcasting, reproduction on microfilms or in any other physical way, and transmission or information storage and retrieval, electronic adaptation, computer software, or by similar or dissimilar methodology now known or hereafter developed.
The use of general descriptive names, registered names, trademarks, service marks, etc. in this publication does not imply, even in the absence of a specific statement, that such names are exempt from the relevant protective laws and regulations and therefore free for general use.
The publisher, the authors and the editors are safe to assume that the advice and information in this book are believed to be true and accurate at the date of publication. Neither the publisher nor the authors or the editors give a warranty, expressed or implied, with respect to the material contained herein or for any errors or omissions that may have been made. The publisher remains neutral with regard to jurisdictional claims in published maps and institutional affiliations.

This Springer imprint is published by the registered company Springer Nature Switzerland AG
The registered company address is: Gewerbestrasse 11, 6330 Cham, Switzerland

If disposing of this product, please recycle the paper.

Preface

The annual Text, Speech and Dialogue Conference (TSD), which originated in 1998, is continuing its third decade. In the course of this time, thousands of authors from all over the world have contributed to the proceedings. TSD constitutes a recognized platform for the presentation and discussion of state-of-the-art technology and recent achievements in the field of natural language processing. It has become an interdisciplinary forum, interweaving the themes of speech technology and language processing. The conference attracts researchers not only from Central and Eastern Europe but also from other parts of the world. Indeed, one of its goals has always been to bring together NLP researchers with different interests from different parts of the world and to promote their mutual cooperation.

One of the declared goals of the conference has always been, as its title says, twofold: not only to deal with language processing and dialogue systems as such, but also to stimulate dialogue between researchers in the two areas of NLP, i.e., between text and speech people. In our view, the TSD Conference was again successful in this respect in 2024. We had the pleasure to welcome two prominent invited speakers this year: Hynek Hermansky from Johns Hopkins University, USA, and Preslav Nakov from Mohamed bin Zayed University of Artificial Intelligence, Abu Dhabi, UAE.

This volume contains the proceedings of the 27th TSD Conference, held in Brno, Czech Republic, in September 2024. TSD 2024 was organized by the Faculty of Informatics, Masaryk University, in cooperation with the Faculty of Applied Sciences, University of West Bohemia in Plzeň. The conference webpage is located at https://www.tsdconference.org/tsd2024/. In the review process, 50 papers were accepted out of 103 submitted, each based on three reviews, with an acceptance rate of 48.5%.

We would like to thank all the authors for the efforts they put into their submissions and the members of the Program Committee and reviewers who did a wonderful job selecting the best papers. We are also grateful to the invited speakers for their contributions. Their talks provided insight into important current issues, applications, and techniques related to the conference topics.

Special thanks are due to the members of the Local Organizing Committee for their tireless effort in organizing the conference. The TeXpertise of Petr Sojka resulted in the production of the volume that you are holding in your hands.

We hope that readers will benefit from the results of this event and disseminate the ideas of the TSD Conference all over the world. Enjoy the proceedings!

July 2024

Elmar Nöth
Aleš Horák
Petr Sojka

Organization

General Chair

Nöth, Elmar   Friedrich-Alexander University
                                   Erlangen-Nürnberg, Germany

Program Committee

Agerri, Rodrigo   University of the Basque Country, Spain
Agirre, Eneko   University of the Basque Country, Spain
Benko, Vladimír   Slovak Academy of Sciences, Slovakia
Bhatia, Archna   Carnegie Mellon University, USA
Černocký, Jan   Brno University of Technology, Czech Republic
Dobrišek, Simon   University of Ljubljana, Slovenia
Ekštein, Kamil   University of West Bohemia, Czech Republic
Evgrafova, Karina   St. Petersburg State University, Russia
Fedorov, Yevhen   Cherkasy State Technological University, Ukraine
Fischer, Volker   EML Speech Technology GmbH, Germany
Fišer, Darja   Institute of Contemporary History, Slovenia
Flek, Lucie   Philipps-Universität Marburg, Germany
Gambäck, Björn   Norwegian University of Science and Technology,
                            Norway
Garabík, Radovan   Slovak Academy of Sciences, Slovakia
Gelbukh, Alexander   Instituto Politécnico Nacional, Mexico
Guthrie, Louise   University of Texas at El Paso, USA
Hajič, Jan   Charles University, Czech Republic
Hajičová, Eva   Charles University, Czech Republic
Haralambous, Yannis   IMT Atlantique, France
Hermansky, Hynek   Johns Hopkins University, USA
Horák, Aleš   Masaryk University, Czech Republic
Hovy, Eduard   Carnegie Mellon University, USA
Jakubíček, Miloš   Lexical Computing, Czech Republic
Khokhlova, Maria   St. Petersburg State University, Russia
Khusainov, Aidar   Tatarstan Academy of Sciences, Russia
Kocharov, Daniil   Saint Petersburg State University, Russia
Konopík, Miloslav   University of West Bohemia, Czech Republic
Kordoni, Valia   Humboldt University Berlin, Germany

Kotelnikov, Evgeny	Vyatka State University, Russia
Král, Pavel	University of West Bohemia, Czech Republic
Kunzmann, Siegfried	Amazon Alexa Machine Learning, Germany
Ljubešić, Nikola	Jožef Stefan Institute, Croatia
Loukachevitch, Natalija	Lomonosov Moscow State University, Russia
Magnini, Bernardo	Fondazone Bruno Kessler, Italy
Matoušek, Václav	University of West Bohemia, Czech Republic
Mouček, Roman	University of West Bohemia, Czech Republic
Mykowiecka, Agnieszka	Polish Academy of Sciences, Poland
Ney, Hermann	RWTH Aachen University, Germany
Nivre, Joakim	Uppsala University, Sweden
Orozco-Arroyave, Juan Rafael	University of Antioquia, Colombia
Piasecki, Maciej	Wrocław University of Science and Technology, Poland
Psutka, Josef	University of West Bohemia, Czech Republic
Pustejovsky, James	Brandeis University, USA
Rigau, German	University of the Basque Country, Spain
Rosso, Paolo	Universitat Politècnica de València, Spain
Rothkrantz, Leon	Delft University of Technology, The Netherlands
Rumshisky, Anna	UMass Lowell, USA
Rusko, Milan	Slovak Academy of Sciences, Slovakia
Rychlý, Pavel	Masaryk University, Czech Republic
Sazhok, Mykola	International Research and Training Center for Information Technologies and Systems, Ukraine
Skrelin, Pavel	Saint Petersburg State University, Russia
Smrž, Pavel	Brno University of Technology, Czech Republic
Sojka, Petr	Masaryk University, Czech Republic
Stemmer, Georg	Intel Corp., Germany
Šikonja, Marko Robnik	University of Ljubljana, Slovenia
Tadić, Marko	University of Zagreb, Croatia
Trmal, Jan	Johns Hopkins University, USA
Varadi, Tamas	Hungarian Academy of Sciences, Hungary
Vetulani, Zygmunt	Adam Mickiewicz University, Poland
Wawer, Aleksander	Polish Academy of Sciences, Poland
Wiggers, Pascal	Amsterdam University of Applied Sciences, The Netherlands
Wróblewska, Alina	Polish Academy of Sciences, Poland
Žganec Gros, Jerneja	Alpineon, Slovenia

Additional Referees

Martin Božič
Evangelia Gogolou
Valeriya Goloviznina
Enikő Héja
Matej Klemen
Zdeněk Krnoul
Jan Lehečka
Noémi Ligeti-Nagy
Jacinto Mata
Jindřich Matoušek
Daša Munková
Zuzana Nevěřilová
Vít Suchomel
Volodymyr Taranukha
Ivor Uhliarik
Tereza Vrabcová
Aleš Žagar
Bartosz Żuk

Organizing Committee

Aleš Horák (Organization Chair)	Faculty of Informatics, Masaryk University, Czech Republic
Adam Rambousek (Web System)	Faculty of Informatics, Masaryk University, Czech Republic
Pavel Rychlý	Faculty of Informatics, Masaryk University, Czech Republic
Petr Sojka (Proceedings Chair)	Faculty of Informatics, Masaryk University, Czech Republic

Sponsors and Support

The TSD conference is regularly supported by the International Speech Communication Association (ISCA). We would like to express our thanks to Lexical Computing Ltd. for their kind sponsoring contribution to TSD 2024.

Contents – Part I

Text

SeqCondenser: Inductive Representation Learning of Sequences
by Sampling Characteristic Functions 3
  Maixent Chenebaux and Tristan Cazenave

Is Prompting What Term Extraction Needs? 17
  *Hanh Thi Hong Tran, Carlos-Emiliano González-Gallardo,
  Julien Delaunay, Antoine Doucet, and Senja Pollak*

Bilingual Lexicon Induction From Comparable and Parallel Data:
A Comparative Analysis 30
  Michaela Denisová and Pavel Rychlý

Explaining Metaphors in the French Language by Solving Analogies
Using a Knowledge Graph 43
  Jérémie Roux, Hani Guenoune, Mathieu Lafourcade, and Richard Moot

The Aranea Corpora Family: Ten+ Years of Processing Web-Crawled Data 55
  Vladimír Benko

Continual Learning Under Language Shift 71
  Evangelia Gogoulou, Timothée Lesort, Magnus Boman, and Joakim Nivre

Neural Spell-Checker: Beyond Words with Synthetic Data Generation 85
  *Matej Klemen, Martin Božič, Špela Arhar Holdt,
  and Marko Robnik-Šikonja*

CoastTerm: A Corpus for Multidisciplinary Term Extraction in Coastal
Scientific Literature 97
  *Julien Delaunay, Hanh Thi Hong Tran, Carlos-Emiliano González-
  Gallardo, Georgeta Bordea, Mathilde Ducos, Nicolas Sidere, Antoine
  Doucet, Senja Pollak, and Olivier De Viron*

New Human-Annotated Dataset of Czech Health Records for Training
Medical Concept Recognition Models 110
  Krištof Anetta and Aleš Horák

Analyzing Biases in Popular Answer Selection Datasets on Neural-Based
QA Models 121
   Chang Nian Chuy, Cherie Ding, and Qinmin Vivian Hu

Using Neural Coherence Models to Assess Discourse Coherence 134
   Lilia Azrou, Houda Oufaida, Philippe Blache, and Israa Hamdine

Named Entity Linking in English-Czech Parallel Corpus 147
   Zuzana Nevěřilová and Hana Žižková

TamSiPara: A Tamil – Sinhala Parallel Corpus 159
   *Randil Pushpananda, Chamila Liyanage, Ashmari Pramodya,
and Ruvan Weerasinghe*

Automatic Ellipsis Reconstruction in Coordinated German Sentences
Based on Text-to-Text Transfer Transformers 171
   Marisa Schmidt, Karin Harbusch, and Denis Memmesheimer

Better Low-Resource Machine Translation with Smaller Vocabularies 184
   Edoardo Signoroni and Pavel Rychlý

Bella Turca: A Large-Scale Dataset of Diverse Text Sources for Turkish
Language Modeling 196
   Duygu Altinok

Evaluation Metrics in LLM Code Generation 214
   Kai Hartung, Sambit Mallick, Sören Gröttrup, and Munir Georges

Kernel Least Squares Transformations for Cross-Lingual Semantic Spaces 227
   Adam Mištera and Tomáš Brychcín

Unsupervised Extraction of Morphological Categories for Morphemes 239
   Abishek Stephen, Vojtěch John, and Zdeněk Žabokrtský

Introducing LCC's NavProc 1.0 Corpus: Annotated Procedural Texts
in the Naval Domain 252
   Michael Mohler, Sandra Lee, Mary Brunson, and David Bracewell

Models and Strategies for Russian Word Sense Disambiguation:
A Comparative Analysis 267
   Anastasiia Aleksandrova and Joakim Nivre

Open-Source Web Service with Morphological Dictionary–Supplemented
Deep Learning for Morphosyntactic Analysis of Czech 279
   Milan Straka and Jana Straková

Mistrík's Readability Metric – an Online Library 291
Mária Pappová and Matúš Valko

Author Index 303

Contents – Part II

Speech

Retrieval Augmented Spoken Language Generation for Transport Domain 3
   Gokul Srinivasagan and Munir Georges

Adapting Audiovisual Speech Synthesis to Estonian 13
   Sven Aller and Mark Fishel

Dysphonia Diagnosis Using Self-supervised Speech Models in Mono
and Cross-Lingual Settings 24
   Dosti Aziz and Dávid Sztahó

Sentences vs Phrases in Neural Speech Synthesis 36
   Daniel Tihelka, Jindřich Matoušek, Zdeněk Hanzlíček, and Lukáš Vladař

Zero-Shot vs. Few-Shot Multi-speaker TTS Using Pre-trained Czech
SpeechT5 Model 46
   Jan Lehečka, Zdeněk Hanzlíček, Jindřich Matoušek, and Daniel Tihelka

Deep Speaker Embeddings for Speaker Verification of Children 58
   Mohammed Hamzah Abed and Dávid Sztahó

Improved Alignment for Score Combination of RNN-T and CTC Decoder
for Online Decoding 70
   Chin Yuen Kwok, Jia Qi Yip, and Eng Siong Chng

Attention to Phonetics: A Visually Informed Explanation of Speech
Transformers 81
   Erfan A. Shams and Julie Carson-Berndsen

Effects of Training Strategies and the Amount of Speech Data
on the Quality of Speech Synthesis 94
   Lukáš Vladař and Jindřich Matoušek

Stream-based Active Learning for Speech Emotion Recognition via Hybrid
Data Selection and Continuous Learning 105
   Santiago A. Moreno-Acevedo, Juan Camilo Vasquez-Correa,
   Juan M. Martín-Doñas, and Aitor Álvarez

Data Alignment and Duration Modelling in VITS 118
   Zdeněk Hanzlíček

Multiword Expressions Resources for Italian: Presenting a Manually
Annotated Spoken Corpus 130
   Ilaria Manfredi

Generating High-Quality F0 Embeddings Using the Vector-Quantized
Variational Autoencoder 139
   David Porteš and Aleš Horák

Anonymizing Dysarthric Speech: Investigating the Effects of Voice
Conversion on Pathological Information Preservation 149
   Abner Hernandez, Paula Andrea Perez-Toro, Tomas Arias-Vergara,
   Juan Camilo Vasquez-Correa, Seung Hee Yang,
   Juan Rafael Orozco-Arroyave, and Andreas Maier

X-Vector-Based Speaker Diarization Using Bi-LSTM and Interim
Voting-Driven Post-processing 161
   J. B. Mala, S. M. Alex Raj, and Rajeev Rajan

A Paradigm for Interpreting Metrics and Measuring Error Severity
in Automatic Speech Recognition 174
   Thibault Bañeras-Roux, Mickael Rouvier, Jane Wottawa,
   and Richard Dufour

Enhancing Speech Emotion Recognition Using Transfer Learning
from Speaker Embeddings 184
   Maroš Jakubec, Roman Jarina, Eva Lieskovská, Peter Kasák,
   and Michal Spišiak

Dialogue

Investigating Low-Cost LLM Annotation for Spoken Dialogue
Understanding Datasets 199
   Lucas Druart, Valentin Vielzeuf, and Yannick Estève

PiCo-VITS: Leveraging Pitch Contours for Fine-Grained Emotional
Speech Synthesis 210
   Kwan-yeung Wong and Fu-lai Chung

Improving and Understanding Clarifying Question Generation
in Conversational Search 222
   Daniel Ortega, Steven Söhnel, and Ngoc Thang Vu

Explainable Multimodal Fusion for Dementia Detection From Text
and Speech 236
   Duygu Altinok

Robust Classification of Parkinson's Speech: an Approximation
to a Scenario With Non-controlled Acoustic Conditions 252
   Diego Alexander Lopez-Santander, Cristian David Rios-Urrego,
   Christian Bergler, Elmar Nöth, and Juan Rafael Orozco-Arroyave

Leveraging Conceptual Similarities to Enhance Modeling of Factors
Affecting Adolescents' Well-Being 263
   Ondřej Sotolář, Jaromír Plhák, and David Šmahel

Joint-Average Mean and Variance Feature Matching (JAMVFM)
Semi-supervised GAN with Additional-Objective Training Function
for Intent Detection 275
   Ankit Kumar and Munir Georges

Capturing Task-Related Information for Text-Based Grasp Classification
Using Fine-Tuned Embeddings 288
   Niko Kleer, Leon Weyand, Michael Feld, and Klaus Berberich

StepDP: A Step Towards Expressive and Pervasive Dialogue Platforms 300
   Julian Wolter, Niko Kleer, and Michael Feld

Automatic Classification of Parkinson's Disease Using Wav2vec
Embeddings at Phoneme, Syllable, and Word Levels 313
   Jeferson David Gallo-Aristizábal, Daniel Escobar-Grisales,
   Cristian David Ríos-Urrego, Elmar Nöth,
   and Juan Rafael Orozco-Arroyave

Author Index 325

Text

"**Text:** a book or other written or printed work, regarded in terms of its content rather than its physical form: *a text which explores pain and grief.*"
 NODE (The New Oxford Dictionary of English), Oxford, OUP, 1998, page 1998, meaning 1.

SeqCondenser: Inductive Representation Learning of Sequences by Sampling Characteristic Functions

Maixent Chenebaux[1] and Tristan Cazenave[2](✉)

[1] Vectors Group, Paris, France
[2] LAMSADE, Université Paris Dauphine - PSL, CNRS, Paris, France
Tristan.Cazenave@dauphine.fr

Abstract. In this work, we introduce SeqCondenser, a neural network layer that compresses a variable-length input sequence into a fixed-size vector representation. The SeqCondenser layer samples the empirical characteristic function and its derivatives for each input dimension, and uses an attention mechanism to determine the associated probability distribution. We argue that the features extracted through this process effectively represent the entire sequence and that the SeqCondenser layer is particularly well-suited for inductive sequence classification tasks, such as text and time series classification. Our experiments show that SCoMo, a SeqCondenser-based architecture, outperforms the state-of-the-art inductive methods on nearly all examined text classification datasets and also outperforms the current best transductive method on one dataset.

1 Introduction

Text classification is a crucial task in natural language processing (NLP). It can be used for various purposes, such as opinion mining, sentiment analysis, fact checking and detecting fake news. One challenge in classifying text is the variability in the length of the sequences. There are several approaches to representing text for classification, including representing it as a sequence of words or characters, as a fixed-size vector, or as a graph [18].

Transductive learning techniques are currently the top performers in text classification. These techniques generally use a network containing both the training set and the testing set, without labels for the latter. Patterns observed in unlabeled data also contribute to the classification process, leading to better performance. ROBERTaGNN [12], a transductive document classification algorithm, is currently the state-of-the-art on many reference datasets. It fine-tunes BERT [6] and uses the generated vectors as node features, then builds a graph where nodes are text units and edges are based on semantic similarity. While this method leverages both the transformer architecture and graph neural networks, it has the same limitations as other transductive techniques, including longer training time, limited generalization, and deployment challenges. Transductive methods also require retraining when dealing with new data points, making them difficult to use in a production environment where documents are classified individually or in batches asynchronously.

Inductive learning is a powerful and versatile approach to machine learning, where the model is trained on a labeled dataset and then makes predictions on unseen data. Unlike transductive learning, inductive learning does not rely on access to the test set during training, making it more suitable for real-world applications where new data continuously arrives. Inductive methods are diverse, with different algorithms achieving better results on different datasets. In recent years, text classification has typically been done by fine-tuning a pre-trained model, such as a transformer-based model, and using aggregation methods, such as average pooling, max over time, or attentive pooling. Alternatively, the BERT model can perform prediction directly through the special token [CLS]. These methods have been successful, but their limitation lies in the ability of the pooling techniques to retain essential information for classification while keeping the dimensionality of the final representation as low as possible.

Present Work: We propose SeqCondenser, a layer that transforms a varying-size sequence of vectors into a compact, fixed-size vector representation that can be used for sequence classification and regression. The benefits of SeqCondenser are as follows: (1) it is fast to train; (2) it is inductive by design; (3) it can easily be added to any TensorFlow model [1] with a single line of code, making all the results presented in this work easily reproducible; (4) it does not need to transform the input data into a graph or another datastructure, and can work on any sequence. The compact representation of the sequence is efficient at classification tasks, and achieves new state-of-the-art performances when compared to both inductive and transductive methods. On all the datasets studied, the direct replacement of any pooling layer with SeqCondenser leads to an immediate performance increase.

Main Contributions: The main contributions of our work are as follows:

1. We introduce a new sequence-to-vector method that relies on a family of characteristic functions and their two subsequent derivatives, where the affiliation probabilities are predicted by a differentiable attention mechanism.
2. We show that dimensionality reduction through random projections can be applied to the SeqCondenser's aggregation vector to improve accuracy.
3. We experimentally demonstrate that this method can be successfully applied to text classification tasks and improve the current state-of-the-art for both inductive and transductive techniques.
4. Our model is implemented using TensorFlow, and an easy-to-use SeqCondenser layer is made publicly available for Keras [4].

The paper is structured as follows: Sect. 2 reviews related literature. The SeqCondenser layer and its theoretical foundations are discussed in Sect. 3. The design and experimental results of SCoMo, a sequence classification architecture utilizing the SeqCondenser layer, are outlined in Sects. 4 and 5 respectively. The paper concludes in Sect. 6.

2 Related Work

Pooling Strategies: In text classification, the input is typically a variable-length sequence of words or characters. Summarizing its features through pooling strategies is an essential step for the task. Pooling, in this context, aims at extracting features from the input and creating a fixed-length representation of it. One approach is to use convolutional filters to extract features from the input text, and apply average or max over time pooling to summarize semantic features [3]. However, average and max pooling have the limitation of not being able to efficiently select salient features that may be particularly useful for the classification task.

To address this limitation, several approaches have been proposed. One technique is to create document representations by summing word embeddings weighted by their corresponding TF-IDF scores, as proposed in [11]. Another approach is to use attentive pooling, which allows the neural network to learn to distinguish important features and create a summary that focuses on relevant information, while ignoring irrelevant information. One example of attentive pooling is the DeepMoji attention layer [8], which creates a dense representation of a sequence of vectors by using attention scores as weights for a weighted summation over all the time steps. These weights are calculated by taking the scalar product of each input vector with a single trainable vector and then applying the softmax function along the time dimension. Another example of an attentive pooling technique is APLN [17], which uses a similar approach to DeepMoji but interprets the weighted sum as an $\mathcal{L}^1$ norm. APLN allows the model to learn the most appropriate $\mathcal{L}^p$ norm for the classification task.

Characteristic Functions: Characteristic functions are mathematical functions that describe the probability distribution of random variables. In the field of graph representation learning, characteristic functions have been used for summarizing information and creating node embeddings. Two notable approaches that utilize characteristic functions are GraphWave [7] and FEATHER [15].

GraphWave is a method for capturing structural roles in graphs using heat diffusion wavelets. Nodes with the same structural role have similar heat diffusion patterns, and the characteristic function is used to summarize the histogram of heat distribution across the network. The node embeddings created through this method are independent of the size of the network, and roles are predicted through the use of PCA and a clustering algorithm applied to the embeddings.

FEATHER is an approach for creating node embeddings in a network that encodes the distribution of features among the neighbors of a given node. It does this by evaluating the characteristic function at differentiable evaluation points, where the probability distribution of the features is deterministically defined in terms of random walks on the graph. This allows FEATHER to capture the salient features in the neighborhood of a given node and create a fixed-length representation of it.

Overall, characteristic functions have proven to be a powerful tool for summarizing information and creating fixed-length representations of data in the context of graph representation learning. In our work, we aim to leverage the power of characteristic functions by incorporating them into an end-to-end trainable sequence classification pipeline.

3 SeqCondenser

In this section, we present the mathematical foundations and rationale for the SeqCondenser layer. A visual representation of the method is provided in Fig. 2.

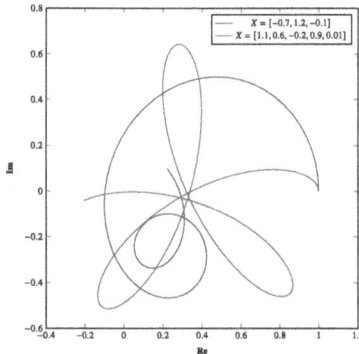

Fig. 1. Graph of two characteristic functions computed from a variable-length list of numbers, where each outcome is assigned an equal probability, showing a partial view of the complex plane representation

3.1 Characteristic Functions on Sequences

The characteristic function is a widely-used concept in probability theory that provides a comprehensive description of the probability distribution of any discrete or continuous random variable X [9]. The kth derivative of the characteristic function, ϕ, at zero is proportional to the kth moment of X: $\phi^{(k)}(0) = i^k \mathbf{E}\left[X^k\right]$. The curve described by the characteristic function lives in the complex plane and theoretically contains all the information needed to describe X. In the discrete case, it can map a random variable with any finite number of outcomes to a curve in the complex plane, as long as a probability is assigned to each outcome. This concept is illustrated in Fig. 1. One way to construct a fixed-size representation of the curve of ϕ is to sample it at different points, however this can result in a loss of important information about the curve's shape. For example, the GraphWave and FEATHER graph algorithms only consider $\phi^{(0)}$. We argue that the use of higher-order derivatives of the characteristic function provides crucial information for the learning process.

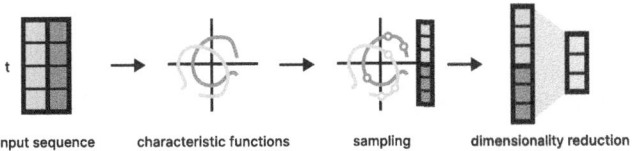

Fig. 2. Main steps of SeqCondenser

3.2 Sequence Feature Distribution Characterization

Let **x** be a sequence vector of length l where x_t is the feature value at time step t. We interpret **x** as the outcomes of a discrete random variable X. The characteristic function of X is defined as:

$$\phi(\theta) = \mathbf{E}[e^{iX\theta}] = \sum_t p_t e^{i\theta x_t} \qquad (1)$$

In Eq. 1, the affiliation probability p_t describes the probability of observing the outcome x_t. The Euler identity applied to Eq. 1 yields: $\mathbf{Re}(\phi(\theta)) = \sum_t p_t \cos(\theta x_t)$ and $\mathbf{Im}(\phi(\theta)) = \sum_t p_t \sin(\theta x_t)$

We are also interested in the first and second derivatives of the characteristic function, as they provide the model with more information to fully characterize the sequence feature distribution. In this way, the model not only uses the position of the characteristic function in the complex plane at specific points, but also the direction, speed, and curvature of its trajectory.

It is worth noting that the characteristic function and its derivatives are independent of the length of the observed sequence, but can still uniquely describe it. This is the main idea behind SeqCondenser's ability to compactly represent variable-length sequences. It is also important to note that the characteristic function does not consider the order of the sequence. It is advisable to include a position-aware layer before SeqCondenser for effective classification.

3.3 Characterization of Multiple Features

In the case of a sequence with multiple features, we require a set of independent characteristic functions and affiliation probabilities for each feature k. This leads to the following definitions:

- $\mathbf{X} \in \mathbb{R}^{(l,d)}$ is the observed sequence of vectors, where l is the length of the sequence and d is the number of features.
- $\mathbf{x}^{(k)} \in \mathbb{R}^l$ is the observed sequence of values for the kth feature, i.e. the kth column of $\mathbf{X}$.
- $p_t^{(k)}$ is the affiliation probability associated with the random variable $X^{(k)}$.
- $x_t^{(k)}$ is the value of the kth feature at time step t.
- $\phi^{(k)}$ is the complex-valued characteristic function associated with the random variable $X^{(k)}$.
- $\phi'^{(k)}$ and $\phi'^{(k)}$ its two derivatives.

3.4 Sampling of the Characteristic Functions and Their Derivatives

The characteristic function is a complex-valued function that takes a real input, and therefore cannot be directly used in a neural model. In order to incorporate the characteristic function into our model, we need to discretize it. Similar to the FEATHER approach, we allow the model to choose the evaluation points of the characteristic functions that are the most discriminative for the downstream classification task.

The evaluation points for the d characteristic functions are represented by a matrix $\Theta \in \mathbb{R}^{(d,n_\theta)}$, where n_θ is the number of sampling points.

3.5 Attention Mechanism to Compute Probabilities

In order to compute the affiliation probabilities $\mathbf{P}$ for the input sequence, we employ an attention mechanism using a two-layered perceptron. The perceptron takes the entire sequence of vectors $\mathbf{X}$ as input and returns a matrix with the same shape, with the softmax function applied column-wise. This results in a matrix with positive columns that sum to 1, representing the probabilities of all observed features. $\mathbf{P}$ is calculated using the following formula: $\mathbf{P} = \text{softmax}(\mathbf{t} \odot \text{ReLU}(\text{ReLU}(\mathbf{X}W_1 + \mathbf{b})W_2 + \mathbf{c}))$ where the softmax is applied column-wise, $W_1, W_2 \in \mathbb{R}^{(d,d)}$ are two parameter matrices, $\mathbf{b}, \mathbf{c} \in \mathbb{R}^d$ are two bias vectors, and $\mathbf{t} \in \mathbb{R}^d$ is a temperature vector that allows the model to independently and quickly adjust the softmax discriminatory power for each dimension. $\odot$ denotes element-wise multiplication. In our notation, $\mathbf{P}_{tk} = p_t^{(k)}$.

3.6 Aggregation of the Evaluations

We define $\psi^{(k)}$ as a weighted average of the kth characteristic function and its two derivatives applied at its corresponding evaluation points:

$$\psi^{(k)} = \alpha_1 \phi^{(k)}(\Theta^{(k)}) + \alpha_2 \phi'^{(k)}(\Theta^{(k)}) + \alpha_3 \phi''^{(k)}(\Theta^{(k)})$$

where α is a softmaxed trainable 3-dimensional vector. Combining the results of the three functions has two advantages: (1) it reduces the memory footprint of the method; (2) it allows the model to balance between the various descriptions of the complex trajectory depending on their discriminative power for the task.

This gives the following formulas for the real and imaginary parts of $\psi^{(k)}$:

$$\mathbf{Re}(\psi^{(k)}(\theta)) = \sum_{x_t} p_t^{(k)} \left[(\alpha_1 - \alpha_3 x_t^2) \cos(x_t \theta) - \alpha_2 x_t \sin(x_t \theta) \right]$$

$$\mathbf{Im}(\psi^{(k)}(\theta)) = \sum_{x_t} p_t^{(k)} \left[(\alpha_1 - \alpha_3 x_t^2) \sin(x_t \theta) + \alpha_2 x_t \cos(x_t \theta) \right]$$

The real and imaginary parts of the $\psi^{(k)}$ are then concatenated into a compact vector $\mathbf{h}$ of size $2 \cdot d \cdot n_\theta$: $\mathbf{h} = \bigoplus_{k=1}^{d} \left[\mathbf{Re}(\psi^{(k)}) \oplus \mathbf{Im}(\psi^{(k)}) \right]$ with $\oplus$ the concatenation operation.

3.7 Dimensionality Reduction Using Random Projections

Using **h** as the final vector of the layer can be challenging due to its large size. For example, with 10 evaluation points and a sequence of 500-dimensional embeddings, **h** has a size of 10,000. One approach to addressing this dimensionality issue is to use techniques such as Principal Component Analysis (PCA), as seen in GraphWave. However, this approach is not suitable for use in an end-to-end training task. Our solution is to multiply **h** with a fixed, non-trainable, randomly initialized, and column-orthogonal matrix $O \in \mathbb{R}^{(2 \cdot d \cdot n_\theta, r)}$, where r is the desired reduced dimensionality. This approach significantly improves the accuracy of the model (Table 3).

Choosing r to be equal to the input dimension d generally works well, although we find that lower values do not significantly penalize the model (Fig. 4). As a final step, we apply the ReLU activation function after adding a scaler and a bias vector to the aggregation, giving: $\mathbf{h} \leftarrow \text{ReLU}\left(\beta(\mathbf{h}^T O) + \gamma\right)$.

3.8 Mathematical Properties

Considering the first derivative of the characteristic function has an interesting mathematical property, summarized by the following proposition:

Proposition 1. *Let* **x** *and* **x'** *be two vectors representing two sequences. Assume that the second sequence differs from the first at a single element at time step* u, *where* $x'_u = -x_u$, *and that the associated probability is kept unchanged, i.e.* $p'_u = p_u$. *The modulus of the difference between their two characteristic functions is bounded by* $2p_u$.

Proof. We start by expressing the difference between the two characteristic functions as follows:

$$\Delta m_\phi = \left|\sum_t p_t e^{ix_t\theta} - \sum_t p'_t e^{ix'_t\theta}\right| = \left|p_u e^{ix_u\theta} - p'_u e^{i(x'_u)\theta}\right|$$

$$= \left|p_u e^{ix_u\theta} - p_u e^{-ix_u\theta}\right| = \left|p_u(e^{ix_u\theta} - e^{-ix_u\theta})\right| = \left|p_u(2i \cdot sin(x_u\theta))\right|$$

Using the fact that the absolute value of $sin(x)$ is always less than or equal to 1, we get:

$$\Delta m_\phi \leq 2p_u$$

This implies that the maximum absolute difference between the two characteristic functions does not depend on the magnitude of x_u, leading to the inability of the model to distinguish opposite features when their probability weights have similar values. To address this issue, the first derivative of ϕ can be used. Carrying out a similar calculation on the derivative of the characteristic function gives us:

$$max(\Delta m_{\phi'}) = max\left|2p_u x_u cos(x_u\theta)\right| = 2p_u|x_u|$$

This result shows that the maximum absolute difference between the derivative of the characteristic functions is dependent on the magnitude of x_u. This property improves the model's expressivity by enabling it to better differentiate between positive and negative values.

4 Text Classification Model Architecture

In this section, we describe the neural network models used for text classification. We introduce two models: the SeqCondenser Model for Classification (referred to as SCoMo) and a simplified model called SCoMo-Bare, which is mainly used for comparison. Before discussing their architectures, we first introduce the units that make up these models.

4.1 Embeddings

As with most models that process text sequences, the input tokens are converted into d-dimensional embeddings. To perform this conversion and learn the representations automatically, the Keras Embedding layer is used as the first unit in the SeqCondenser models.

4.2 Self Attention

SCoMo uses a simplified version of the popular Self Attention layer [16]. Instead of using three different sets of projection matrices (Key, Query, and Value), we only use one. For a sequence X with shape (l, d), we compute the attention score matrix of shape (l, l) using the following formula: $Scores(X) = \text{softmax}(t \cdot \text{ReLU}(XWX^T + b))$ where the softmax function is applied along the rows, $W \in \mathbb{R}^{(d,d)}$ is a trainable matrix, b is a bias scalar, and t is a temperature scalar that controls the strength of the softmax.

The output of the simplified Self Attention layer is then given by: $SA(X) = Scores(X)X$

4.3 Positional Encoding

The characteristic function is insensitive to the order of the elements in the sequence. To make the model position-aware, we let the model learn its own positional embeddings. $PE(X) = X + \text{SWISH}(W)$ where X is the input sequence of shape (l, d), $W \in \mathbb{R}^{(l,d)}$ is a trainable parameter matrix, and *SWISH* is an activation function used to avoid vanishing gradients [14]. The *SWISH* function was chosen early on in our experiments because it provided better classification performance.

4.4 Model Architecture

The SCoMo model consists of a Positional Encoding layer, Self Attention layers, a SeqCondenser unit and a final dense layer for classificaton.

To buildS SCoMo-Bare, we follow the same structure as SCoMo, but remove the Self Attention layers and the Positional Encoding unit. The output of the Embedding layer is directly passed through the SeqCondenser layer, and the resulting fixed-size vector is used for classification. Despite its simplicity and disregard for the order of the sequence, SCoMo-Bare performs very well in comparison to other models, demonstrating the effectiveness of the SeqCondenser layer in capturing relevant information from the input sequence.

5 Experiments

In this section, we present experimental results to demonstrate the efficacy of our model.

5.1 Experiments Setups

For the text classification task, we run our experiments on five well-established text classification datasets, namely 20 Newsgroups (20NG), R8, R52, Ohsumed and Movie Review (MR). The 20 Newsgroups dataset consists of approximately 20,000 newsgroup posts from 20 different categories, and is widely used for multi-label classification tasks. The R8 and R52 datasets are collections of Reuters news articles, with 8 and 52 categories, respectively. The Ohsumed dataset is a collection of medical abstracts with 23 categories. The Movie Review dataset consists of movie reviews with binary labels indicating positive or negative sentiment. We use the standard splits for all datasets. Table 1 provides a numerical summary of the benchmarks.

Preprocessing was kept to a minimum. Documents were tokenized and truncated at 500 tokens if they exceeded that length. All tokens and punctuation were retained and no stopwords were removed. No pretrained word embeddings or language models were used.

We compare the performance of the proposed SeqCondenser architecture with two state-of-the-art inductive models: GraphStar [13] and SparseTensorClassifier (referred to as STC hereafter) [10]. As baselines, we used four popular algorithms implemented in Scikit-Learn [2]: Support Vector Machine (SVM), Multinomial Naive Bayes (MNB), Random Forest (RF), and K-Nearest Neighbors (KNN). Text vectorization for these models was performed using TF-IDF, a widely used approach for representing text data.

To compare the effectiveness of the SeqCondenser unit to other pooling methods, three models were built: SAt-avg, SAt-max, and SAt-weighted. The prefix 'SAt' stands for 'Self-Attention'. These models have identical architecture as SCoMo, with the only difference being the method of aggregation they use:

- **SAt-avg** aggregates the sequence by taking its average over the temporal dimension.
- **SAt-max** uses max over time pooling.
- **SAt-weighted** computes a weighted summation of the sequence, where the weights are predicted using an attention mechanism similar to DeepMoji.

Since no pretraining was utilized, both SCoMo and SAt models were kept small, incorporating only two Self-Attention layers for the experiments.

Table 1. Summary of Text Classification Datasets.

Dataset	N_{train}	N_{test}	Classes
20Newsgroups	11,314	7,532	20
R8	4,937	2,189	8
R52	5,879	2,568	52
Ohsumed	3,021	4,043	23
MR	7,108	3,554	2

5.2 Settings

We kept the SeqCondenser model's architecture and parameter sizes consistent across all datasets to showcase its 'drop-in replacement' capabilities. Both the token embedding dimension d and the reduction dimension r were set to 500. To maintain a low model size, we used only 2 self-attention layers and 10 evaluation points. The matrix Θ was initialized by linearly interpolating values between -10 and 100. While fine-tuning the trainable weights or increasing the model size might have improved accuracies, this was not explored in this paper.

All models were trained using the RMSProp optimizer [5] and minimizing cross-entropy loss, with a learning rate of 0.001 and a batch size of 30 for 10 epochs.

Table 2. Mean accuracies over 10 runs for different models. The best accuracies for inductive methods are in bold, and the best overall accuracies are in italic.

Model	20NG	R8	R52	OH	MR
SVM	77.7	94.6	88.9	48.6	75.5
MNB	73.6	83.6	71.2	29.8	**77.8**
RF	74.2	92.8	84.8	57.3	68.7
KNN	52.4	87.0	83.5	54.5	70.2
SAt-max	77.6	96.6	92.1	65.1	73.2
SAt-avg	82.7	97.1	92.7	65.0	75.7
SAt-weighted	81.5	97.0	92.5	63.5	75.5
STC	86.3	95.1	90.9	67.4	75.7
GraphStar	86.9	97.4	95.0	64.2	76.6
SCoMo-Bare	85.6	97.6	94.8	68.7	76.9
SCoMo	**87.4**	**98.5**	**95.5**	**70.1**	76.6
RoBERTaGCN	*89.5*	98.2	*96.1*	*72.8*	*89.7*

5.3 Main Results

Table 2 shows that SCoMo outperforms all other inductive algorithms on all datasets except one. SeqCondenser also performs significantly better than all other pooling strategies, despite all models having otherwise the same architecture. This demonstrates that replacing a traditional pooling layer with SeqCondenser can significantly improve model performance. Additionally, using a SeqCondenser layer immediately after an embedding layer without self-attention or positional embedding (SCoMo-Bare) still outperforms more sophisticated models. At the time of writing, SCoMo achieves the highest reported accuracy on the R8 dataset according to the Papers with Code leaderboard, outperforming the transductive text classification algorithm RoBERTaGCN.

5.4 Importance of Dimensionality Reduction

As explained earlier, we introduce an additional step after the computation of the embeddings of the characteristic functions in order to reduce the size of the output of the aggregation step. This is achieved by multiplying the aggregation vector by a randomly initialized column-orthogonal matrix O with r columns, where r is less than $2dn_\theta$.

We used $r = 500$ in our experiments and evaluated testing accuracies according to the following three modalities:

- **non-trainable reduction:** the aggregation vector is multiplied by a randomly initialized but fixed matrix O.
- **trainable reduction:** the agregation vector is multiplied by O, and O is trained alongside the rest of the neural network.
- **no reduction:** the aggregation vector is used as the output of the layer without being multiplied by O.

According to the results presented in Table 3, using a non-trainable O yields the best performance on the three datasets. These results suggest that using a non-trainable matrix is an effective approach for reducing the size of the output of the aggregation step and improving the test accuracy of the neural network.

Table 3. Mean accuracies over 10 runs for different dimensionality reduction schemes.

Dimensionality reduction	R8	R52	OH
non-trainable reduction	**98.5**	**95.5**	**70.1**
trainable reduction	97.8	95.2	66.6
no reduction	97.9	95.0	67.3

5.5 Parameters Sensitivity

In this section, we investigate the influence of the two main model parameters on the quality of learning: the number of evaluation points and the size of the dimensionality reduction. To do this, we conducted an analysis of the model's accuracy by varying the number of evaluation points from 1 to 14 on three datasets: Ohsumed, R8, and R52. The numbers presented in Fig. 3 are the average accuracy over 10 runs. It is worth noting that error bars were not included in the graph as they are too small to be legible: the standard deviation of the accuracies is never more than 0.1 and typically falls within the range of 0.03 to 0.06.

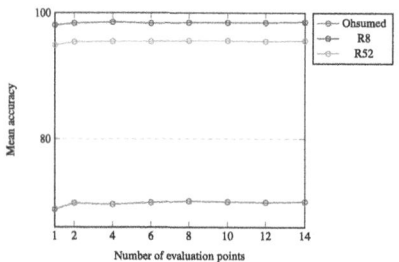

Fig. 3. Impact of the number of evaluation points on accuracy

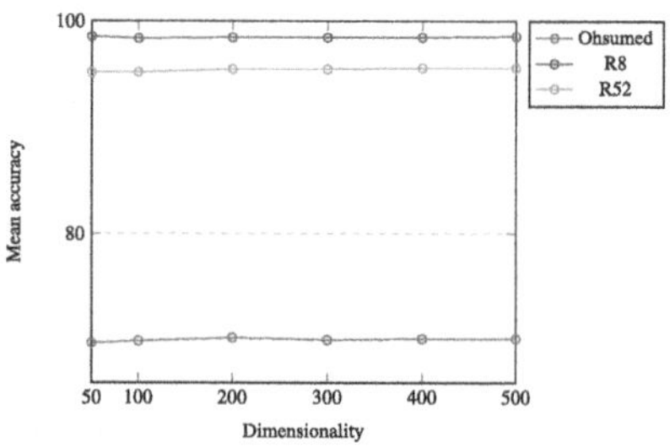

Fig. 4. Impact of the dimensionality reduction on accuracy

Based on the calculations, increasing the number of sampling points significantly improves accuracy from 1 to around 8 points. However, beyond 8 points, the improvement becomes negligible, indicating that the model's accuracy is stable across a range of sampling point values. Thus, using more than 8 evaluation points is unnecessary.

It is important to note that using a sampling point is not the same as using a single parameter. Instead, it represents a sampling point for each input feature, resulting in a $2d$-dimensional vector before reduction, contributing to the robustness of the results in this situation. Moreover, if the accuracies with a sampling point appear similar to larger values on Fig. 3, it is primarily a visual artifact of the compression of the y-axis.

In addition, an analysis of the impact of the size of the low-dimensional space that the aggregation vector is projected onto reveals that the model is able to effectively compress information across a range of dimensionality values from 50 to 500. The observed high stability in accuracy across this range demonstrates the effectiveness of the SeqCondenser layer in producing a rich representation for the classification task.

6 Conclusion and Future Work

In this work, we introduced SeqCondenser, a novel TensorFlow layer that transforms any sequence into a fixed-size vector, serving as an alternative to traditional pooling layers. We presented SCoMo, a neural model for classification that incorporates SeqCondenser and achieved state-of-the-art results compared to both inductive and transductive models. We also introduced SCoMo-Bare, a simple model that is unaware of sequence order, yet still outperforms more sophisticated, position-aware models using standard architectures.

The SeqCondenser unit can be integrated into any Keras model for classification or regression tasks. Our experiments demonstrate the effectiveness of the SeqCondenser layer in providing a concise and accurate summary of a sequence. In addition, we found that using a dimensionality reduction step, which involves multiplying the aggregation vector with a randomly initialized but fixed column-orthogonal matrix, significantly improves the training accuracy. Overall, the use of this dimensionality reduction step was a key factor in achieving the high accuracy and stability observed in our experiments.

Future work could explore the use of SeqCondenser to fine-tune language models such as BERT for classification purposes. We also plan to test SCoMo on classification tasks involving time series data, and preliminary results have been encouraging. We believe that the layer may also be successful in regression tasks, although we have not yet conducted any tests in this area.

References

1. Abadi, M., et al.: TensorFlow: large-scale machine learning on heterogeneous systems (2015)
2. Buitinck, L., et al.: API design for machine learning software: experiences from the scikit-learn project. In: ECML PKDD Workshop: Languages for Data Mining and Machine Learning, pp. 108–122 (2013)
3. Chen, Y.: Convolutional neural network for sentence classification. Master's thesis, University of Waterloo (2015)
4. Chollet, F., et al.: Keras (2015). https://keras.io
5. Dauphin, Y.N., de Vries, H., Chung, J., Bengio, Y.: RMSprop and equilibrated adaptive learning rates for non-convex optimization (2015)
6. Devlin, J., Chang, M., Lee, K., Toutanova, K.: BERT: pre-training of deep bidirectional transformers for language understanding. CoRR abs/1810.04805 (2018)
7. Donnat, C., Zitnik, M., Hallac, D., Leskovec, J.: Spectral graph wavelets for structural role similarity in networks. CoRR abs/1710.10321 (2017)
8. Felbo, B., Mislove, A., Søgaard, A., Rahwan, I., Lehmann, S.: Using millions of emoji occurrences to learn any-domain representations for detecting sentiment, emotion and sarcasm. In: Empirical Methods in Natural Language Processing, pp. 1615–1625 (2017)
9. Feller, W.: An Introduction to Probability Theory and Its Applications, vol. 1, 3rd edn. Wiley, New York (1968)
10. Guidotti, E., Ferrara, A.: An explainable probabilistic classifier for categorical data inspired to quantum physics (2021). https://arxiv.org/abs/2105.13988

11. Huang, E., Socher, R., Manning, C., Ng, A.: Improving word representations via global context and multiple word prototypes. In: Proceedings of the 50th Annual Meeting of the Association for Computational Linguistics, pp. 873–882 (2012)
12. Lin, Y., et al.: BERTGCN: transductive text classification by combining GCN and BERT. arXiv preprint arXiv:2105.05727 (2021)
13. Lu, H., Huang, S.H., Ye, T., Guo, X.: Graph star net for generalized multi-task learning. CoRR abs/1906.12330 (2019)
14. Ramachandran, P., Zoph, B., Le, Q.V.: Swish: a self-gated activation function. arXiv: Neural and Evolutionary Computing (2017)
15. Rozemberczki, B., Sarkar, R.: Characteristic functions on graphs: birds of a feather, from statistical descriptors to parametric models. In: CIKM, [p. 1325–1334. ACM (2020)
16. Vaswani, A., et al.: Attention is all you need. In: Advances in Neural Information Processing Systems, vol. 30. Curran Associates, Inc. (2017)
17. Wu, C., Wu, F., Qi, T., Cui, X., Huang, Y.: Attentive pooling with learnable norms for text representation, pp. 2961–2970 (2020). https://doi.org/10.18653/v1/2020.acl-main.267
18. Yao, L., Mao, C., Luo, Y.: Graph convolutional networks for text classification. CoRR abs/1809.05679 (2018). http://arxiv.org/abs/1809.05679

Is Prompting What Term Extraction Needs?

Hanh Thi Hong Tran[1,2,3], Carlos-Emiliano González-Gallardo[1(✉)],
Julien Delaunay[1(✉)], Antoine Doucet[1(✉)], and Senja Pollak[3(✉)]

[1] University of La Rochelle, L3i, La Rochelle, France
{thi.tran,carlos.gonzalez_gallardo,julien.delaunay,
antoine.doucet}@univ-lr.fr
[2] Jožef Stefan International Postgraduate School, Ljubljana, Slovenia
[3] Jožef Stefan Institute, Ljubljana, Slovenia
senja.pollak@ijs.si

Abstract. Automatic term extraction (ATE) is a natural language processing (NLP) task that reduces the effort of manually identifying terms from domain-specific corpora by providing a list of candidate terms. This paper summarizes our research on the applicability of open and closed-sourced large language models (LLMs) on the ATE task compared to two benchmarks where we consider ATE as sequence-labeling (*iobATE*) and seq2seq ranking (*templATE*) tasks, respectively. We propose three forms of prompting designs, including (1) sequence-labeling response; (2) text-extractive response; and (3) filling the gap of both types by text-generative response. We conduct experiments on the ACTER corpora in three languages and four domains with two different gold standards: one includes only terms (ANN) and the other covers both terms and entities (NES). Our empirical inquiry unveils that above all the prompting formats, text-extractive responses, and text-generative responses exhibit a greater ability in the few-shot setups when the amount of training data is scarce, and surpasses the performance of the templATE classifier in all scenarios. The performance of LLMs is close to fully supervised sequence-labeling ones, and it offers a valuable trade-off by eliminating the need for extensive data annotation efforts to a certain degree. This demonstrates LLMs' potential use within pragmatic, real-world applications characterized by the constricted availability of labeled examples.

Keywords: Term extraction · LLMs · prompting · in-context learning

1 Introduction

Terms are "*the designation of a defined concept in a special language by a linguistic expression.*" (ISO 1087). They are beneficial not only for several terminographical tasks by linguists (e.g., specialized dictionary construction [14]) but also for several downstream tasks (e.g., topic detection [5] and information retrieval [15]). To minimize the effort needed to extract terms from domain-specific corpora, automatic term extraction (ATE) approaches have been proposed.

TermEval 2020: Shared Task on Automatic Term Extraction, organized as part of the CompuTerm workshop [17], presented an important step forward in systematic comparison among several ATE systems with the introduction of a new manually-annotated corpus, namely ACTER corpora [17]. The corpora contain domain-specific texts from four different fields in three languages with two versions of gold standards (with or without named entities). This is also the dataset on which we conduct our experiments.

After the evolution of transformer-based token classifiers toward term extraction (e.g., XLMR [21–24]), recent years witnessed the blossoming of large-scale generative models with the advent of prompt engineering [16]. Despite several works with state-of-the-art (SOTA) performance on downstream tasks [8,11], no application has been found on ATE tasks yet, and the performance of sequence-labeling tasks is still significantly below supervised baselines.

The main contribution of our work is threefold:

1. We conduct an empirical evaluation of term extraction using three distinct approaches, where we treat the task as (1) a sequence labeling task, (2) a seq2seq ranking task, and (3) a generative task using LLMs prompting;
2. We investigate the potential of LLMs' prompting for our ATE tasks to highlight their valuable insights in both rich- and low-resourced language niches;
3. We experiment with open and closed-sourced LLMs with comprehensive error analysis. This allows a task-oriented comparison among models and enriches the debate concerning the importance and utility of open LLMs.

This paper is organized as follows: Sect. 2 presents the related work while Sect. 3 describes the methods with the experimental setup, the datasets, and the evaluation metrics. The results with error analysis are discussed in Sect. 4 before we conclude with future work in Sect. 5.

2 Related Work

2.1 Automatic Term Extraction

Traced back to the 1990s, term extraction was first proposed under the research of [4] with the two-step procedure: (1) extracting a list of candidate terms, and (2) determining their correctness. Traditional methods primarily relied on either linguistic or statistical aspects [6] or combined both [10]. The advancement of representation learning and neural networks has led to the exploration of various text embedding techniques for term extraction (e.g., local-global [1], non-contextual [26], and contextual [12] or their combinations [7]). Language models have also been applied to the task, as demonstrated in the *TermEval 2020* [17], e.g., feeding GloVe embeddings into a Bi-LSTM [17], feeding all possible extracted n-gram combinations into a BERT binary classifier [9]. In recent years, the evolution of term extraction has seen a shift towards treating the task as a sequence-labeling problem, extending beyond monolingual learning [12] to include cross-lingual and multilingual learning [13,21]. A systematic review of the tasks can be found in [20].

2.2 In-context Learning with LLMs

The emergence of LLMs has significantly improved performance across several downstream tasks [25]. Two strategies for incorporating LLMs into these tasks include fine-tuning and in-context learning (ICL). While the fine-tuning involves initializing a pre-trained model and conducting additional training epochs on task-specific supervised data, ICL leverages the LLM ability to generate texts with only a few task-specific examples as demonstrations. The concept of prompts with few-shot demonstrations was first introduced by [16], followed by an empirical analysis of the ICL paradigm with GPT-3 [2] and PaLM [3], in specific. With the release of ChatGPT[1] by OpenAI and the blossom of open-sourced LLMs, recent research focuses on evaluating its performance in various NLP tasks. We evaluate the performance of ChatGPT's reinforcement learning with human feedback (RLHF) model *gpt-3.5-turbo* and the open-sourced *Llama 2-Chat* model family with the ICL paradigm and compare it to the traditional sequence-labeling and fine-tuned seq2seq classifiers. To our knowledge, none of these two directions (seq2seq and LLMs) had been previously explored in the ATE task. Therefore, we would like to provide a comprehensive view of the ATE task from three perspectives: (1) as a token classification task; (2) as a seq2seq ranking task; and (3) for prompting.

3 Methodology

This section investigates the impact of semantically ambiguous and complex terms on prompt-based methods compared to the sequence-labeling baseline.

3.1 Sequence-Labeling ATE

A common approach is to consider ATE as a token classification task, which assigns a label $y \in Y = \{B, I, O\}$ to each word x in a given sentence $X = \{x_1, ..., x_n\}$, where Y denotes the set of labels in IOB annotation regimes, and n denotes the length of the given sentence. We refer to this approach as *iobATE*. Inspired by [21]'s work, our baseline is the XLM-R token classifier with a standard hyperparameter configuration, which is also the SOTA in term extraction.

---

[1] https://openai.com/chatgpt.

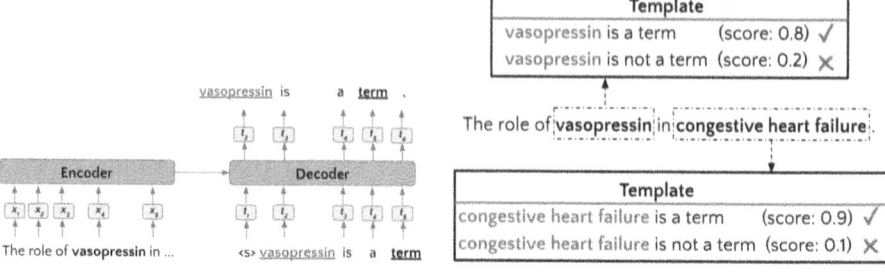

Fig. 1. *templATE* training architecture. **Fig. 2.** *templATE* inference.

3.2 Template-Based ATE

The task is formulated as a template-based (seq2seq) ranking problem for ATE (*templATE*), where the original sentence $X = \{x_1, \ldots, x_n\}$ is the source sequence, and the templates $\{t_1, \ldots, t_m\}$ filled by the candidate term span $x_{i:j}$ are the target sequence during training. The method contains the following steps:

1. Identify the gold standard terms in a sentence (e.g., *The role of* **vasopressin** *in* **congestive heart failure**...).
2. Create a positive template for gold standard terms: $<MASK>$ is a term. (e.g., **vasopressin** *is a term*; **congestive heart failure** *is a term*; ...).
3. Create a negative template for the rest: $<MASK>$ is not a term. (e.g., **The** *is not a term*; **role** *is not a term*; **of** *is not a term*; **in** *is not a term*;...).
4. **Training**: Feed into the mBART[2] [18] the terms with their related positive and only the other 30% of negative ones to reduce imbalance. For example:
   – Sentence: *The role of* **vasopressin** *in* **congestive heart failure**.
   – Output: *The* is not a term; *role* is not a term; *of* is not a term; *sentence* is not a term; *in* is not a term; *vasopressin* is a term; ...
5. **Inference**: Calculate the term score for each n-gram ($n = \{1, 2, 3, 4\}$). If the positive score is higher, consider it as a term.

We used mBART with 5 epochs, a batch size of 32, and a max sequence length of 70. The training and inference steps of the templATE architecture are visualized in Figs. 1 and 2, respectively.

3.3 LLMs Prompting

We propose *promptATE*, which uses the close-sourced ChatGPT's *gpt-3.5-turbo*[3] and the open-sourced *Llama 2-Chat* (i.e., *Llama 2-Chat-7B*[4], *Llama 2-Chat-13B*[5], and *Llama 2-Chat-70B*[6]) RLHF models to address the ATE task. The

---

[2] https://huggingface.co/facebook/mbart-large-50-many-to-one-mmt.
[3] https://platform.openai.com/.
[4] https://huggingface.co/meta-llama/Llama-2-7b-chat-hf.
[5] https://huggingface.co/meta-llama/Llama-2-13b-chat-hf.
[6] https://huggingface.co/meta-llama/Llama-2-70b-chat-hf.

approach follows the general paradigm of in-context (few-shot) learning with a three-step procedure as in Fig. 3 where (1) **Task Description** instructs *promptATE* to detect the candidate terms using terminological knowledge; (2) **Few-shot Demonstrations** gives the model a few examples; and (3) **Input Sentence** indicates the input sentence while *promptATE*'s output is highlighted in green.

Task Description. Given input sentence X, construct a $Prompt(X)$ to give a descriptive overview of the task with the following steps:

1. ***SYSTEM_PROMPT***: First, *"You are ... extraction (ATE) system."* tells *promptATE* to produce the output using terminological knowledge. Second, *"I will provide you ... extract the terms"* indicates the input information, including the domain and sentence having domain-specific terms while *"and the output ... with examples."* shows the position of few-shot demonstrations and marks the end of the description.
2. ***USER_PROMPT_1***: *"Are you clear about your role?"*. triggers a response by the assistant explicitly asking for confirmation of the task comprehension.
3. ***ASSISTANT_PROMPT_1***: *"Sure. I am ready to ... get started."* is the acknowledgment by *promptATE* but designed by the user only.
4. ***PROMPT***: This guideline prompt defines how *promptATE* should perform the ATE task. In the guidelines, we provided the requirements and the output format to guide *promptATE*'s responses for further processing.

SYSTEM_PROMPT
You are an excellent automatic term extraction (ATE) system. I will provide you the domain of the terms you need to extract and the sentence from which you need to extract the terms and the output in given format with examples.
USER_PROMPT_1
Are you clear about your role?
ASSISTANT_PROMPT_1
Sure, I'm ready to help you with your ATE task. Please provide me with the necessary information to get started.
PROMPT
What are the terms in the following text? Terms should not include named entities.
Output Format: [list of terms present]
If no terms are presented, keep it empty list: [ ]   **Task Description**

EXAMPLES:
Sentence: Treatment of anemia in patients with heart disease : a clinical practice guideline from the American College of Physicians .
Domain: Heart failure
Output: ['anemia', 'patients', 'heart disease', 'clinical practice guideline', 'Physicians']
Sentence: Recommendation 2 : ACP recommends against the use of erythropoiesis-stimulating agents in patients with mild to moderate anemia and congestive heart failure or coronary heart disease .
Domain: Heart failure
Output: ['erythropoiesis-stimulating agents', 'patients', 'anemia', 'congestive heart failure', 'coronary heart disease']
Sentence: Moreover , there is yet to be established a common consensus being used in current assays .
Domain: Heart failure
Output: [ ]   **Few-shot Demonstration**

Sentence: The role of vasopressin in congestive heart failure .
Domain: Heart failure
Output: ['vasopressin', 'congestive heart failure']   **Input sentence**

Fig. 3. A complete prompt with the output format #2 for the ANN version (Color figure online)

Few-Shot Prompting. We focus on the few-shot demonstrations where we provide examples that are appended to the task description phase to regulate the format of outputs for each test input, as *promptATE* will generate outputs that mimic the demonstration format. For example, in the **Few-shot Demonstrations** rectangle of Fig. 3, the demonstration sequentially packs a list of examples, each consisting of both the input and output sequences. The demonstration is set up as follows: The first two examples contain terms while the last one is without terms inside the input sequence. All the examples of sequences are only from the test domain (Heart Failure) without any further information from the other three domains from ACTER corpora.

The following three output formats (OF) are tested: (1) *Sequence-labeling output (OF#1)*, where the output contains the information for each word label in the IOB annotation regime; (2) *List of candidate terms output (OF#2)*, which is the same format as our original gold standard; (3) *Generative output (OF#3)*, where we use unique tokens "@@" and "##" to surround the candidate terms.

ChatGPT vs. Llama 2-Chat. We delved into the capabilities of few-shot demonstrations, employing both the close-sourced ChatGPT (*gpt-3.5-turbo*) [2] and the open-sourced *Llama 2-Chat* [19] RLHF models. While both exhibit remarkable language understanding and generation abilities, they employ divergent training methods and prompting mechanisms. Thus, slight modifications are required in the prompt structure while preserving the overarching concepts. By doing so, we aimed to evaluate how each model adapts to varying input cues and assess their respective adaptability in handling the same set of instructions. This study not only sheds light on the comparative performance of these RLHF models but also underscores their flexibility and versatility in comprehending and generating content, even when their underlying architectures differ significantly.

3.4 Data

The experiments have been conducted on ACTER v1.5 [17], a manually annotated collection of 12 comparable corpora (same domains in different languages) covering four domains (Corruption - Corp, Dressage - Equi, Wind energy - Wind, Heart failure - Htfl) in English, French, and Dutch. The corpora have two versions of gold standard annotations: one containing both terms and named entities (NES), and the other containing only terms (ANN). We apply the same configuration as in the TermEval 2020 shared task and related works [9,13,21] where Htfl domain of each language is considered the test set.

3.5 Evaluation Metrics

The performance of each term extractor is assessed by strictly comparing the aggregated list of candidate terms identified across the entire test set against the manually designated gold standard list of terms, using precision, recall, and F1-score [9,13,21,22].

4 Results

Table 1 presents the comprehensive evaluation of three different ATE approaches on the Htfl domain in the ACTER dataset. For all our experiments, we fixed the

Table 1. Evaluation of different approaches on Htfl test set.

Settings	English			French			Dutch		
	Precision	Recall	F1-score	Precision	Recall	F1-score	Precision	Recall	F1-score
				ANN versions					
				Benchmarks					
iobATE	58.4 ± 0.216	46.1 ± 3.861	51.4 ± 2.439	70.0 ± 0.852	39.9 ± 4.241	50.7 ± 3.691	72.4 ± 1.464	58.7 ± 3.387	64.7 ± 1.699
templATE	29.1 ± 3.445	25.1 ± 3.232	27.0 ± 3.240	33.0 ± 5.587	29.9 ± 5.025	30.6 ± 1.219	31.5 ± 1.359	42.3 ± 3.534	36.1 ± 2.175
				$promptATE_{Llama2-Chat-7B}$					
OF#1	12.4	4.8	6.9	7.5	9.3	8.3	19.2	14.4	16.5
OF#2	↓ $40.4_{-18.0}$	↑ $62.6_{+16.5}$	↓ $49.1_{-2.3}$	36.3	↑ $59.2_{+19.3}$	↓ $45.0_{-5.7}$	40.4	↑ $73.1_{+14.4}$	↓ $52.0_{-12.7}$
OF#3	40.3	26.8	32.2	↓ $58.5_{-11.5}$	23.4	33.4	↓ $53.8_{-18.6}$	41.6	46.9
				$promptATE_{Llama2-Chat-13B}$					
OF#1	12.1	1.7	3.0	11.2	6.6	8.3	25.6	5.9	9.6
OF#2	35.0	↑ $63.4_{+17.3}$	↓ $45.1_{-6.3}$	38.4	↑ $59.2_{+19.3}$	↓ $46.6_{-4.1}$	43.3	↑ $75.0_{+16.3}$	↓ $54.9_{-9.8}$
OF#3	↓ $40.0_{-18.4}$	36.9	38.4	↓ $41.0_{-29.4}$	48.7	44.5	↓ $46.1_{-26.3}$	56.2	50.7
				$promptATE_{Llama2-Chat-70B}$					
OF#1	15.6	5.7	8.3	4.6	3.9	4.2	23.7	8.2	12.2
OF#2	36.8	↑ $65.9_{+19.8}$	47.2	38.0	↑ $64.8_{+44.9}$	47.9	42.3	↑ $74.8_{+16.1}$	54.0
OF#3	↓ $46.4_{-12.0}$	50.0	↓ $48.1_{-3.3}$	↓ $47.1_{-22.9}$	51.4	↓ $49.2_{-1.5}$	↓ $50.5_{-21.9}$	67.3	↓ $57.7_{-7.0}$
				$promptATE_{gpt-3.5-turbo}$					
OF#1	10.8	14.4	12.3	11.3	11.6	11.4	18.3	14.1	15.9
OF#2	26.6	↑ $67.6_{+21.5}$	38.2	28.5	↑ $67.0_{+27.1}$	40.0	36.8	↑ $79.6_{+20.9}$	50.3
OF#3	↓ $39.6_{-18.8}$	48.3	↓ $43.5_{-7.9}$	↓ $45.5_{-24.5}$	50.8	↓ $48.0_{-2.7}$	↓ $61.1_{-11.3}$	56.6	↓ $58.8_{-5.9}$
				NES versions					
				Benchmarks					
iobATE	63.0 ± 0.735	49.1 ± 3.014	55.2 ± 1.893	71.3 ± 1.330	45.4 ± 3.555	55.4 ± 3.074	74.0 ± 1.330	59.6 ± 1.322	66.0 ± 0.736
templATE	30.7 ± 3.122	31.2± 1.203	31.0 ± 2.164	36.1 ± 3.926	32.2 ± 6.193	33.8 ± 4.720	34.6 ± 4.678	43.4 ± 4.075	38.0 ± 1.239
				In-domain $promptATE_{Llama2-Chat-7B}$					
OF#1	17.3	7.3	10.3	8.4	11.0	9.5	16.6	23.8	19.6
OF#2	42.9	↑ $63.4_{+14.3}$	↓ $51.2_{-4.0}$	36.0	↑ $61.6_{+16.2}$	↓ $45.4_{-10.0}$	40.3	↑ $75.6_{+16.0}$	↓ $52.6_{-13.4}$
OF#3	↓ $45.0_{-18.0}$	32.5	37.7	↓ $52.1_{-19.2}$	34.5	41.5	↓ $48.8_{-25.2}$	52.3	50.5
				In-domain $promptATE_{Llama2-Chat-13B}$					
OF#1	25.9	2.4	4.4	8.4	5.3	6.5	23.5	5.9	9.4
OF#2	38.4	↑ $66.1_{+17.0}$	↓ $48.6_{-6.6}$	33.8	↑ $60.2_{+14.8}$	↓ $43.3_{-12.1}$	41.4	↑ $73.6_{+16.0}$	↓ $53.0_{-13.4}$
OF#3	↓ $40.3_{-22.7}$	47.5	43.6	↓ $35.7_{-35.6}$	49.4	41.4	↓ $47.9_{-26.1}$	49.1	48.5
				In-domain $promptATE_{Llama2-Chat-70B}$					
OF#1	21.4	4.7	7.7	7.9	9.5	8.6	18.7	17.5	18.1
OF#2	39.9	↑ $67.2_{+18.1}$	50.1	33.2	↑ $61.8_{+16.4}$	43.2	41.1	↑ $74.8_{+15.2}$	53.1
OF#3	↓ $48.3_{-14.7}$	54.9	↓ $51.4_{-3.8}$	↓ $40.8_{-30.5}$	57.7	↓ $47.8_{-7.6}$	↓ $53.1_{-20.9}$	57.9	↓ $55.4_{-10.6}$
				In-domain $promptATE_{gpt-3.5-turbo}$					
OF#1	10.3	13.1	11.5	10.8	12.0	11.4	14.8	13.2	14.0
OF#2	29.2	↑ $69.2_{+20.1}$	41.1	27.9	↑ $66.8_{+21.4}$	39.4	39.8	↑ $78.5_{+18.9}$	52.8
OF#3	↓ $39.8_{-23.2}$	53.1	↓ $45.5_{-9.7}$	↓ $44.7_{-26.6}$	54.4	↓ $49.1_{-6.3}$	↓ $63.6_{-10.4}$	60.6	↓ $62.1_{-3.9}$

Htfl domain as the test dataset, *iobATE* and *templATE* classifiers were trained and validated with all possible combinations of the other three domains always having two domains for training and one domain for validation.

We present these combinations' average scores and standard deviations for both benchmarks. We emphasize the settings yielding the most favorable outcomes for each of the three approach types of *promptATE* by rendering them in bold. Additionally, we identify in grey the best-performing model across all approach types, as measured by the F1-score, for each version of the gold standards. The arrows are used to compare our proposed methods and best benchmark for each setting, where ↑ is used to show the better performance of our approaches compared to the benchmark, while ↓ denotes the lower performance.

4.1 General Observations

As shown in Table 1, the *iobATE* approach consistently demonstrates a competitive balance between precision and recall, achieving a stable F1-score. This indicates the reliability of the fully supervised token classifier in terms of providing accurate predictions, but the approach requires a manually annotated training set. Comparatively, the *templATE* method showcases a mixed performance. While it can achieve high precision in certain scenarios, its recall lags, implying that it might struggle to identify all relevant examples, potentially resulting in missing important information. Compared to other approaches, there was a significant gap in F1-score performance.

The *promptATE* approach with in-domain few-shot demonstrations exhibits a considerable performance gap depending on the output format. It struggles with low precision and recall for sequence labeling (*OF#1*) compared to the others. This suggests the gap between the semantic labeling task and the text generation one, which open-sourced LLMs (i.e., *Llama 2-Chat*) and close-sourced LLMs (i.e., *gpt-3.5-turbo*) have been trained for. *OF#2* and *OF#3* show much higher scores compared to *OF#1*, even surpassing the *templATE*, and achieving competitive results to *iobATE* classifiers.

Results show variations given the language and model size, however, *Llama 2-Chat* with fewer parameters demonstrates to be better suited for listing candidate terms (*OF#2*) while *gpt-3.5-turbo* and the largest version of *Llama 2-Chat* show to be to a good option for generative output with specific delimiting tokens (*OF#3*). An interesting behavior is present for the *OF#2*, independently of the model, all recall scores are equal or higher than 59.2%, even reaching 79.6% in the case of *gpt-3.5-turbo* for Dutch. These scores outperform *iobATE* in the recall by an important margin, nevertheless in terms of precision, *promptATE* present a limited performance. We explain this by the complexity of the ATE task regarding the definition of a "term" inside a sentence, which is closely related to a specific domain. Generative models (i.e., *Llama 2-Chat* and *gpt-3.5-turbo*) can retrieve what can behave like a term but this leads to a big amount of false positives reflecting a high recall but limited precision.

4.2 Error Analysis

Impact of Output Formats. As ATE is a sequence-labeling rather than a generative task, it is not readily suited for the ICL paradigm by default. Additionally, the expected candidate terms to be extracted depend not only on the role that words occupy inside the sentence but also on a specific domain. *gpt-3.5-turbo* tends to generate outputs that exhibit a broader range of variability, often producing results that can be less predictable. However, *Llama 2-Chat* stands out for its remarkable ability to consistently adhere to the desired output structure and maintain a high level of reliability in generating content, especially *Llama 2-Chat-70B*. This contrast underscores the importance of choosing the right model for specific applications, where predictability and adherence are critical factors in decision-making processes and content generation.

IOB Format (OF#1) contains the information for each word label and can be easily transformed into the term sequence. However, three main obstacles led to the poor performance in this format for all tested LLMs: (1) The model needs to learn the alignment between each position in the input sequence and the output labels, which naturally adds to the difficulty of the generation task; (2) It is difficult for the model to generate the output with the same length as the input sentence, especially when the input sentence is long, a case where the model is more likely to exhibit *hallucinations*; (3) The model either added an extra explanation per label of the input sequence or failed to provide the labels.

Despite reducing the obstacles from the previous format design, *List of candidate terms format (OF#2)* faces the following challenges: (1) The model failed to finish their predictions for elongated sentences containing multiple terms due to their limited amount of tokens as inputs and outputs by default; and (2) The model generated the predictions for candidate terms that do not appear in the original sentence (*hallucination*), which is mostly found in the Dutch corpus.

Text Generation Format (OF#3) solves to a certain degree the obstacles faced by the two previous formats. As the model only needs to mark the position of the terms and make copies for the rest, it can (1) significantly decrease the difficulty in generating text that fully encodes label information (as in *OF#1*) of the input sequence, (2) avoiding self-explanation and repetition of the few-shot demonstration, and (3) preventing the wrong output formats.

Impact of Term Length Variants. To determine whether the term length affects the models' performance, we calculate the precision, recall, and F1-score of *promptATE* for terms of length $k = \{1, 2, 3, 4, \geq 5\}$ over the Htfl domain with both ANN and NES gold standards. Generally, as the term length increases, precision and recall decrease across most output formats and languages. This suggests that longer terms are more challenging to predict accurately, since they may have more variability and complexity, making them harder for the models to capture effectively. Independently of the approach, the highest F1-scores are achieved on the Dutch dataset. The proportion of different word lengths for the gold standard terms is shown in Table 2. It can be seen that, for the Htfl domain, the number of terms with more than one word (k ≥ 2) is considerably smaller compared to French and English, which facilitates their extraction.

Besides, different output formats have also varying effects on model performance across term lengths and languages. *OF#1* consistently exhibits lower precision and recall compared to the other formats across all the term lengths, indicating that it might not be as effective for ATE. For the English ANN version, *OF#2* has higher recall but lower precision in comparison with other formats across most term lengths. *OF#3* shows the best F1-score, striking a better balance between precision and recall. In the French and Dutch corpora, precision and recall are generally lower compared to the English ones, suggesting potential challenges in term extraction for these languages.

Table 2. Term proportion of different word lengths in each domain and language

Language	Domain	ANN version					NES version				
		k = 1	k = 2	k = 3	k = 4	k ≥ 5	k = 1	k = 2	k = 3	k = 4	k ≥ 5
English	Corp	389	377	117	30	14	502	419	146	52	54
	Equi	646	418	69	18	4	884	540	100	36	15
	Wind	319	527	198	39	8	565	639	245	58	24
	Htfl	1,064	767	358	118	54	1,170	801	377	142	91
French	Corp	440	326	131	51	31	550	356	158	75	68
	Equi	579	203	111	49	19	712	253	137	58	21
	Wind	315	232	122	65	39	446	265	128	74	55
	Htfl	1,207	604	264	79	74	1,309	620	266	91	88
Dutch	Corp	682	246	67	30	22	803	287	96	44	65
	Equi	1,091	185	65	37	15	1,181	224	82	40	17
	Wind	701	186	35	9	9	881	263	67	17	17
	Htfl	1,587	368	87	20	12	1,687	391	108	35	33

Impact of Language Distribution in Pretraining. The study by [19] pointed out that having a training dataset predominantly in English could potentially limit the model's effectiveness when used in languages other than English[7]. Despite French and Dutch not conventionally falling into the category of low-resourced languages, they are relatively under-resourced in the context of training LLMs, where English dominates (the pretraining distribution of French and Dutch accounts for 0.16% and 0.12% in *Llama 2-Chat*, 1.82% and 0.34% in *gpt-3.5-turbo* while English accounts for 89.70% in *Llama 2-Chat* and 92.65% in *gpt-3.5-turbo*. Our results indicate that our LLM prompting can indeed potentially enhance term extraction performance for under-represented languages.

---
[7] *"A training corpus with a majority in English means that the model may not be suitable for use in other languages."* [19].

5 Conclusions

In this paper, we evaluated the applicability of RLHF models toward ATE through an empirical study on different prompt designs in comparison with classical sequence labeling and the seq2seq approach. Although the RLHF models have achieved SOTA performances on various NLP tasks, there is still a gap between their performance in ATE and the fully supervised sequence-labeling baselines. We bridge the gap between the text generation and the sequence labeling task inherent in the ATE task by guiding the RLHF models to produce predictions with three designed formats.

Our empirical inquiry unveils that RLHF models exhibit a greater ability in the few-shot setups when the amount of training data is scarce and surpasses the performance of *templATE* in all scenarios with the last two output-designed formats: (1) as a list of candidates terms, (2) encapsulating the candidate terms using specialized tokens. Its performance is not only close to fully supervised sequence-labeling baselines, but it offers a valuable trade-off by eliminating the need for extensive data annotation efforts as well. These findings demonstrate the capabilities of RLHF models' prompting to ATE tasks within pragmatic, real-world applications characterized by the constricted availability of labeled examples. Nevertheless, RLHF models, which are built upon LLMs, are pre-trained with an enormous amount of general data, making them agnostic to the specific domain of a term. This leads to an over-extraction of terms, resulting in good coverage but poor precision. In consequence, when a complete training dataset is accessible, opting for a fully-supervised ATE system remains the optimal choice.

Acknowledgments. The work was partially supported by the Slovenian Research and Innovation Agency (ARIS) core research program Knowledge Technologies (P2-0103) and projects Linguistic Accessibility of Social Assistance Rights in Slovenia (J5-50169) and Embeddings-based techniques for Media Monitoring Applications (L2-50070). The work has also been supported by the ANNA (2019-1R40226) and TERMITRAD (2020-2019-8510010) projects funded by the Nouvelle-Aquitaine Region, France. Besides, the work was supported by the project Cross-lingual and Cross-domain Methods for Terminology Extraction and Alignment, a bilateral project funded by the program PROTEUS under the grant number BI-FR/23-24-PROTEUS006.

References

1. Amjadian, E., Inkpen, D., Paribakht, T., Faez, F.: Local-global vectors to improve unigram terminology extraction. In: Proceedings of the 5th International Workshop on Computational Terminology (Computerm2016), pp. 2–11 (2016)
2. Brown, T., et al.: Language models are few-shot learners. In: Advances in Neural Information Processing Systems, vol. 33, pp. 1877–1901 (2020)
3. Chowdhery, A., et al.: Palm: scaling language modeling with pathways. arXiv preprint arXiv:2204.02311 (2022)
4. Damerau, F.J.: Evaluating computer-generated domain-oriented vocabularies. Inf. Process. Manag. **26**(6), 791–801 (1990)

5. El-Kishky, A., Song, Y., Wang, C., Voss, C.R., Han, J.: Scalable topical phrase mining from text corpora. Proc. VLDB Endow. **8**(3), 305–316 (2014). https://doi.org/10.14778/2735508.2735519
6. Frantzi, K.T., Ananiadou, S., Tsujii, J.: The *C-value/NC-value* method of automatic recognition for multi-word terms. In: Nikolaou, C., Stephanidis, C. (eds.) ECDL 1998. LNCS, vol. 1513, pp. 585–604. Springer, Heidelberg (1998). https://doi.org/10.1007/3-540-49653-X_35
7. Gao, Y., Yuan, Yu.: Feature-less end-to-end nested term extraction. In: Tang, J., Kan, M.-Y., Zhao, D., Li, S., Zan, H. (eds.) NLPCC 2019. LNCS (LNAI), vol. 11839, pp. 607–616. Springer, Cham (2019). https://doi.org/10.1007/978-3-030-32236-6_55
8. Guo, B., et al.: How close is ChatGPT to human experts? Comparison corpus, evaluation, and detection (2023)
9. Hazem, A., Bouhandi, M., Boudin, F., Daille, B.: TermEval 2020: TALN-LS2N system for automatic term extraction. In: Proceedings of the 6th International Workshop on Computational Terminology, pp. 95–100 (2020)
10. Kessler, R., Béchet, N., Berio, G.: Extraction of terminology in the field of construction. In: 2019 First International Conference on Digital Data Processing (DDP), pp. 22–26. IEEE (2019)
11. Kocoń, J., et al.: ChatGPT: jack of all trades, master of none (2023)
12. Kucza, M., Niehues, J., Zenkel, T., Waibel, A., Stüker, S.: Term extraction via neural sequence labeling a comparative evaluation of strategies using recurrent neural networks. In: INTERSPEECH, pp. 2072–2076 (2018)
13. Lang, C., Wachowiak, L., Heinisch, B., Gromann, D.: Transforming term extraction: transformer-based approaches to multilingual term extraction across domains. In: Findings of the Association for Computational Linguistics: ACL-IJCNLP 2021, pp. 3607–3620 (2021)
14. Le Serrec, A., L'Homme, M.C., Drouin, P., Kraif, O.: Automating the compilation of specialized dictionaries: use and analysis of term extraction and lexical alignment. Terminology. Int. J. Theor. Applied Issues Spec. Commun. **16**(1), 77–106 (2010)
15. Lingpeng, Y., Donghong, J., Guodong, Z., Yu, N.: Improving retrieval effectiveness by using key terms in top retrieved documents. In: Losada, D.E., Fernández-Luna, J.M. (eds.) ECIR 2005. LNCS, vol. 3408, pp. 169–184. Springer, Heidelberg (2005). https://doi.org/10.1007/978-3-540-31865-1_13
16. Radford, A., et al.: Language models are unsupervised multitask learners. OpenAI Blog **1**(8), 9 (2019)
17. Rigouts Terryn, A., Hoste, V., Drouin, P., Lefever, E.: TermEval 2020: shared task on automatic term extraction using the annotated corpora for term extraction research (ACTER) dataset. In: 6th International Workshop on Computational Terminology (COMPUTERM 2020), pp. 85–94. European Language Resources Association (ELRA) (2020)
18. Tang, Y., et al.: Multilingual translation with extensible multilingual pretraining and finetuning. arXiv preprint arXiv:2008.00401 (2020)
19. Touvron, H., et al.: Llama 2: open foundation and fine-tuned chat models. arXiv preprint arXiv:2307.09288 (2023)
20. Tran, H.T.H., Martinc, M., Caporusso, J., Doucet, A., Pollak, S.: The recent advances in automatic term extraction: a survey. arXiv preprint arXiv:2301.06767 (2023)

21. Tran, H.T.H., Martinc, M., Doucet, A., Pollak, S.: Can cross-domain term extraction benefit from cross-lingual transfer? In: Pascal, P., Ienco, D. (eds.) DS 2022. LNCS, vol. 13601, pp. 363–378. Springer, Cham (2022). https://doi.org/10.1007/978-3-031-18840-4_26
22. Tran, H.T.H., Martinc, M., Pelicon, A., Doucet, A., Pollak, S.: Ensembling transformers for cross-domain automatic term extraction. In: Tseng, Y.H., Katsurai, M., Nguyen, H.N. (eds.) ICADL 2022. LNCS, vol. 13636, pp. 90–100. Springer, Cham (2022). https://doi.org/10.1007/978-3-031-21756-2_7
23. Tran, H.T.H., Martinc, M., Repar, A., Ljubešić, N., Doucet, A., Pollak, S.: Can cross-domain term extraction benefit from cross-lingual transfer and nested term labeling? Mach. Learn. 1–30 (2024)
24. Tran, H., Martinc, M., Doucet, A., Pollak, S.: A transformer-based sequence-labeling approach to the Slovenian cross-domain automatic term extraction. In: Slovenian Conference on Language Technologies and Digital Humanities (2022)
25. Vilar, D., Freitag, M., Cherry, C., Luo, J., Ratnakar, V., Foster, G.: Prompting palm for translation: assessing strategies and performance. arXiv preprint arXiv:2211.09102 (2022)
26. Zhang, Z., Gao, J., Ciravegna, F.: SemRe-rank: improving automatic term extraction by incorporating semantic relatedness with personalised PageRank. ACM Trans. Knowl. Discov. Data (TKDD) **12**(5), 1–41 (2018)

Bilingual Lexicon Induction From Comparable and Parallel Data: A Comparative Analysis

Michaela Denisová[1(✉)] and Pavel Rychlý[1,2(✉)]

[1] Natural Language Processing Centre, Masaryk University, Brno, Czech Republic
{x449884,pary}@fi.muni.cz
[2] Lexical Computing, Brno, Czech Republic

Abstract. Bilingual lexicon induction (BLI) from comparable data has become a common way of evaluating cross-lingual word embeddings (CWEs). These models have drawn much attention, mainly due to their availability for rare and low-resource language pairs. An alternative offers systems exploiting parallel data, such as popular neural machine translation systems (NMTSs), which are effective and yield state-of-the-art results. Despite the significant advancements in NMTSs, their effectiveness in the BLI task compared to the models using comparable data remains underexplored. In this paper, we provide a comparative study of the NMTS and CWE models evaluated on the BLI task and demonstrate the results across three diverse language pairs: distant (Estonian-English) and close (Estonian-Finnish) language pair and language pair with different scripts (Estonian-Russian). Our study reveals the differences, strengths, and limitations of both approaches. We show that while NMTSs achieve impressive results for languages with a great amount of training data available, CWEs emerge as a better option when faced less resources.

Keywords: Bilingual lexicon induction · Cross-lingual word embeddings · Neural machine translation systems

1 Introduction

Bilingual Lexicon Induction (BLI) is an intrinsic evaluation task focusing on retrieving translations of individual words. This task has been widely adopted for evaluating cross-lingual word embeddings (CWEs). The advantage of CWEs lies in their ability to align two sets of monolingual word embeddings (MWEs) into a shared cross-lingual space while exploiting comparable data and only a few or no bilingual supervision signals. [18]

Leveraging this property, they have proven to be useful in many natural language processing (NLP) applications, including machine translation [4,8], cross-lingual information retrieval [21], language acquisition and learning [22].

The BLI task from comparable data offers a promising alternative for low-resource or rare language pairs with insufficient parallel data. Traditionally, in lexicography, exploiting parallel data for retrieving translations has been a preferred method for many years. However, in the NLP field, while word-level extraction from parallel data was central during the era of statistical machine translation [13], the specific task of BLI has not received as much attention.

In NLP, neural machine translation systems (NMTSs) that utilise parallel data present another solution for retrieving translations. Although they have been proven effective for translating sentences or texts, yielding state-of-the-art results, their potential in the BLI task has not been fully explored yet, and to our knowledge, there are no experiments using NMTSs for the BLI task.

In this paper, we comparatively study the BLI task from comparable and parallel data. We select MARIANMT [20] to represent NMTSs using parallel data, and the three most cited state-of-the-art CWE methods using comparable data, i.e., MUSE [5], VECMAP [2,3], and RCLS [11]. We evaluate all models across three diverse language pairs: distant language pair (Estonian-English), close language pair (Estonian-Finnish), and language pair with different scripts (Estonian-Russian). Our motivation is to study the differences, similarities, strengths, and limitations of both approaches. Moreover, the discrepancy in training data volumes between these two approaches motivated us to understand how models perform under such different conditions on the same task. On top of that, we aim to investigate whether recent trends favouring comparable data for the BLI task can compete with standard, widely used parallel data.

Our contribution is threefold.

1. We provide a thorough comparison of the advantages and disadvantages of the BLI task from comparable and parallel data.
2. We comprehensively evaluate three CWE models and one NMTS across diverse and rare language pairs.
3. We make our code and datasets publicly available.[1]

This paper is structured as follows. In Sect. 2, we explain the background behind the BLI task from comparable and parallel data. In Sect. 3, we present the metrics, data and training details. In Sect. 4, we evaluate the baseline models exploiting comparable and parallel data and discuss the results. In Sect. 5, we offer concluding remarks.

2 Background

The objective behind the BLI task is to find the most suitable target word (or words) w_i^t for each source word w_i^s, given a list of P source words, where $P = \{w_1^s, w_2^s \ldots, w_n^s | n \in \mathbb{N}\}$. Afterwards, the output L, i.e., the list of the source and target words $L = \{(w_1^s, w_1^t), (w_2^s, w_2^t), \ldots, (w_l^s, w_l^t) | l \in \mathbb{N}\}$, is compared to a gold-standard evaluation dataset.

---
[1] https://github.com/x-mia/marianmt-bli.

To achieve this objective, various approaches are available, often leveraging the two most common data types: comparable and parallel data. In the following Subsects. 2.1 and 2.2, we outline the background of methods exploiting both data types, focusing on their advantages and limitations.

2.1 Comparable Data

Comparable data or comparable corpus consists of texts in two or more languages that share a common domain or were collected under identical conditions. It is characterised as non-aligned and, most importantly, similar in genre. Additionally, these texts can be similar in size. [14]

The advantage of the models using comparable data lies in their availability for low-resource or untypical language combinations. By contrast, parallel corpora also often skew the actual distributions of lexical items in the target language, artificially elevating the occurrence of frequent words and cognates while disproportionately diminishing the presence of other, potentially more natural equivalents. Additionally, the texts in the parallel corpora are typically limited to the legislative or public domain, while comparable corpus tends to be more diverse.

In NLP, the BLI task from comparable corpora typically evaluates CWE models, where the aim is to find the closest target word vector to the source word vector in the aligned cross-lingual space, usually by computing cosine similarity between the source and target word vectors. The seminal study by Mikolov et al. (2013) [16] introduced this trend to NLP, followed by a plethora of research papers ranging from most cited baseline methods [1,5,11], comprehensive evaluation studies and surveys [9,18] to recent experiments with dynamic embeddings [15]. In this paper, we demonstrate the results across three CWE models, which are cited as baseline models in many research papers: MUSE [5], VECMAP [2], and RCLS [11]. We selected these models as they are publicly available and straightforward to use, and the performance gap compared to newer methods is not substantial. On top of that, they are more accessible and computationally less demanding to train than NMTSs.

MUSE was released as a strong baseline model along with the evaluation and training datasets for over 110 languages. They employed a two-step process: adversarial training that develops a linear mapping between the source and target embedding spaces, challenging a discriminator to distinguish between them, and Procrustes refinement that optimises this mapping, leveraging a synthetic dictionary derived from the initial alignment. The introduction of the Cross-Domain Similarity Local Scaling (CSLS) metric aimed to address the high-dimensional space's hubness issue[2], significantly improving nearest-neighbour searches.

VECMAP presented a multi-step framework for learning bilingual word embeddings while generalising and refining a wide array of previous approaches.

---

[2] Hubness is an issue observed in high-dimensional space where some points are the nearest neighbours of many other points. [17].

Central to this framework is an orthogonal transformation, which allows for a detailed reinterpretation and improvement alongside additional steps such as normalisation, whitening, re-weighting, de-whitening, and dimensionality reduction. They also proposed a method in an unsupervised mode [3], relying on an unsupervised initialisation that exploits structural similarities between monolingual embeddings, coupled with a self-learning algorithm that iteratively refines the mappings.

RCLS aligned word embeddings from different languages by optimising the CSLS criterion, using convex relaxations for efficient optimisation, in contrast to traditional approaches that typically solve a quadratic problem. It incorporated unsupervised data to enhance alignment, addressing the hubness problem by ensuring consistency between the loss used in training and inference.

2.2 Parallel Data

The opposite of comparable data is parallel data or parallel corpora. It is a type of corpus that comprises two or more monolingual text collections that are aligned at the word, phrase, or sentence level. [14]

Exploiting parallel corpus for retrieving translations has been a preferred method mainly in lexicography, for instance, in the statistical-based method that computes the probabilities of word pair candidates based on their occurrences and co-occurrences presented in Kovář et al. (2016) [14]. In NLP, the parallel-data-based methods are often represented by NMTSs, which are not typically used for the BLI task. However, NMTSs could be used for compiling gold-standard dictionaries, as in the case of the widely used evaluation datasets MUSE for evaluating CWEs.

The main advantage of the parallel corpus is that it contains rich context information while offering many target word candidates and performing well for polysemous words and multi-word expressions in contrast to the CWE models, which focus on pure word-to-word alignment. Moreover, the NMTSs can translate words that were unseen in training data, whereas CWEs are limited to the vocabulary in MWEs.

In this work, we opted for NMTS called MARIANMT [20] for evaluation. MARIANMT was trained using the Marian C++ library[3] on OPUS parallel corpora[4] [19], which have various domains, such as subtitles, public texts, web texts, etc. It contains over 1,000 models, of which all are transformer encoders-decoders with six layers in each component. Additionally, it supports a wide diversity of languages and language combinations, including European, non-European, endangered languages, etc.

3 Experimental Setup

In this Section, we introduce four key aspects of this experiment: the evaluation and training datasets used, training details of the CWE models, the setup details

---

[3] https://marian-nmt.github.io/.
[4] https://opus.nlpl.eu/.

of MarianMT, and the evaluation metrics and procedures we employed during the evaluation.

3.1 Data

Training Data. To train CWEs, we utilised pre-trained fastText MWEs for English, Estonian, Finnish, and Russian. These were trained on Wikipedia with dimension 300 and contain over 9.2 billion words in English and under 10 million tokens for the other languages [10]. For supervised mode, we selected the training datasets MUSE [5] for Estonian-English. For Estonian-Finnish and Estonian-Russian, we compiled new training datasets by aligning Estonian-English with English-Finnish and English-Russian MUSE training datasets while using English as the pivot language. All training datasets contain 5K source words.

Regarding the training data for MarianMT, OPUS parallel corpora contain 115,564,910 sentences for Estonian-English, 42,353,565 sentences for Estonian-Finnish, and 29,699,112 sentences for Estonian-Russian.

Evaluation Data. In the evaluation part, we exploited the Estonian-English evaluation dataset MUSE. Since the evaluation datasets MUSE are often criticised for uneven part of speech distribution [12] and containing errors in translations [7], we included Estonian-English[5], Estonian-Finnish[6], and Estonian-Russian[7] dictionaries that were manually post-edited by lexicographers from the Institute of the Estonian Language (EKI). All of these dictionaries are published under a CC BY 4.0 Deed licence.[8]

3.2 Training Details of CWEs

For our comparison, we selected three state-of-the-art CWE methods, MUSE, VecMap (VM), and RCLS. All three models are trained in a supervised mode (MUSE-S, VM-S, RCLS), while only MUSE and VM in an unsupervised (MUSE-U, VM-U) mode and mode that relies on identical strings (MUSE-I, VM-I).

The default settings closely followed the MUSE training described in [5], RCLS setting in [11], and VM-S and VM-I in [2], and VM-U settings in [3]. The results are computed from the first 200K aligned embeddings.

3.3 MarianMT

We experimented with three pre-trained MarianMT models: Helsinki-NLP/opus-mt-et-en, Helsinki-NLP/opus-mt-et-fi, and Helsinki-NLP/opus-mt-et-ru, using Python programming language with PyTorch framework and HuggingFace library.[9]

---

[5] http://www.eki.ee/dict/ies/.
[6] http://www.eki.ee/dict/efi/.
[7] https://portaal.eki.ee/dict/evs/.
[8] https://creativecommons.org/licenses/by/4.0/.
[9] https://huggingface.co/.

We employed a series of parameters during the translation generation. We set the beam search to 20, disabled random sampling, specified to return ten target words, restricted the maximum number of new tokens that the model can generate in the response to 10, and enabled the output of scores.

3.4 Metrics

The most common evaluation metric in the BLI task is precision@k where k represents the number of target words retrieved for a single source word. In this paper, we report, in addition to precision, also recall and F1 scores using fixed and dynamic k.

We calculate the precision (P) as the ratio of the positive target words to the number of all target words that the model found (positive and negative). The recall (R) and F1 score representing the balance between precision and recall are computed using the standard formula.

In the case of the fixed k, we set it to 1, i.e., we report P@1, R@1, and F1@1. When evaluating CWEs using dynamic k, instead of limiting the retrieved target words based on top-k nearest neighbours, we restrict cosine similarity scores with the following formula adopted from Denisová (2022) [6]:

$$limit = S_C(x_i^s, x_j^t) + j * 0.01,$$

where $S_C(x_i^s, x_i^t)$ represents cosine similarity between the source word vector x_i^s and target word vector x_i^t, and j denotes the position of the target word, i.e., the target word with the closest target word vector has a position 0, etc. The value of $S_C(x_i^s, x_j^t)$ was adjusted for each model and language pair individually.

When evaluating MARIANMT using dynamic k, we retrieved scores stored in the model to determine the reliability of each target word candidate. Then, we excluded each target word candidate with a score < 0.05.

4 Evaluation

Overall results for Estonian-English (et-en) are displayed in Tables 1 and 2, where Table 1 presents dynamic k and Table 2 fixed k, both using two different evaluation datasets. General results for Estonian-Finnish (et-fi) are outlined in Table 4 and for Estonian-Russian (et-ru) in Table 5.

When looking at Tables 1 and 2, MARIANMT outperformed CWE models in almost all metrics measured across Estonian-English language pair by a margin approximately ranging from 3% to 51%. Generally, the results for EKI evaluation dataset were worse than MUSE by around up to 20%.

We examined some examples from both evaluation datasets, and the reason behind such a decrease in performance is that the MUSE dataset is polluted by English to English equivalents, such as *act - act, ever - ever, girls - girls*, etc. Moreover, it contains a lot of English proper nouns which are identical in both languages, e.g., *adelaide - adelaide, hannah - hannah, selma - selma*, etc. We naturally get better outcomes when generating the target words in English for

Table 1. The results for the Estonian-English language pair evaluated using dynamic k.

et-en (%)	MUSE dataset			EKI dataset		
	P	R	F1	P	R	F1
MARIANMT	**50.39**	49.64	**50.01**	**32.17**	29.70	**30.89**
MUSE-S	17.42	45.02	25.13	9.78	32.85	15.07
MUSE-I	17.60	38.75	24.20	9.17	23.43	13.18
MUSE-U	0.00	0.00	0.00	0.00	0.00	0.00
VM-S	21.60	50.36	30.23	8.31	**40.11**	13.77
VM-I	17.67	50.96	26.24	6.90	37.29	11.65
VM-U	15.15	46.95	22.91	6.74	36.84	11.40
RCLS	20.84	**53.82**	30.04	19.12	30.79	23.58

Table 2. The results for the Estonian-English language pair evaluated using fixed $k = 1$.

et-en (%)	MUSE dataset			EKI dataset		
	P@1	R@1	F1@1	P@1	R@1	F1@1
MARIANMT	**56.33**	**46.51**	**50.95**	**33.85**	**33.83**	**33.84**
MUSE-S	40.33	33.30	36.48	25.11	23.49	24.28
MUSE-I	36.80	30.38	33.28	19.42	18.17	18.77
MUSE-U	0.00	0.00	0.00	0.00	0.00	0.00
VM-S	49.20	40.62	44.50	28.60	26.76	27.65
VM-I	49.07	40.51	44.38	24.26	22.70	23.46
VM-U	42.13	34.78	38.11	23.75	22.23	22.96
RCLS	45.53	37.59	41.18	30.74	28.76	29.71

an English word. The same applies to fastText embeddings that are often noisy and contain English words.

Furthermore, we compared examples from MARIANMT and RCLS models evaluated with the Estonian-English EKI evaluation dataset. Table 3 exemplifies the main findings. The performance of the model RCLS was worse than that of the model MARIANMT, which confirmed the examination of their outputs. The main error which we discovered in RCLS model was that it aligned mainly word pairs with similar lexical-semantic relationships instead of translations. In MARIANMT, the errors were various. Firstly, it often generated a target word with a capital letter (Type A), but the evaluation datasets were lowercase. The model appended extra numbers or punctuation to some target words (Types A and E) and, in some cases, generated complete sentences (Type C). Additionally, some target word candidates exhibited part-of-speech mismatches with the source word, for instance, a verb (jutustama) was translated as an adjective (narrated) (Type B). Finally, the model demonstrated the capability

Table 3. Examples of the source (SRC) and target (TGT) words from the Estonian-English evaluation dataset EKI compared to the output from MarianMT and RCLS models. The target words in bold are correct.

Type	SRC	TGT	MarianMT	RCLS	Explanation
A	aasta	year	Year 4 Year 3 Year **year**	**year** month summer autumn	capital letter, numbers
	mänguasi	toy	Toy **toy**	game play boardgame	
B	jutustama	narrate	**narrate** narrated	tale story tell	part-of-speech mismatch
C	aluspüksid	panties	Terry towelling and similar woven terry fabrics	trousers pants shirts	nonsense/ sentence
	loodetavasti	hopefully	I hope so. Hopefully.	probably possibly hope greatly	
D	ristsõna	crossword	crossword puzzles crossword word Crossword Puzzle	-	multi-word expression
E	au	honor	(au) - (au)	**honor** honour ain	punctuacion, symbols, English
	hõlmama	encompass	cover:	covers covering **encompass**	

to handle multi-word expressions (Type D), offering a distinct advantage over CWE models that typically map single-word units to corresponding single-word units.

On the other hand, Tables 4 and 5 indicate a significant decrease in MarianMT performance. Although MarianMT still surpassed nearly all CWE mod-

Table 4. The results for the Estonian-Finnish language pair evaluated using dynamic and fixed $k = 1$.

et-fi (%)	P	R	F1	P@1	R@1	F1@1
MarianMT	19.40	**21.21**	**20.27**	**50.57**	**13.62**	**21.46**
Muse-S	17.74	11.25	13.77	37.57	10.12	15.95
Muse-I	16.11	11.38	13.34	36.22	9.76	15.37
Muse-U	14.70	12.03	13.24	36.93	9.95	15.67
VM-S	15.43	15.80	15.62	38.28	10.31	16.25
VM-I	13.65	16.59	14.97	42.33	11.40	17.97
VM-U	13.66	16.59	14.98	42.26	11.38	17.94
RCLS	**28.76**	12.11	17.04	39.49	10.64	16.76

Table 5. The results for the Estonian-Russian language pair evaluated using dynamic and fixed $k = 1$.

et-ru (%)	P	R	F1	P@1	R@1	F1@1
MarianMT	9.97	4.05	5.76	17.73	2.91	5.01
Muse-S	22.77	6.40	10.0	35.17	5.79	9.94
Muse-I	21.92	5.51	8.81	31.98	5.26	9.04
Muse-U	0.00	0.00	0.00	0.00	0.00	0.00
VM-S	18.54	10.04	13.03	35.96	5.92	10.16
VM-I	18.04	11.09	13.73	39.58	6.51	11.19
VM-U	19.21	**11.93**	**14.72**	**44.21**	**7.28**	**12.5**
RCLS	**30.59**	8.29	13.04	35.89	5.90	10.14

els in the Estonian-Finnish evaluation by a margin of around 3% to 14%, it did not perform as well as the CWE model VM-U in the Estonian-Russian evaluation, where VM-U achieved the best performance.

The reason behind this is twofold. Firstly, the Estonian-Russian evaluation dataset contains a lot of target word variants for each source word, which influences the result, especially for models that are not performing well for polysemous words. Table 6 shows the number of target words in each dataset. We can observe that both Estonian-English evaluation datasets contain a lot of targets with one equivalent, whereas Estonian-Finnish and Estonian-Russian are more spread out. This is consistent with the performance of our models, i.e., the recall is high for Estonian-English but significantly decreases for Estonian-Finnish and Estonian-Russian.

NMTSs are known for generating only one output, which is reflected in the recall performance. Some models, including MarianMT, are able to offer more than one output, but access to the model is necessary. Table 7 displays a few examples from the models MarianMT and VM-U trained across Estonian-Russian. It can be observed that the MarianMT model typically generates

Table 6. The number of target words (TGW) in the evaluation datasets. [1] MUSE dataset. [2] EKI dataset.

TGW	et-en[1]	et-en[2]	et-fi	et-ru
1	1240	3150	381	343
2	211	2	275	230
3	43	0	226	186
4	4	0	152	160
5	2	0	118	93
6+	0	0	256	488

fewer target words, and a majority of these words include symbols, punctuation, and capital letters, i.e., the output is in the form of a sentence since the NMTSs are trained to translate sentences and not words in isolation.

Secondly, the amount of training data available in parallel corpus OPUS is larger and of better quality for Estonian-English than for Estonian-Finnish and Estonian-Russian. The Estonian-English corpora include over 115,500,000 sentences sourced from top-tier resources such as ParaCrawl, Europarl, DGT, and open subtitles. In contrast, the Estonian-Finnish corpora comprise over 42,300,000 sentences from similar high-quality sources like MultiParaCrawl,

Table 7. Examples of the source (SRC) and target (TGT) words from the Estonian-Russian evaluation dataset compared to the output from MARIANMT and VM-U models. The target words in bold are correct.

SRC	TGT	MARIANMT	VM-U	Explanation
inglane	англичанин англичанка британец бритт	Англичанин. Англичанин	**англичанин** американец шотландец **британец** француз	capital letters, sentence-form of the output
ületama	переходить проходить пересекать переезжать переправляться преодолевать превышать превосходить перекрывать	Преодолеть	**преодолевать** пересекать доходить подниматься **переправляться**	capital letters
patarei	батарейка куча батарейный батареи	& Батарея Батарея... Батарея:	**батарея** **батареи** батарее **батарейный**	symbols, punctuation

Europarl, DGT, and open subtitles. However, the Estonian-Russian corpora contain approximately 29,700,000 sentences, primarily from open subtitles and lower-quality sources like KDE4. Although over 29,000,000 sentences of training data for a language pair are not considered under-resourced, it has a major impact on the resulting quality of the translations. In the lower-data scenario and despite the training data volume discrepancy, the CWE models yield better results and prove to be a good supplement to the NMTSs.

5 Conclusion

In this paper, we have conducted a comparative analysis of the BLI task from comparable and parallel data. We have thoroughly discussed both data types and compared their advantages and limitations. From each group, we have selected models representing the specific data type, i.e., popular CWE models MUSE, VECMAP, and RCLS for the BLI task from comparable data, and NMTS MARIANMT for the parallel data. We have evaluated these models across three diverse language pairs: distant (Estonian-English), close (Estonian-Finnish), and language pair with different scripts (Estonian-Russian), and we have analysed the results rigorously.

In conclusion, although NMTSs are still a competition to the CWE models due to their ability to capture context and handle multi-word expressions, their outcomes heavily depend on the amount of training data available. The CWE models represent a good alternative or can serve as a supplement data, especially for languages with fewer resources or when recall is favour over precision.

References

1. Artetxe, M., Labaka, G., Agirre, E.: Learning principled bilingual mappings of word embeddings while preserving monolingual invariance. In: Proceedings of the 2016 Conference on Empirical Methods in Natural Language Processing, pp. 2289–2294. Association for Computational Linguistics (2016). https://doi.org/10.18653/v1/D16-1250
2. Artetxe, M., Labaka, G., Agirre, E.: Generalizing and improving bilingual word embedding mappings with a multi-step framework of linear transformations. In: Proceedings of the Thirty-Second AAAI Conference on Artificial Intelligence, pp. 5012–5019 (2018). https://doi.org/10.1609/aaai.v32i1.11992
3. Artetxe, M., Labaka, G., Agirre, E.: A robust self-learning method for fully unsupervised cross-lingual mappings of word embeddings. In: Proceedings of the 56th Annual Meeting of the Association for Computational Linguistics (Volume 1: Long Papers), pp. 789–798. Association for Computational Linguistics (2018). https://doi.org/10.18653/v1/P18-1073
4. Artetxe, M., Labaka, G., Agirre, E.: Unsupervised statistical machine translation. In: Proceedings of the 2018 Conference on Empirical Methods in Natural Language Processing, pp. 3632–3642. Association for Computational Linguistics (2018). https://doi.org/10.18653/v1/D18-1399

5. Conneau, A., Lample, G., Ranzato, M., Denoyer, L., J'egou, H.: Word translation without parallel data. arXiv **abs/1710.04087** (2017). https://doi.org/10.48550/arXiv.1710.04087
6. Denisová, M.: Parallel, or comparable? That is the question: the comparison of parallel and comparable data-based methods for bilingual lexicon induction. In: Proceedings of the Sixteenth Workshop on Recent Advances in Slavonic Natural Languages Processing. RASLAN 2022, pp. 3–13. Tribun EU (2022)
7. Denisová, M., Rychlý, P.: When word pairs matter: analysis of the English-Slovak evaluation dataset. In: Recent Advances in Slavonic Natural Language Processing (RASLAN 2021), pp. 141–149. Tribun EU, Brno (2021)
8. Duan, X., et al.: Bilingual dictionary based neural machine translation without using parallel sentences. In: Proceedings of the 58th Annual Meeting of the Association for Computational Linguistics, pp. 1570–1579. Association for Computational Linguistics (2020). https://doi.org/10.18653/v1/2020.acl-main.143
9. Glavaš, G., Litschko, R., Ruder, S., Vulić, I.: How to (properly) evaluate cross-lingual word embeddings: on strong baselines, comparative analyses, and some misconceptions. In: Korhonen, A., Traum, D., Màrquez, L. (eds.) Proceedings of the 57th Annual Meeting of the Association for Computational Linguistics, pp. 710–721. Association for Computational Linguistics (2019). https://doi.org/10.18653/v1/P19-1070
10. Grave, E., Bojanowski, P., Gupta, P., Joulin, A., Mikolov, T.: Learning word vectors for 157 languages. In: Proceedings of the International Conference on Language Resources and Evaluation (LREC 2018) (2018)
11. Joulin, A., Bojanowski, P., Mikolov, T., Jégou, H., Grave, E.: Loss in translation: learning bilingual word mapping with a retrieval criterion. In: Proceedings of the 2018 Conference on Empirical Methods in Natural Language Processing, pp. 2979–2984. Association for Computational Linguistics (2018). https://doi.org/10.18653/v1/D18-1330
12. Kementchedjhieva, Y., Hartmann, M., Søgaard, A.: Lost in evaluation: misleading benchmarks for bilingual dictionary induction. In: Proceedings of the 2019 Conference on Empirical Methods in Natural Language Processing and the 9th International Joint Conference on Natural Language Processing (EMNLP-IJCNLP), pp. 3336–3341. Association for Computational Linguistics (2019). https://doi.org/10.18653/v1/D19-1328
13. Koehn, P., Knight, K.: Learning a translation lexicon from monolingual corpora. In: Proceedings of the ACL-02 Workshop on Unsupervised Lexical Acquisition, pp. 9–16. Association for Computational Linguistics (2002). https://doi.org/10.3115/1118627.1118629
14. Kovář, V., Baisa, V., Jakubíček, M.: Sketch engine for bilingual lexicography. Int. J. Lexicogr. **29**(3), 339–352 (2016). https://doi.org/10.1093/ijl/ecw029
15. Li, Y., Korhonen, A., Vulić, I.: On bilingual lexicon induction with large language models. In: Bouamor, H., Pino, J., Bali, K. (eds.) Proceedings of the 2023 Conference on Empirical Methods in Natural Language Processing, pp. 9577–9599. Association for Computational Linguistics (2023). https://doi.org/10.18653/v1/2023.emnlp-main.595
16. Mikolov, T., Le, Q.V., Sutskever, I.: Exploiting similarities among languages for machine translation. arXiv preprint arXiv:1309.4168 (2013)
17. Radovanović, M., Nanopoulos, A., Ivanović, M.: Hubs in space: popular nearest neighbors in high-dimensional data. J. Mach. Learn. Res. **11**, 2487–2531 (2010). https://doi.org/10.5555/1756006.1953015

18. Ruder, S., Vulić, I., Søgaard, A.: A survey of cross-lingual word embedding models. J. Artif. Intell. Res. **65**, 569–631 (2019). https://doi.org/10.1613/jair.1.11640
19. Tiedemann, J.: News from OPUS - A Collection of Multilingual Parallel Corpora with Tools and Interfaces, vol. V, pp. 237–248. Recent Advances in Natural Language Processing (2009)
20. Tiedemann, J., Thottingal, S.: OPUS-MT – building open translation services for the world. In: Proceedings of the 22nd Annual Conference of the European Association for Machine Translation, pp. 479–480. European Association for Machine Translation, Lisboa, Portugal (2020)
21. Vulić, I., Moens, M.F.: Monolingual and cross-lingual information retrieval models based on (bilingual) word embeddings. Proceedings of the 38th International ACM SIGIR Conference on Research and Development in Information Retrieval, pp. 363–372 (2015). https://doi.org/10.1145/2766462.2767752
22. Yuan, M., Zhang, M., Van Durme, B., Findlater, L., Boyd-Graber, J.: Interactive refinement of cross-lingual word embeddings. In: Proceedings of the 2020 Conference on Empirical Methods in Natural Language Processing (EMNLP), pp. 5984–5996. Association for Computational Linguistics (2020). https://doi.org/10.18653/v1/2020.emnlp-main.482

Explaining Metaphors in the French Language by Solving Analogies Using a Knowledge Graph

Jérémie Roux(✉), Hani Guenoune, Mathieu Lafourcade, and Richard Moot

LIRMM, Univ Montpellier, CNRS, Montpellier, France
{jeremie.roux,hani.guenoune,mathieu.lafourcade,richard.moot}@lirmm.fr

Abstract. An analogy is a relation which operates between two pairs of terms representing two distant domains. It operates by transferring meaning from a concept that is known to another that one would like to clarify or define. In this report, we address analogy both from the aspect of modeling and by automatically explaining it. We will then propose a system of resolution of analogical equations in their notation in symbol chains. The model, based on the common sense knowledge base *JeuxDeMots* (a semantic network), operates by generating a list of potential candidates from which it chooses the most suitable solution. We conclude by evaluating our model on a collection of equations, and reflecting upon future work.

Keywords: Analogy · Metaphor · Figurative language · Natural language processing · Knowledge base

1 Introduction and State of the Art

From more or less complex ideas and reasoning, eloquent and persuasive expressions can emerge in a non-trivial way, carrying clear or nuanced meanings. Language production and understanding are accepted as faculties specific to humans, being capable of high-level semantic interpretation. A speaker seems to refer, subconsciously and effortlessly, to a complex and hierarchical apparatus built from his knowledge of the world. We seek here to "achieve [...] the formalization of an operation that everyone recognizes as being at work in language" [8] with the aim of automatizing it. Our main challenge is to carry out such refined semantic reasoning automatically.

We can describe analogy as a fundamental way of expression used in languages across the world [2]. Beyond this universality, the omnipresence of analogies and metaphors in written language justifies the interest of further research in the field. Manual annotation of metaphorical figures from the British National Corpus revealed that 241 out of 761 sentences contained this type of language [9]. The study, although marked by an idiomatic dimension specific to the English language, represents an indicator of the prevalence of analogies in natural language. However, these figures cannot *a priori* be generalized to other languages.

It is important to underline the central character of the operation of analogy in the human cognitive apparatus, whether in its fundamental natural operations, or in more complex and methodical demonstrations and reasoning (*system 1/system 2* of Kahneman [5]). We are used to approaching the resolution of a difficult or new problem by trying to reduce it to another for which the solution is known. Hofstadter [3] [4] believes that thought and analogy are inseparable, he argues that analogy is the core of the cognitive functioning of human beings, and that every problem we encounter is nothing other than an assembly of analogies that we navigate more or less fluidly through our reasoning. He thus bases his theories on this hypothesis of a close connection between reasoning and analogy.

In this perspective, the interpretation of analogies represents a crucial step forward in the development of work in Natural Language Processing (NLP). In the ambitious perspective of identifying and then analyzing this mechanism, it is crucial to scrutinize the subtleties of language, and to analyze their functioning from cognitive and linguistic points of view. A large part of the work in NLP still focuses on elementary, or first level, linguistic tasks (morpho-syntactic labeling, syntactic analysis, coreference resolution, recognition of named entities, etc.), while another part of the research aims to improve automatic inference mechanisms and the extraction of new knowledge from textual corpora. Ultimately, even less work focuses on bringing together the lessons and progress provided by each of these scientific directions, in order to get closer to human linguistic capacity and thus simulate high-level linguistic reasoning such as the understanding of expressions in all their creativity.

In this work, we present a prototype for interpreting analogies by exploiting a common sense knowledge base in the form of a graph, the *JeuxDeMots* network (*JDM*) [7]. We will formalize statements in analogical squares (see Sect. 2) and present methods for evaluating the quality of these analogies (see Sect. 3). For an analogy, it is a question of bringing out correspondences between terms which are provided to the system by the user. The demonstrator of a first proof of concept associated with this article is available at the following address: https://analogie.demo.lirmm.fr.

2 Analogy Square

An analogy (when formalized in an analogy square) is a set of 4 terms linked by similarity relationships (see Fig. 1). The strength of an analogy comes from the similarity of the terms that compose it, its *explainability* is based on the multiple relationships between these terms.

2.1 Analogy and Similarity

Generally speaking, figurative speech aims to integrate, into the description of one concept, attributes of a second, chosen on the basis of its semantic similarity

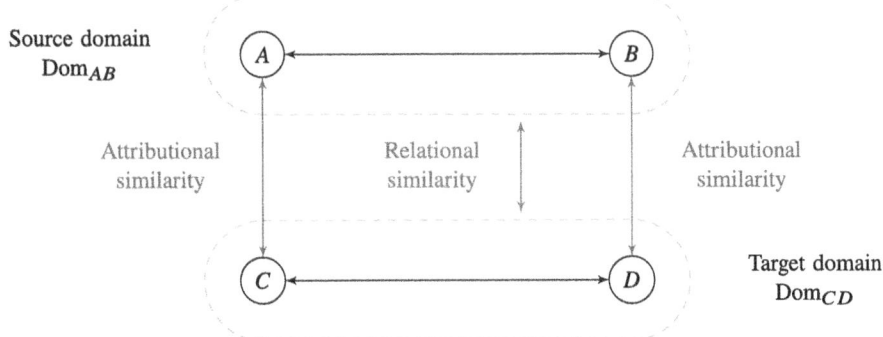

Fig. 1. Simplified diagram of the analogy square (example: A=eye, B=sight, C=hand, D=touch)

with the concept described. Here, we tackle the distinction between the two notions of similarity emanating from the work of Gentner [1], who argues that there are at least two types of similarities:

- **Relational similarity**, which consists of the correspondence between the relations of two pairs of concepts.
- **Attributional similarity**, which is the correspondence between the attributes of two concepts.

The notions of attribute and relation are accepted in the sense of first-order logic, where an attribute is a single variable predicate, while a relation is a predicate with two variables. We qualify two terms, each designating a concept, as synonyms, when their **attributional similarity** is sufficiently high, while we designate two pairs of terms as analogous if their **relational similarity** is high [10].

We can, following this relational similarity, argue that it would be possible to generate a correspondence (analogy) $A \rightarrow B$ transferring knowledge from a so-called source concept A to a target concept B. The source concept is generally abstract, uncertain or unknown, the one on which we wish to establish an easy to understand and concrete target. Here are examples of pairs of concepts constituting analogies:

$$(confidence \text{ and } success) \text{ with } (sun \text{ and } flower) \quad \text{abstract-concrete} \quad (1)$$
$$(electron \text{ and } plasma) \text{ with } (person \text{ and } crowd) \quad \text{unknown-known} \quad (2)$$
$$(carpenter \text{ and } wood) \text{ with } (mason \text{ and } stone) \quad (3)$$

The analogy (3) taken from the work of Turney and Pantel [11] could be formulated: "the *carpenter* is to *wood* what the *mason* is to *stone*". The meanings of the relationships between the concepts *mason* and *carpenter* respectively with *stone* and *wood* are indeed similar. On the one hand, these are professions, and on the other hand, materials very closely linked to these respective professions. The introduction of a characteristic specific to the distant concept (*source*) therefore makes available to the speaker all the contextual knowledge of the concept and operates, in the case of an explanatory speech or reasoning, as a familiar and edifying support, or, in the case of a poetic intention, as an evocative agent helping to color the language and offering a complementary lyrical richness.

2.2 Symbol Strings and Analogy Equation

When it comes to the notion of equality of relationships (proportionality), between two pairs of terms (A, B) and (C, D), the statement (3) can be written more concisely with a symbol chain notation "$A : B :: C : D$". The notation illustrated by the example of the proportional analogy between the pairs (carpenter, wood) and (mason, stone) previously mentioned then becomes (4).

$$carpenter : wood :: mason : stone \qquad (4)$$

The operator ":" indicates the existence of relational relationships between its operands, in this case the terms *carpenter* and *wood*. The "::" operator then transfers this relationship **from the *source* to the *target*** by asserting the existence of a relationship of the same semantic nature between the terms of the second pair (terms *mason* and *stone*). This notation is adopted for its clear separation between the *source* (*carpenter* and/or *wood*) and the *target* (*mason* and/or *stone*) in an analogy.

2.3 Metaphors and Comparisons: Gap Analogies

In the context of Aristotelian analogies, we find the case where one (or more) of the four symbols is missing, raising what is called an analogical equation of the form "$A : B :: C :?$". The interpretation of the analogy comes down to solving this equation and consists of deducing the possible values of the missing term(s). In a desire for formalism, similar to King and Gentner [6], we start from the postulate that metaphors and comparisons can be formalized as manifestations of gap analogies (analogical equations with 1 or 2 unknowns). Figure 2 illustrates the different cases observed.

3 Strength of an Analogy and Election of Candidates

Let us now see the methods proposed for the evaluation of similarities with a view to electing the best candidate term(s) for resolving a gap analogy.

3.1 Relationships Between Words

We denote "$a\ r\_t\ b$", the relation r of type t from a to b and its weight "$p(a, r\_t, b)$". In our knowledge base, we have information about a term (or node in the context of a knowledge graph): its relationships with other words.

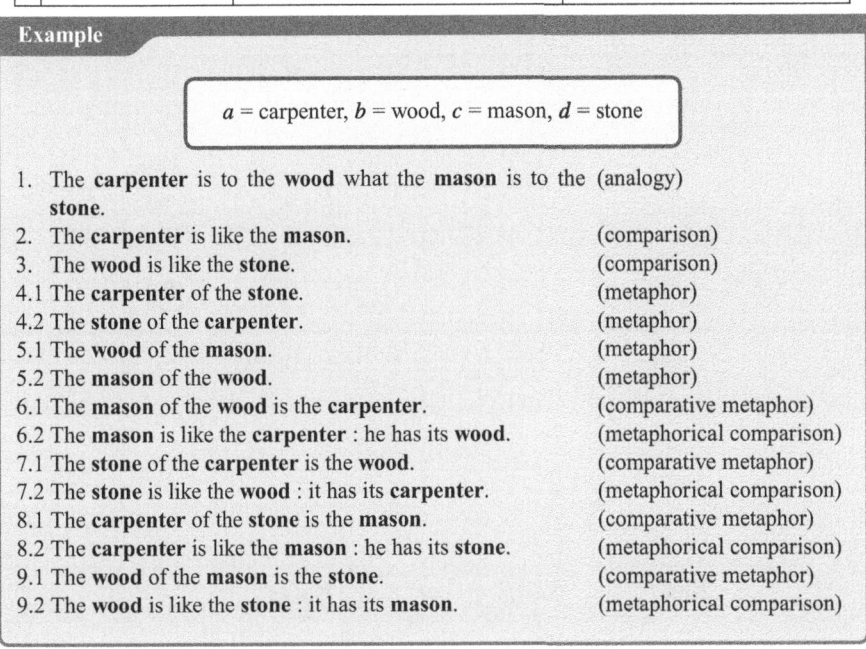

Fig. 2. Behavior observed depending on the positioning of unknowns in an analogy

We exploit the types of relationships present in JDM^1 and we limit ourselves to those mainly relating to semantics: $r\_associated$, r_domain, r_isa, r_anto, r_hypo, r_has_part, r_holo, r_agent, $r\_patient$, r_lieu, r_instr, r_carac^2.

We can group the relationships between terms/nodes A, B, as presented in Fig. 3, depending on whether they are direct or indirect (with iAB nodes) as well as oriented in one direction or the other. For a given relationship, it is possible to calculate a normalized weight p_{norm}: the ratio of the weight of the relationship

---

[1] https://www.jeuxdemots.org/jdm-about-detail-relations.php.
[2] Some relations are conversive, that is to say that $a \, r\_t \, b \Leftrightarrow b \, r\_t_{-1} \, a$ has with r_t_{-1} the conversive relation to r_t (example: r_isa and r_hypo).

$a\ r\_t\ b$ by the maximum weight of all relationships of type t from node a. The p_{norm} value makes it possible to classify the relationships of a node and to select those considered to be the most strongly associated, i.e. the first[3] n relationships with the best normalized weights with regard to the type t of the relationship. This allows us to control the combinational cost when calculating the strength of similarities (see Sect. 3.2). The detailed formula for the normalized weight for fixed a, r_t and b is defined in (5).

$$p\_norm(a,r,b) = \frac{p(a,r\_t,b)}{max\_val\_for\_type(a,t)} \qquad (5)$$

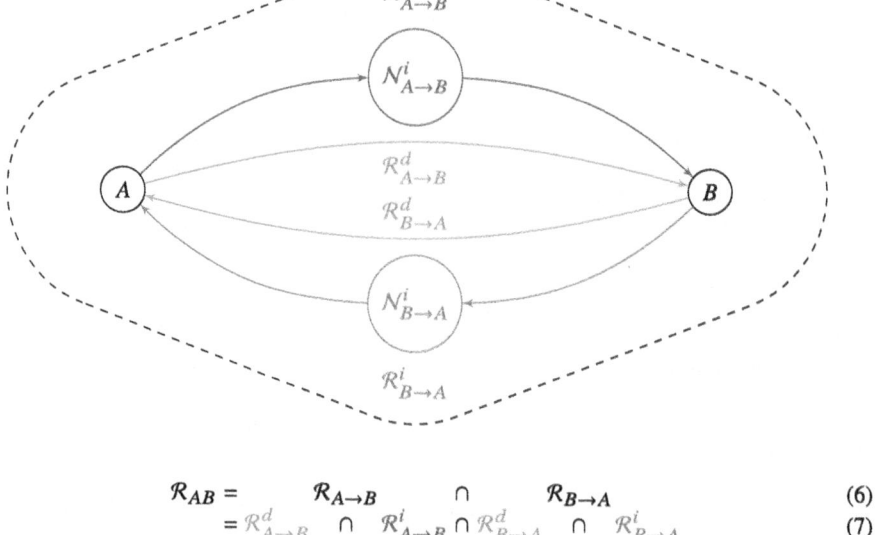

$$\mathcal{R}_{AB} = \qquad \mathcal{R}_{A \to B} \qquad \cap \qquad \mathcal{R}_{B \to A} \qquad (6)$$
$$= \mathcal{R}^d_{A \to B} \quad \cap \quad \mathcal{R}^i_{A \to B} \cap \mathcal{R}^d_{B \to A} \quad \cap \quad \mathcal{R}^i_{B \to A} \qquad (7)$$

Fig. 3. Notation of sets of direct and indirect relationships between A and B

3.2 Relational Similarity

When analyzing an analogy $A : B :: C : D$, we want to recover all the direct and indirect relationships between A and B (in both directions) as well as between C and D (in both directions). The goal is then to perform an intersection between these types of relationships between A and B on the one hand and C and D on the other hand in order to evaluate the strength of the relational similarity.

---

[3] We have arbitrarily chosen in our demonstrator the 2 most relevant relationships (if there are any) for reasons of simplification. Note that the intermediate node may be the same for these 2 relationships.

In the example presented in Fig. 4, we see that we have the same types of relationships between *electron* and *nucleus/stone*[4] as between *planet* and *star*. In this case, the relationships are indirect, because no direct relations of the same type were found. We establish the strength of the relational similarity observed when exploring from A to B[5] by averaging the normalized weights of the relations from A to B combined with those from A to the intermediate nodes iAB. For the relations from B to A, we operate the same way. In practice, the results are satisfactory in the sense that the values obtained correspond to intuition, with the values approaching 1 for a similarity considered strong. Similarly, the similarity is considered weak when the calculated value approaches 0. It is equal to 0 when no relationship (direct or indirect) is observed. Note that this measurement is taken as a baseline and can be refined to improve precision.

We also implement some weighting according to the type of relationship, which allows, for example, to consider a relationship of the type r_isa, r_lieu or r_has_part as more important with regard to its semantic contribution, relative to an $r\_associated$ relationship which is more vague[6]. We hypothesize that establishing a precise link between electron and nucleus then finding this same link on the other side of the analogy between planet and star (with other intermediate nodes) makes it possible to ensure the strength of this **relational similarity** more than if it were a simple relation of association of ideas or even if one did not exist.

3.3 Attributional Similarity

We now analyze the relationships of the same types between each of the terms A, B and a certain intermediate node iAB. The desired pattern therefore corresponds to $A\ r\_t\ iAB$ and $B\ r\_t\ iAB$. We proceed in the same way and independently with C and D. In Fig. 4, we see that there is no such intermediate node between electron and planet in *JDM*, the **attributional similarity** is therefore equal to 0. This is not enough to assert that electron and planet have nothing to do with each other, this rather means an absence of attributional similarity from the point of view of the knowledge base in its current state. On the other hand, we note between *nucleus/stone* and *star* the presence of the *swallow* node via a relationship $r\_patient_{-1}$, undoubtedly because a star can for example "be swallowed by a black hole" and the *stone* of a fruit can "be swallowed by a child" (*stone* and *nucleus* in french (*noyau*) being homonyms). A human speaker will immediately notice that two different meanings of the word *noyau* are involved. It will therefore be necessary to carry out semantic refinements by

---

[4] The word 'noyau' in French has a translation of 'stone' as in the hard core of stone fruits.
[5] Note that the strength of relational similarity from A to B is not necessarily the same as that from B to A since there are relationships oriented in both directions and of different weights.
[6] In the future, we will be able to use a *TF-IDF* type approach which consists of seeing how an ArB is as specific as possible to A, such an approach would however be computationally intensive.

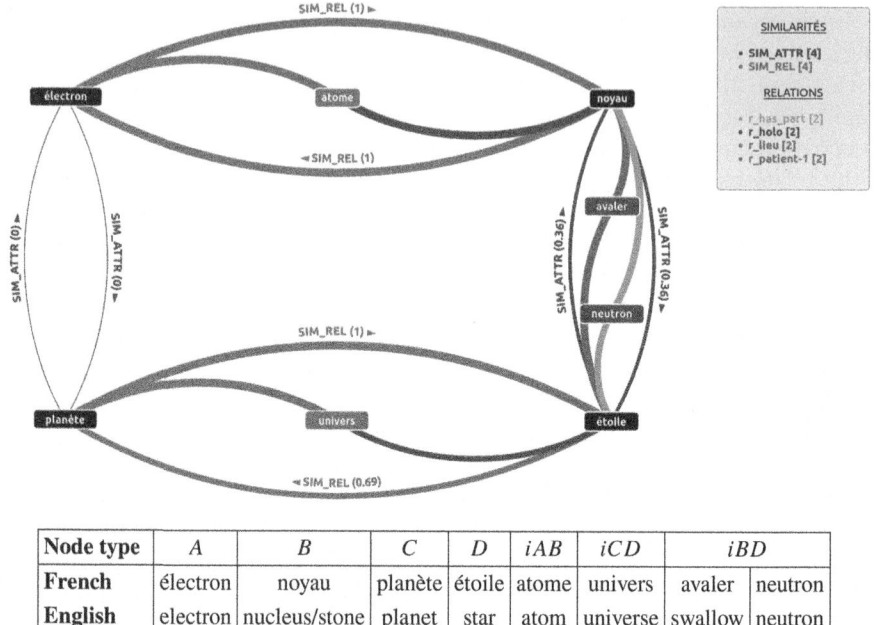

Node type	A	B	C	D	iAB	iCD	iBD	
French	électron	noyau	planète	étoile	atome	univers	avaler	neutron
English translation	electron	nucleus/stone	planet	star	atom	universe	swallow	neutron

Fig. 4. Execution of our demonstrator (https://analogie.demo.lirmm.fr) on the analogy square *electron : nucleus :: planet : star*

distinguishing the different meanings of each word to obtain better results; *JDM* is equipped to handle this type of word sense disambiguation.

3.4 Overall Strength and Election of Candidates in the Case of Metaphor

An analogy is considered more meaningful when the relational and attributional similarities it contains are sufficient, or from a statistical point of view if their score is high. By taking an average[7] between all of these similarities we obtain a value which we will consider as a strength of association in the form of a probability. This score takes on its meaning when it comes to classifying different candidates in the case of a gap analogy[8]. We can therefore evaluate the relevance of these candidates by calculating the score for each of them with the known nodes of the analogy with which they have sufficiently high similarity ratios. Take for example the hole analogy *leg : knee :: arm : ?*, the proposed candidate is *elbow* which corresponds to a response that comes naturally. In our implementation,

---

[7] Arithmetic and geometric means produce similar results.
[8] We have only discussed metaphors so far, comparisons being a broader subject given that their two unknowns in the context of the analogical square constitute a combinatorial challenge.

the relations $a\ r\_t\ b$ displayed and currently used for calculating the weight are the first 2 in order of decreasing weight for each pair (a, b); this makes the calculation times reasonable and gives satisfactory results for a proof of concept.

3.5 Evaluation of Analogy Equation Resolution

We have a list of analogies with a distribution of responses provided by *JDM* players (control)[9]. We check if the top 4 predicted candidates include the best control term (see Fig. 5). This preliminary work results in an accuracy of around 37%. A perspective for improvement consists, firstly, of taking into account a more covering set of semantic relationships, and then calibrating the methods for calculating similarities and their aggregations. It will be necessary to put in place more in-depth procedures for exploring the knowledge base such as inference and reasoning mechanisms. It is also possible to carry out data preparation, such as morpho-syntactic normalization[10] or even semantic disambiguation.

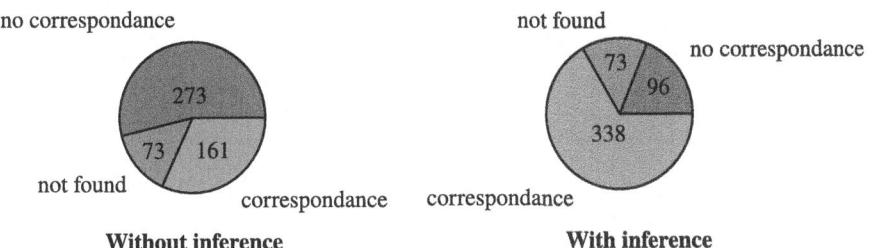

Fig. 5. Evaluation of the relevance of the candidates (first 4) compared with the most played control in *JDM* for the same metaphor (on a corpus of 507 metaphors) with and without inference

4 Synthesis

We consider the Aristotelian analogy in its symbol chain notation **A : B :: C : D**, which means "*A* **is to** *B* **what** *C* **is to** *D*" and we say that there is (see Fig. 6):

---

[9] By simulating the intersections for a list of just over 2000 metaphors played in JDM (http://jeuxdemots.org/analogies.php → "exporter données"), we note that around 87% of failure cases are due to missing relationships. Deductive inference processes can overcome this problem in 65% of cases. Example of inference: *child r_has_part leg → child r_isa human r_has_part leg*.

[10] Passage through lemmatization by observing all the relationships of close words such as *petit, petite, petits, petites* (which means *small* in feminine and masculine and in singular and plural in french) when one of them is concerned.

- **relational similarity** (correspondence between the relations of 2 pairs of concepts):
  - $R_{AB} = A : B$ (with A and $B \in$ *source* domain denoted Dom$_{AB}$)
  - $R_{CD} = C : D$ (with C and $D \in$ *target* domain denoted Dom$_{CD}$)
  - SimRel = $R_{AB} \cup R_{CD}$
  - It is possible to have a correspondence $A \rightarrow B$ (respectively $C \rightarrow D$) transferring knowledge from a generally familiar and concrete *source* concept A (respectively C) to a generally unknown and abstract *target* concept B (respectively D)
- **attributional similarity** (correspondence between the attributes of 2 concepts):
  - SimAttr$_{AC}$ between A and C (co-P)[11]
  - SimAttr$_{BD}$ between B and D (co-P)
- **analogy** when there is a non-empty intersection of the relations of R_{AB} and R_{CD}, its understanding is improved with the presence of attributional similarities

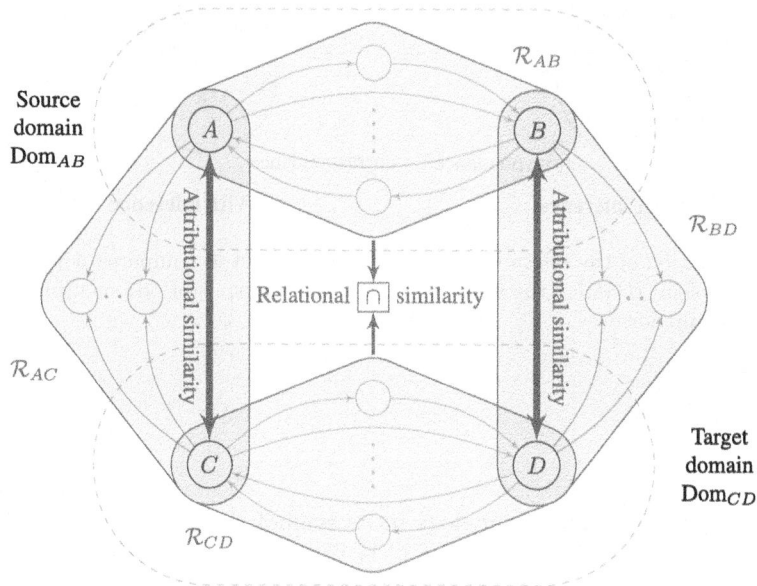

Fig. 6. Detailed diagram of the analogy square

We believe that the strength of an analogy is defined through several aspects. First of all, it should be noted that the approach does not take place, with regard to the knowledge base, in a closed world context, in the sense that a missing relation ar_tb does not mean that it's not possible. Gaps in the knowledge base can

---
[11] P being the semantic relation seen as a unary predicate of the same value.

result in mistakes in the system's analysis of the analogy. These gaps, highlighted by mistakes, need to be identified and corrected. The ongoing improvements of this proof of concept include, among other things, more efficient reasoning mechanisms, and approaches to resolving the polysemy of input terms.

In an analogical square, the relationships between words in the *source* domain are of the same type (one could say analogous) to those in the *target* domain establishing the strength of the relational similarity. Equivalent terms from opposite domains must have a relationship of the same type to an intermediate node (attribute). This configuration is referred to as *attributional similarity*, which is key to a better explanation of the analogy.

In any case, the more semantically precise the relation type, the more useful that type is in explaining the analogy as a whole. The overall strength of the analogy could correspond to the average of all the relational and attributional similarities weights. This currently makes it possible to elect the candidate(s) proving to be the most appropriate for a given metaphor (a hole analogy with a single unknown) in 37% of cases with a simple algorithm (constituting our baseline). However, a more precise aggregation of the different similarity scores would make it possible to take into account the nuances carried by each type of relationship; reason why the definition of this aggregation method represents one of the central aspects of our research. By definition, figurative language is subject to interpretation; the same analogy or its explanation can turn out to be more or less telling depending on the angle from which the explanation is approached. The goal of an automatic analyzer is to highlight what can most be considered a convincing or satisfactory interpretation (subjective concepts) for a given analogy. In the state of the art, it should be noted that the subject is often treated through theoretical and linguistic points of view rather than in computational and applied work.

This work aims to produce a resolution model but will also serve as a tool for identifying anomalies in the knowledge base (imperfect and incomplete by nature). Any shortcomings will be highlighted by cases of algorithm failure and will make it possible to consolidate the appropriate types of relationships. One avenue for future research could then be to design a game (with a purpose) on the intersections of associations to broaden the knowledge base in this regard.

5 Conclusions and Future Work

We have presented our to the automated treatment of metaphors and analogies in natural language, using the knowledge graph of JeuxDeMots to generate plausible solution, and providing a prototype solver. We have also presented some promising preliminary results of our approach.

In the future, we will first extend our approach to incorporate word sense disambiguation, then look at different methods for detecting and resolving metaphors in their linguistic context.

References

1. Gentner, D.: Structure-mapping: a theoretical framework for analogy. Cognitive science **7**(2), 155–170 (1983). https://www.sciencedirect.com/science/article/pii/S0364021383800093, publisher: Elsevier
2. Gentner, D., Holyoak, K.J.: Reasoning and learning by analogy: Introduction. Am. Psychol. **52**(1), 32 (1997). https://psycnet.apa.org/record/1997-02239-004, publisher: American Psychological Association
3. Hofstadter, D.R.: Fluid concepts and creative analogies: computer models of the fundamental mechanisms of thought. Basic books (1995). https://psycnet.apa.org/record/1995-98269-000
4. Hofstadter, D.R.: Analogy as the core of cognition. The analogical mind: Perspectives from cognitive science (2001)
5. Kahneman, D.: Thinking, fast and slow. Farrar, Straus and Giroux (2017)
6. King, D., Gentner, D.: Verb Metaphors are Processed as Analogies (2023)
7. Lafourcade, M.: Making people play for Lexical Acquisition with the JeuxDeMots prototype. In: SNLP'07: 7th International Symposium on Natural Language Processing, p. 7 (2007). https://hal-lirmm.ccsd.cnrs.fr/lirmm-00200883/
8. Lepage, Y.: De l'analogie rendant compte de la commutation en linguistique. Ph.D. thesis, Université Joseph-Fourier-Grenoble I (2003)
9. Shutova, E.: Automatic metaphor interpretation as a paraphrasing task. In: Human Language Technologies: The 2010 Annual Conference of the North American Chapter of the Association for Computational Linguistics, pp. 1029–1037 (2010). https://aclanthology.org/N10-1147.pdf
10. Turney, P.D.: Measuring Semantic Similarity by Latent Relational Analysis (2005). http://arxiv.org/abs/cs/0508053, arXiv:cs/0508053
11. Turney, P.D., Pantel, P.: From frequency to meaning: Vector space models of semantics. J. Artif. Intell. Res. **37**, 141–188 (2010). https://www.jair.org/index.php/jair/article/view/10640

The Aranea Corpora Family: Ten+ Years of Processing Web-Crawled Data

Vladimír Benko[1,2(✉)]

[1] Slovak Academy of Sciences, Ľ. Štúr Institute of Linguistics, Panská 26, 811 011 Bratislava, Slovakia
[2] Comenius University Science Park, UNESCO Chair in Plurilingual and Multicultural Communications, Ilkovičova 8, 814 04 Bratislava, Slovakia
vladimir.benko@juls.savba.sk
https://juls.savba.sk/

Abstract. *Aranea* is a project to create a family of web-crawled corpora for languages taught at Slovak Universities. Since 2013, more than two dozen languages have been added to the project: they are often represented with Gigaword+ size corpora and sometimes have subcorpora for their territorial varieties. Our paper summarizes the development of the *Aranea* project in the past decade. We describe the step-by-step optimization of the processing pipeline, highlight existing issues, and discuss the linguistic rationale behind some engineering decisions associated with the idiosyncrasies of individual languages.

Keywords: web-crawled corpora · 'Brno pipeline' · FLOSS tools · ensemble lemmatization and PoS tagging

1 Introduction

Aranea[1] is a multilingual corpus-building and corpus-managing framework designed to build web-corpora for languages taught at Slovak universities and primarily intended for pedagogical purposes. As the corpora are created using the same technology and have the same size(s), they can be considered "comparable" to a considerable extent[2].

While the project was officially released at the TSD conference in 2014 [5], this jubilee contribution is focused on the aspects of *Aranea* development that did not receive any or enough coverage elsewhere.

2 Related Work

Though the methodology and technology of building web-crawled corpora have been available for more than two decades, there have been just a few projects that used them to produce and maintain large multilingual collections.

---

[1] araneum *pl.* aranea *n.* (https://www.latin-dictionary.net/search/latin/araneum).
[2] The question of comparability with respect to web-crawled corpora is discussed in [7].

The *Web as Corpus* (*WaC*) project was a pioneer in this area, producing Gigaword corpora for four languages, i.e., *ukWaC* (British English), *deWaC*, *frWaC*, and *itWaC* [2,3], respectively.

The *TenTen* corpora collection is another notable example of a large-scale multilingual project that has created web-corpora for more than 40 languages and is making them available through the *Sketch Engine*[3] portal [11].

The {*bs,hr,sr*}*WaC* corpora resulted from using the same methodology and tools [14], while the *COW* (*Corpora of the Web*) yielded corpora for six languages using a different set of tools [22].

And finally, there is an impressive collection of corpora[4] created and maintained by Mark Davies from the Brigham Young University [9].

3 The *Aranea* Language Collection

The original intention of the *Aranea* project was to create corpora for Slovak, Czech and the four "large" languages (i.e. English, French, German, and Russian). However, over time our ambition first extended to covering all the languages taught at the Slovak Universities. Then, for various reasons, we have included other languages with available PoS taggers, even though these corpora are unlikely to be used by Slovak students or teachers in the foreseeable future.

4 Crawling and Pre-tokenization Filtration

Since its publication in 2013, we have been using *SpiderLing*[5] [5, 8, 27] as the tool for crawling the web to obtain and pre-process the corpus data. From the initial Version 0.72 up to today's Version 2.2, *SpiderLing* has gradually developed into a mature, stable, and extremely efficient tool capable (within our IT infrastructure and setup) of gathering as much as two billion tokens of corpus data in 24 h. During these recent years, important improvements have included reducing the RAM necessary to store the internal data structures, refining the language identification procedure, and providing more metadata for the downloaded documents (for example, the web page title, original and detected character encoding, language detected through the *CLD2*[6] procedure, etc.).

After the data for a language is downloaded, we apply the following stack of operations: deduplication, secondary language detection, and deletion of texts with encoding issues.

---

[3] https://www.sketchengine.eu/.
[4] https://www.english-corpora.org/.
[5] https://corpus.tools/wiki/SpiderLing.
[6] https://github.com/CLD2Owners/cld2.

4.1 Secondary and Tertiary Deduplication

Though *SpiderLing* is performing basic detection and removal of 100%-duplicate documents on the fly, there are situations when duplicate documents do appear in the output, for example, when the new crawling session had to be (for various reasons) launched without taking into consideration the "context" of the previous one. The deduplication procedure is based on Rabin's fingerprint method [20] which compares the cryptographic MD4 Message-Digest Algorithm[7] hashes calculated for every document.

We also delete identical subsequent paragraphs that often appear in the documents. Though it can be argued that we lose some interesting information by deleting those copies, we decided to delete them as we suspect that in most cases duplicate subsequent paragraphs are the result of search engine optimization (SEO) efforts. The removal is performed using the *uniq* program. An excerpt from a deduplication log (of the Norwegian data) is shown in Fig. 1. The left-hand column indicates the number of repeating paragraphs.

```
 3  <p heading="yes">Skolelaboratoriet</p>
 2  <p>Er du sikker på at du vil slette?</p>
 2  <p heading="yes">Vi inkluderer personer med
       utviklingshemming i det ordinære arbeidslivet</p>
 2  <p>Varen er bestilt fra leverandør, men leveringsdato er
       ikke bekreftet.</p>
 2  <p heading="yes">Løse Dine Sinus problemer</p>
10  <p>Læs, hvordan kunder verden over drager nytte af vores
       alsidige løsninger.</p>
```

Fig. 1. Secondary (subsequent paragraph) deduplication in *Araneum Norvegicum*.

4.2 Secondary Language Detection

Although the language identification procedure works fairly well for detecting "sufficiently different" languages, it often fails for the closely related ones. We, therefore, apply a more sensitive procedure that uses character frequencies: for each language, we count characters appearing in that language and not in any of the others which are similar. For example, Czech can be identified by the presence of the ě, ř, and ů characters; Slovak by ä, ĺ, ľ, ô, and ŕ; Hungarian by ő, and ű, etc. The respective thresholds for removing documents in languages similar to the expected language of the crawled country domain are set experimentally.

---

[7] https://datatracker.ietf.org/doc/html/rfc1186.

4.3 Character Encoding Issues

It sometimes happens that a document is garbled, either as a whole by incorrect detection of the character encoding by *SpiderLing*, or partially, when the issue is associated only with some turns in online discussions (most likely due to a misunderstanding of encoding between the portal and the respective user browser). As a result, the text can contain artifacts like this: *"You're no piker!"*. Here, the UTF-8 encoding was incorrectly identified as 8-bit, resulting in the garbled single quote signs. A series of regular expressions has been written to identify these issues.

5 Tokenization and Post-Tokenization Processing

Tokenization is a technical process, yet some linguistically motivated decisions have to be made here. We prefer having the same "tokenization policy" for all languages, even in situations when the tagger expects a slightly different input format.

We use the *Unitok*[8] [16] universal tokenizer with a custom parameter file provided for each language. The tokenization strategy mostly adheres to that defined by the default parameter files, with some additional modifications related to the definition of period-final abbreviations, rare clitics, and other idiosyncrasies of the respective languages.

5.1 Sentence Segmentation

A rule-based algorithm was written to segment the tokenized text into sentences. The algorithm relies on punctuation and period-final abbreviations already recognized by the tokenizer. The outcomes are reasonably high quality for languages using scripts that distinguish between upper and lowercase letters (Latin and Cyrillic); for the other languages, there is more noise in the results.

5.2 Detection and Removal of Partially Duplicate Documents

At this step in the corpus preprocessing, we again use a tool from the "Brno" pipeline - a program referred to as *Onion*[9] [19]. As it expects the corpus text to be already tokenized, it is best to use *Onion* at this step in the workflow.

The description of the process and the discussion related to the optimal parameters can be found in [4]. We would only mention here that (for all corpora) we use the same input parameters: an n-gram length of 5 and a similarity threshold value of 0.9.

Depending on the language, the proportion of deleted material can vary from 30% to as much as 60% – the higher value is usually due to unnecessary long crawling sessions that return too many similar downloaded documents.

---

[8] https://corpus.tools/wiki/Unitok.
[9] https://corpus.tools/wiki/Onion.

5.3 Comparing the Corpora

After the pre-processing, tokenization and deduplication are completed, the sizes of the corpus data for the *Aranea* languages can be compared. For most languages, we performed two crawls - the initial one and the second one in 2021 in an attempt to cover the COVID pandemic discourse. For some languages, the more recent downloads are still being processed. Table 1 reflects the corpora sizes after preprocessing by language and indicates the years of the available crawls.

Table 1. Available crawls and corpora sizes (in millions of tokens) after preprocessing by language (as of June 2024).

Lang	Year(s) crawled	Tokenization & depuplication	
		Last year processed	Size
ar	2017, 2021	2017	172
be	2021, 2023	2023	483
bg	2016, 2021	2016	2,325
cs	2013, 2015-2018, 2021, 2022, 2024	2024	12,248
cre da	2023	2023	2,759
de	2013-2015, 2018, 2021	2018	9,555
en	2013-2015, 2017, 2021, 2024	2021	21,655
es	2013, 2014, 2021	2021	5,620
et	2017-2019, 2021	2019	3,553
fa	2022	2022	3,892
fi	2014, 2021	2014	1,629
fr	2013, 2015-2019, 2021, 2024	2024	20,915
hu	2014, 2021	2023	3,352
hr	2021	2021	1,886
it	2014, 2021	2014	2,001
ja	2021	-	n/a
ka	2014, 2018, 2021, 2023, 2024	2024	1,568
kk	2021	2021	340
ko	2021	2021	1,113
la	2018, 2020	2020	126
lv	2017, 2018, 2021	2018	864
mt	2017, 2021	2021	24
nl	2013, 2021	2015	1,592
no	2019, 2021-2023	2023	3,532
pl	2013, 2021-2023	2024	7,867
pt	2015, 2021	2021	1,683
ro	2017, 2021	2018	1,874
ru	2013-2018, 2019-2023	2021	43,800
sk	2013-2024	2024	8,711
sl	2021	2021	1,211
sq	2021	2021	294
sr	2021	2024	2,423
sv	2017, 2021	2017	1,955
tr	2023	2023	1,933
tt	2018, 2021	2021	152
uk	2014, 2015, 2021, 2022	2022	5,249
uz	2020, 2024	2024	1,712
zh	2015, 2021, 2024	2015	1,254

6 Some Language Idiosyncrasies

6.1 Chinese, Clitics and Sub-Word Tokenization

Unlike most other languages, Chinese does not separate words by spaces, and a whole paragraph appears as a single string in the text. "Tokenization" is therefore performed by a heuristic algorithm that tries to guess the word boundaries. We used the program designed by Serge Sharoff[10].

Correct tokenization of clitics is crucial for languages such as English and French: it helps the tagger recognize expressions like *I haven't been* or *J'en parle*. This, however, means that the tokenization of clitics has to be compatible with what is expected by the tagger.

However, there are situations when the tagger is not capable of recognizing either a split or a joint clitic - this has to be treated for each language individually.

Sub-word tokenization is traditionally considered in some languages, notably Polish and Georgian. We, however, decided not to perform it within the framework of our project[11].

6.2 Digraphia

Two languages in our collection use two scripts (Cyrillic and Latin) in parallel. While in Serbia, the existence of digraphia is an official state policy, in Uzbekistan, the situation of a de facto digraphia may be considered as a result of the partially unsuccessful transition from Cyrillic to Latin at the beginning of the 1990s.

As in both cases, the respective taggers expect the texts to use the Latin script only; corpus linguistics projects usually convert everything to Latin, thus hiding the information about the original appearance of the respective texts. Contrary to that, we consider the presence of Cyrillic script an interesting sociolinguistic attribute and keep all the texts in their original scripts. We introduced a "normalized word form" as an additional attribute to be used for lemmatization and PoS tagging. Using the appropriate attribute, queries can be performed in Latin script (to get all occurrences) or in Cyrillic script (to get only the relevant ones).

6.3 Romanization

We expect users of languages with Latin and/or Cyrillic scripts to have a keyboard with the respective layout. For other scripts it may be useful to be able to perform queries transliterated to Latin. In our corpus collection, the latter applies to Georgian, Persian, and Korean, where a solution similar to the digraphia languages is used. Unlike Serbian and Uzbek, however, "Romanization" is applied not only to word forms but also to lemmas, so that all queries

---

[10] Serge's web page containing his language resources is currently not accessible.

[11] Sub-word tokenization is a rather complex topic that is beyond the scope of this paper.

can be performed in both native and Romanized forms. The result of such a query in the Georgian corpus is shown in Fig. 2.

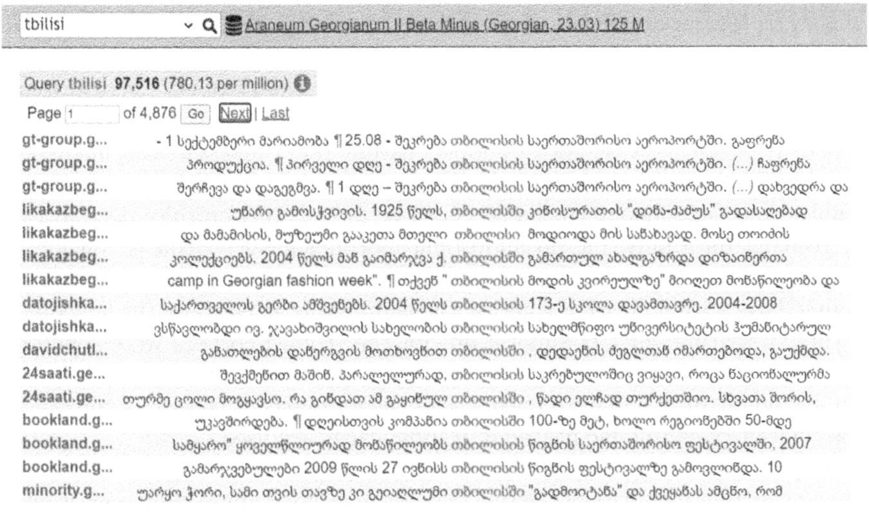

Fig. 2. Querying for "tbilisi" in *Araneum Georgianum*.

6.4 "Half-Spaces"

One of the peculiarities of the standard Persian orthography is that some affixes are to be separated from their respective main word by a "zero-width non-joiner" character (U+200C) which prevents the neighboring characters from being joined into a ligature (as they would normally be). While this character can be entered on standard Persian keyboard, (apparently), in situations when writing on a different keyboard (such as Arabic), those affixes are either joined with the respective word or separated by a regular whitespace character. We provide a special normalized attribute for "half-spaces", which makes querying more deterministic.

7 Lemmatization and PoS Tagging

7.1 The Baseline Annotation

For basic processing, we prefer tools with multilingual support, i.e., those equipped with language models for more than one language. Within the context of our project, we mostly use four taggers and one lemmatizer, as follows.

TreeTagger[12] [23] belongs to the oldest tools available and still being supported by its author, providing models for almost 50 languages. It is especially suitable for coarse tagsets, i.e., for languages with 'poor' morphology. Its main advantage is the tagging speed, sometimes more than by an order of magnitude higher than for newer neural networks-based tools. Its main deficiencies include lower precision for languages with rich morphology (Slavic and agglutinative languages) and no lemma guessing for the out-of-vocabulary (OOV) lexical items. The tag guessing, however, works reasonably well.

UDPipe[13] [25] is a tagger developed within the *Universal Dependencies*[14] (*UD*) [15] framework which provides language models for all languages with a suitably-sized dependency treebank. As those models rely on rules derived from the treebank data only, i.e. no additional morphological lexicon is considered, lemma guessing is performed for all lexical items, even for those present in the training data, with no testing whether the lemma was attested by the lexicon. At the time of writing, all *UD* annotations for the *Aranea* corpora were generated using *UDPipe 1* with the latest *UDs 2.5* language models released in December 2019. In the future, we plan to make use of the more recent version of the tagger, whose efficiency can be facilitated if graphic cards are used.

MorphoDiTa[15] [24] is a high-performance tagger combining, from the user perspective, functionalities of the two taggers mentioned above. However, it offers language models for just three languages (Czech, English, and Slovak) and is not supported anymore.

Apertium[16] [13] is an open-source machine translation system with a morphological module that can also be used separately for lemmatization and PoS tagging. Its main advantage is that it has language models for languages missing in other tools.

Unlike other tools mentioned above, *CSTlemma*[17] [12] performs lemmatization only, though it can consider PoS tags provided by other tools during disambiguation.

Two other tools used in our project include *LU MII Tagger*[18] for Latvian [17] and *HunPos*[19] [10] for Hungarian[20].

The behavior of these tools with respect to the OOV lexical items is summarized in Table 2.

Each of the tools has pros and cons, and the decision to use a tool largely depends on the availability of a model for a specific language. In addition to the

---

[12] https://www.cis.uni-muenchen.de/~schmid/tools/TreeTagger/.
[13] https://ufal.mff.cuni.cz/udpipe/1.
[14] https://universaldependencies.org/.
[15] https://ufal.mff.cuni.cz/morphodita.
[16] https://www.apertium.org/.
[17] https://cst.dk/online/lemmatiser/uk/.
[18] https://peteris.rocks/blog/latvian-part-of-speech-tagging/.
[19] http://mokk.bme.hu/en/resources/hunpos/.
[20] Hungarian was the only language with processing being "outsourced" - it has been done for us by the Research Institute of Linguistics at the Hungarian Academy of Sciences.

Table 2. Baseline tools used for lemmatization and PoS tagging.

Tool		Non-OOV lexical items	OOV lexical items
TreeTagger	Lemmas	assigned from the lexicon	marked as 'unknown'
	PoS Tags	assigned from the lexicon	guessed
	Disambiguation	statistical	statistical
MorphoDiTa	Lemmas	assigned from the lexicon	guessed & marked
	PoS Tags	assigned from the lexicon	guessed & marked
	Disambiguation	statistical	statistical
UDPipe	Lemmas	n/a	guessed
	PoS Tags	n/a	guessed
	Disambiguation	n/a	statistical
Apertium	Lemmas	assigned from the lexicon	marked as 'unknown'
	PoS Tags	assigned from the lexicon	marked as 'unknown'
	Disambiguation	rule-based	n/a
CSTlemma	Lemmas	assigned from the lexicon[a]	guessed
	PoS Tags	n/a	n/a
	Disambiguation	statistical	statistical

[a] If no lexicon is available, all lemmas are guessed.

varying extent of language support, the tools also differ in processing speed - *UDPipe*, for example, is typically at least ten times slower than *TreeTagger*.

One of the measures to assess the quality of a language model is the lexical coverage measured by the OOV rate. For languages where the baseline tagger provides such information, the results are summarized in Table 3.

It can be seen that the lexical coverage varies a lot across languages - while being surprisingly high for some languages, it indicates questionable quality for other languages. To counteract this disparity, we applied an "ensemble" approach to lemmatization, and partially also to PoS tagging.

7.2 Ensemble Annotation

In Computational Linguistics, the "ensemble" term is used to describe approaches where several tools are used to (independently) perform the same operation, assuming that aggregation of their outputs could improve the overall success rate of the whole process. In the framework of morphosyntactic annotation, we can speak about "ensemble lemmatization/tagging" if more than one lemmatizer/tagger is available for a particular language, which is the case of several languages in our corpus collection.

If more than two taggers use the same tagset, the aggregation can be performed by simple majority voting. In our case, however, at least three "full-fledged" taggers are unavailable, and the respective tagsets are incompatible.

Table 3. Out-of-vocabulary (OOV) rates.

Lang	Baseline tagger	Year tagged	OOVs (%)	Lang	Baseline tagger	Year tagged	OOVs (%)
ar	(not tagged yet)			la	TreeTagger	2018	7.02
be	TreeTagger	2023	16.96	lv	LU MII Tagger	2018	n/a
bg	TreeTagger	2016	6.41	mt	Apertium	2021	16.28
cs	MophoDiTa	2020	2.50	nl	TreeTagger	2015	7.07
da	TreeTagger	2023	3.64	no	TreeTagger	2021	6.51
de	TreeTagger	2020	5.66	pl	TreeTagger	2021	9.47
en	TreeTagger	2017	2.72	pt	TreeTagger	2015	6.11
es	TreeTagger	2015	5.74	ro	TreeTagger	2019	5.97
et	TreeTagger	2019	17.50	ru	TreeTagger	2023	5.69
fa	TreeTagger	2022	10.48	sk	MorphoDiTa	2024	4.33
fi	TreeTagger	2015	12.38	sl	TreeTagger	2021	3.73
fr	TreeTagger	2021	4.26	sq	Apertium	2021	26.87
hr	UDPipe	2021	n/a	sr	(not tagged yet)		
hu	HunPos	2014	2.46	sv	TreeTagger	2017	19.27
it	TreeTagger	2021	3.89	tr	UDPipe	2023	n/a
ja	(not tagged yet)			tt	Apertium	2021	6.79
ka	TreeTagger	2023	34.06	uk	UDPipe	2022	n/a
kk	Apertium	2021	6.03	uz	Apertium	2024	11.22
ko	TreeTagger	2021	14.16	zh	TreeTagger	2015	6.55

We started to apply the ensemble approach to the newer *Aranea* corpora created after 2022 and to the older corpora re-processed after this year. A typical setup consists of *TreeTagger*, *UDPipe* and *CSTlemma*. The main idea is to improve the annotation of the OOV items by "double guessing" (*UDPipe* + *CSTlemma* for lemmatization, and *TreeTagger* + *UDPipe* for PoS tags). Due to the differences in tagsets, we only consider main word classes encoded using the *Araneum Universal Tagset*[21] (*AUT*). The aggregation is rule-based, and can be described as follows:

1. If PoS can be assigned by means of a regular expression (punctuation, digits, symbols, e-mail addresses, etc.), ignore information provided by the taggers.
2. If the respective token was present in the morphological lexicon of *TreeTagger*, take both lemma and PoS from it.
3. If the token was OOV in *TreeTagger* and *UDPipe* guessed a lemma, take both lemma and PoS from *UDPipe*. If the lemma guessed by *CSTlemma* is different, add it as an alternative. If PoS guessed by the *TreeTagger* is different, add it as an alternative.
4. Otherwise take the lemma from *CSTlemma* and PoS from *TreeTagger* and *UDPipe* (if different).

The result of the aggregation is stored in a special attribute "ztag": the respective value consists of two parts separated by a period - the left part has

---

[21] http://aranea.juls.savba.sk/aranea_about/aut.html.

details about the assignment of lemma, while the right – about the PoS. The uppercase letters indicate success in the morphological lexicon lookup (in the case of *TreeTagger*), the lowercase letters indicate guessing, and the exclamation mark indicates that the respective value differs from that on the left. The actual situation in a 125-Megatoken (Farsi) *Araneum Persicum* is shown in Table 4 [6].

Table 4. Result of aggregation for *Araneum Persicum Minus* (125 million tokens)

ztag	freq	%	ztag	freq	%
T!c.T!u	313,397	0.25	u!c.t!u	734,753	0.59
T!c.Tu	1,744,755	1.40	u!c.tu	2,098,439	1.68
T!u!c.T!u	141,740	0.11	uc.t!u	2,077,455	1.66
T!u!c.Tu	2,082,343	1.67	uc.tu	3,751,262	3.00
T!uc.T!u	644,554	0.52	c.t!u	1,411,739	1.13
T!uc.Tu	4,477,277	3.58	c.tu	3,026,913	2.42
Tc!u.T!u	783,445	0.63	z!	6,629,046	5.30
Tc!u.Tu	1,242,515	0.99	z#	788,956	0.63
Tc.T!u	694,332	0.56	z$	910,190	0.73
Tc.Tu	3,817,999	3.05	z@	1,068	< 0.01
Tu!c.T!u	455,065	0.36	zu	5,987	< 0.01
Tu!c.Tu	13,464,342	10.77	zv	17,675	0.01
Tuc.T!u	7,834,406	6.27	zw	1,972	< 0.01
Tuc.Tu	65,848,758	52.68	Total	125,000,383	100.00

Flags starting with the "z" letter indicate lexical items "tagged" by regular expressions. For example, "z!" denotes punctuation, "z#" numbers, "z$" symbols (special graphic characters, emoji, etc.), "z@" e-mail addresses, and "zu", "zv", and "zw" Internet addresses in different formats. As it can be seen, most items have a "Tuc.Tu" flag, indicating equal values assigned both for lemma and PoS by all tools, followed by a "Tu!c.Tu", where *CSTlemma* assigned a different lemma than *TreeTagger* and *UDPipe*. If we add counts for the "uc.tu" and "uc.t!" ztag values, we will find out that for 4.66% of the word forms both guessers assigned the same lemma, making it probably correct, and improving the overall lemmatization success rate.

8 Post-tagging Processing

Lemmatization and PoS tagging result in corpus data that is minimally sufficient for some applications. However, for lexicographic purposes or language teaching, where the reading of concordances is often involved, the corpora must undergo additional processing steps.

8.1 Paragraph and Sentence-Level Deduplication

Detection and removal of partially duplicate paragraphs is performed by the *Onion* program (with the same settings as for the document deduplication, i.e. n-gram length of 5, and threshold level 0.9), and Rabin's fingerprint method is used for the sentences while ignoring numbers, special characters and punctuation (making use of the PoS annotation already present in the data). Both processes are described in [4].

The paragraph and sentence deduplication procedure typically removes 15–20% and 3–5% tokens, respectively, on top of the document deduplication.

8.2 Sentence-Level Language Identification

Foreign language (mostly English) sentences appear in almost every corpus of the *Aranea* family. They may be undesirable when frequency statistics are needed because frequent foreign words, such as English "the", can rank high in the frequency lists and contaminate the results for the targeted language.

Recently, we started removing foreign sentences using a procedure designed by Vít Suchomel [26] but based on our own implementation and on the statistics derived from the *Aranea* corpora. This process usually removes up to 5% of corpus tokens.

8.3 Removal of Spam and Other Machine-Generated Contents

There is no single method of detecting machine-generated content that often contains nonsensical parts. Until recently, auto-generated texts were relatively easy to identify using heuristic algorithms. With the advent of generative AI, texts produced by a machine are often indistinguishable from those created by a human.

This step appears as the last step in the corpus processing pipeline because it is relatively independent of the previous stages. Given a method for identification of "suspicious" documents, the preprocessed corpus can be put through a filtering procedure to remove spam and machine-generated content.

As spotting problematic texts depends on comprehension of the respective language, we were able to apply the large-scale spam-removing procedures to the Czech, Slovak, and Russian data only. The Czech and Slovak spam pages were identified as "link farms" advertising online bets, while Russian Internet spam was associated with prostitution. In both cases, the activities are illegal in the respective countries.

9 Processing by the Corpus Manager

To keep our corpora as comparable as possible, we sample them into two sizes: 125 million token corpora for teaching purposes and (if possible) 1.25 billion token corpora for general use. For some languages, we also have the as-much-as-can-get size.

Our two *Aranea* corpus portals have several thousand registered users, over two hundred of whom access our corpora online daily. On demand, we also provide processed source corpus data (for research purposes).

A final note on the last processing step in the preparation of a corpus is associated with the corpus manager. We use the *NoSketch Engine* [21], an open-source subset of the commercial *Sketch Engine*. To our knowledge, this is the only system with a FLOSS license capable of processing corpora of several billions of tokens in size.

One of the functionalities less known to users is the possibility of creating user sub-corpora, either using the metadata or by saving the result of any corpus query. Such sub-corpora can subsequently be used to calculate keywords, for example, by comparing frequency lists from different time periods. The Appendix shows keywords calculated from the 2021 *French Araneum Francogalicum* with the previous version of the same corpus used as a reference. Most items indicate the pandemic discourse prevailing in the respective time period. The presence of English words is because sentence-level language filtration has not been performed yet.

10 Conclusions, New Challenges and Further Work

The main *Aranea* corpus portal[22] currently hosts corpora for 27 languages, with four of them also having territorial varieties (based on the top-level domain). The "sandbox" portal (accessible for all registered users) hosts the beta versions of corpora for 7 additional languages. New crawling sessions for all languages performed since 2021 yielded data covering the discourse of the pandemic and the Russian-Ukrainian war conflict. The new crawls yielded huge amounts of data that are still being processed.

Newer versions of *Aranea* corpora are lemmatized and PoS tagged using the ensemble approach, which can significantly improve the quality of annotation. Several other improvements, including sentence-level language filters, have been incorporated into the processing pipeline. We hope to be able to reprocess all languages using this new pipeline in the near future.

New challenges include the option of using datasets from the Common Crawl[23] [18] initiative provided by the OSCAR Project[24] [1]. An important point for deliberation is whether the increasing amount of AI-generated texts means that the web-crawled corpora will lose their importance as natural language resources in the future.

---

[22] http://aranea.juls.savba.sk/guest/, http://unesco.uniba.sk/guest/.
[23] https://commoncrawl.org/.
[24] https://oscar-project.org/.

Acknowledgements. This work has been, in part, supported by the Slovak VEGA Grant Agency for Science, Projects No. 2/0015/14 (2014–2016), 2/0017/17 (2017–2020), and 2/0016/21 (2021–2024), respectively.

Disclosure of Interests. The author has no competing interests to declare relevant to this article's content.

Appendix

lemma	Araneum Francogallicum Novum MMXXI Duplex Minus (Global French French, 21.03) 125 M		Araneum Francogallicum III Minus (Global French, 20.05) 125 M		Score
	document frequency	document frequency/mill	document frequency	document frequency/mill	
Covid-19	2,576	20.6	0	0.0	21.6
coronavirus	1,892	15.1	4	0.0	15.6
confinement	2,893	23.1	129	1.0	11.9
pandémie	2,768	22.1	126	1.0	11.5
COVID-19	1,287	10.3	0	0.0	11.3
Covid	962	7.7	0	0.0	8.7
Coronavirus	563	4.5	0	0.0	5.5
déconfinement	372	3.0	1	0.0	3.9
covid	354	2.8	0	0.0	3.8
COVID	345	2.8	0	0.0	3.8
télétravail	627	5.0	85	0.7	3.6
covid-19	295	2.4	0	0.0	3.4
distanciation	524	4.2	79	0.6	3.2
Biden	286	2.3	20	0.2	2.8
couvre-feu	437	3.5	74	0.6	2.8
épidémie	2,188	17.5	695	5.6	2.8
that	568	4.5	140	1.1	2.6
which	282	2.3	33	0.3	2.6
cookies	1,094	8.8	357	2.9	2.5
RGPD	281	2.2	39	0.3	2.5
visioconférence	400	3.2	89	0.7	2.5
Castex	219	1.8	17	0.1	2.4
LREM	301	2.4	51	0.4	2.4
sanitaire	5,768	46.1	2,373	19.0	2.4
autrice	213	1.7	23	0.2	2.3
présentiel	385	3.1	99	0.8	2.3
this	505	4.0	152	1.2	2.3
reconfinement	158	1.3	0	0.0	2.3
their	243	1.9	40	0.3	2.2
also	219	1.8	30	0.2	2.2
Macron	1,847	14.8	770	6.2	2.2
based	216	1.7	32	0.3	2.2
has	328	2.6	84	0.7	2.2
hydroalcoolique	159	1.3	9	0.1	2.1
Netflix	560	4.5	202	1.6	2.1

References

1. Abadji, J., Suarez, P.O., Romary, L., Sagot, B.: Towards a cleaner document-oriented multilingual crawled corpus. In: Thirteenth Language Resources and Evaluation Conference-LREC 2022 (2022)
2. Baroni, M., Bernardini, S.: BootCaT: bootstrapping corpora and terms from the web. In: LREC, pp. 1313–1316 (2004)
3. Baroni, M., Bernardini, S., Ferraresi, A., Zanchetta, E.: The WaCky wide web: a collection of very large linguistically processed web-crawled corpora. Lang. Resour. Eval. **43**, 209–226 (2009)
4. Benko, V.: Data Deduplication in Slovak Corpora. In: Slovko 2013: Natural Language Processing, Corpus Linguistics, E-learning, pp. 27–39. RAM-Verlag: Lüdenscheid (2013)
5. Benko, V.: Aranea: yet another family of (comparable) web corpora. In: Sojka, P., Horák, A., Kopeček, I., Pala, K. (eds.) TSD 2014. LNCS (LNAI), vol. 8655, pp. 247–256. Springer, Cham (2014). https://doi.org/10.1007/978-3-319-10816-2_31
6. Benko, V.: Aranea Go Middle East: Persicum. In: RASLAN, pp. 113–121 (2022)
7. Benko, V., Kunilovskaya, M.: Comparable web-crawled corpora as a resource for contrastive studies. In: Accepted for presentation at the Between Languages: Methods of Contrastive Research Using Corpora Workshop, Biennial of Czech Linguistics (2024)
8. Benko, V., Zakharov, V.: Very Large Russian Corpora: New Opportunities and New Challenges. In: Kompjuternaja lingvistika i intellektual'nyje technologii, pp. 83–98 (2016)
9. Davies, M.: The best of both worlds: multi-billion word "dynamic" corpora. In: Proceedings of the Workshop on Challenges in the Management of Large Corpora (CMLC-7), pp. 23–28 (2019)
10. Halácsy, P., Kornai, A., Oravecz, C.: HunPos – an open source trigram tagger. In: Proceedings of the 45th Annual Meeting of the Association for Computational Linguistics Companion Volume Proceedings of the Demo and Poster Sessions, pp. 209–212 (2007)
11. Jakubíček, M., Kilgarriff, A., Kovář, V., Rychlý, P., Suchomel, V.: The TenTen corpus family. Corpus. Linguistics **2013**, 125 (2013)
12. Jongejan, B., Dalianis, H.: Automatic training of lemmatization rules that handle morphological changes in pre-, in-and suffixes alike. In: Proceedings of the Joint Conference of the 47th Annual Meeting of the ACL and the 4th International Joint Conference on Natural Language Processing of the AFNLP, pp. 145–153 (2009)
13. Khanna, T., et al.: Recent advances in Apertium, a free/open-source rule-based machine translation platform for low-resource languages. Mach. Translation **35**(4), 475–502 (2021)
14. Ljubešić, N., Klubička, F.: bs, hr, srWaC – Web corpora of Bosnian, Croatian and Serbian. In: Proceedings of the 9th web as corpus workshop (WaC-9), pp. 29–35 (2014)
15. McDonald, R., et al.: Universal dependency annotation for multilingual parsing. In: Proceedings of the 51st Annual Meeting of the Association for Computational Linguistics (Volume 2: Short Papers), pp. 92–97 (2013)
16. Michelfeit, J., Pomikálek, J., Suchomel, V.: Text Tokenisation Using unitok. In: RASLAN, pp. 71–75 (2014)
17. Paikens, P.: Deep neural learning approaches for Latvian morphological tagging. In: Human Language Technologies–The Baltic Perspective, pp. 160–166. IOS Press (2016)

18. Patel, J.M., Patel, J.M.: Introduction to common crawl datasets. Getting structured data from the internet: running web crawlers/scrapers on a big data production scale, pp. 277–324 (2020)
19. Pomikálek, J.: Removing Boilerplate and Duplicate Content from Web Corpora. Ph.D. thesis, Masaryk University, Faculty of Informatics, Brno, Czech Republic (2011)
20. Rabin, M.O.: Fingerprinting by random polynomials. Technical report (1981)
21. Rychlý, P.: Manatee/Bonito – A Modular Corpus Manager. In: Recent Advances in Slavonic Natural Language Processing (RASLAN 2007), pp. 65–70 (2007)
22. Schäfer, R., Bildhauer, F., et al.: Building large corpora from the web using a new efficient tool chain. In: Lrec, pp. 486–493 (2012)
23. Schmid, H.: Probabilistic part-of-speech tagging using decision trees. In: New Methods in Language Processing, pp. 154–164. Routledge (2013)
24. Spoustová, D., Hajič, J., Raab, J., Spousta, M.: Semi-supervised training for the averaged perceptron POS tagger. In: Proceedings of the 12th Conference of the European Chapter of the ACL (EACL 2009), pp. 763–771 (2009)
25. Straka, M., Hajič, J., Straková, J.: UDPipe: trainable pipeline for processing CoNLL-U files performing tokenization, morphological analysis, pos tagging and parsing. In: Proceedings of the Tenth International Conference on Language Resources and Evaluation (LREC'16), pp. 4290–4297 (2016)
26. Suchomel, V.: Discriminating between similar languages using large web corpora. In: Recent Advances in Slavonic Natural Language Processing (RASLAN 2019), p. 129 (2019)
27. Suchomel, V., Pomikálek, J., et al.: Efficient web crawling for large text corpora. In: Proceedings of the seventh Web as Corpus Workshop (WAC7), pp. 39–43 (2012)

Continual Learning Under Language Shift

Evangelia Gogoulou[1,2](✉), Timothée Lesort[3,4], Magnus Boman[5,6], and Joakim Nivre[1,7]

[1] RISE Research Institutes of Sweden, Kista, Sweden
evangelia.gogoulou@ri.se
[2] KTH Royal Institute of Technology, Stockholm, Sweden
[3] Université de Montréal, Montreal, Canada
[4] MILA-Quebec AI Institute, Montreal, Canada
[5] Karolinska Institutet, Solna, Sweden
[6] MedTechLabs, Solna, Sweden
[7] Uppsala University, Uppsala, Sweden

Abstract. The recent increase in data and model scale for language model pre-training has led to huge training costs. In scenarios where new data become available over time, updating a model instead of fully retraining it would therefore provide significant gains. We study the pros and cons of updating a language model when new data comes from new languages – the case of continual learning under language shift. Starting from a monolingual English language model, we incrementally add data from Danish, Icelandic and Norwegian to investigate how forward and backward transfer effects depend on pre-training order and characteristics of languages, for models with 126M, 356M and 1.3B parameters. Our results show that, while forward transfer is largely positive and independent of language order, backward transfer can be positive or negative depending on the order and characteristics of new languages. We explore a number of potentially explanatory factors and find that a combination of language contamination and syntactic similarity best fits our results.

Keywords: Multilingual NLP · Continual Learning · Large Language Models

1 Introduction

Pre-training on large amounts of text is the current paradigm for training language models. While these models have good generalization capabilities for downstream tasks, they will soon become obsolete given the current rate of producing new data that differ from the training data [21]. Continual learning aims to update a model with new data while retaining the previous model knowledge in different scenarios of data distribution shift [31]. In this paper, we focus on the specific case of language shift, where each new data source is in a different language, and study how the model learns and forgets when sequentially trained on multiple languages, in a standard language modeling scenario. From a continual learning perspective, this multilingual scenario is particularly interesting because the distribution shift between two languages is expected to have a big impact on model knowledge. In addition, we can study cross-lingual transfer effects as a monolingual model becomes progressively multilingual.

The main research question of this paper is: How do the characteristics of languages and their order in the task sequence influence model performance? We focus on how the model performs on new languages – *forward transfer* – and on previous languages – *backward transfer*. We do this by training generative language models on data from English, Danish, Icelandic, and Norwegian, in different sequential orders. We also investigate whether increasing the model capacity alleviates catastrophic forgetting of previous languages or changes the transfer effects in other respects. Finally, in an attempt to explain the transfer effect patterns, we explore language similarity metrics and track language contamination in training data.

Our main findings are: **(i)** Forward transfer is consistently positive and language-independent. **(ii)** Backward transfer can be positive or negative depending on the specific languages and the order in which they are added. **(iii)** Increasing model size improves final performance on all languages but does not affect transfer patterns. **(iv)** Backward transfer effects correlate with language contamination as well as syntactic similarity.

2 Related Work

Continual Learning Continual learning in the deep learning literature generally addresses the problem of learning without forgetting on sequences of datasets [12]. The classic setup for this problem is image classification [2,10]. However, recent progress has led to increased interest in continual learning also in NLP [3,27]. Notably, the increased availability of large pre-trained models provides good foundations for continual learning where only slight adaptations are needed for downstream tasks [1]. We study the influence of model size on continual performance as in previous studies [22,34].

Continual Language Pre-training. Continual pre-training of language models has been previously applied to data evolving through time [15,16,26]. In our setup, we assume factual knowledge to be invariant and are interested in incorporating more knowledge while minimizing forgetting as in domain adaptation [19,36]. As the capacity of algorithms to continuously learn depends on data distribution shifts [23], we study the multilingual setup as a corner case with drastic distribution changes.

Multilingual Language Models. Examples of multilingual Transformer encoder models are mBERT [9] and XLM-R [7]. Following the development of Transformer decoder architectures, several multilingual GPT-style [33] models have been proposed, such as BLOOM [35], XGLM [24], Mixtral [17] and GPT-SW3 [11]. Joint pre-training is used as a baseline in our setup, while our work focuses on multilingual models where each new language is added to the model one by one, starting from scratch. Our model architecture, data, and training process closely follow GPT-SW3 [11]. Studies related to continual learning of multilingual models include a method for continual fine-tuning of mBERT on unbalanced language distributions [4], a study of catastrophic forgetting when continually fine-tuning mBERT on two sequence labeling tasks one language at a time [8], a novel continual multilingual fine-tuning method, which opts for both minimum forgetting in the source languages and reduction of the fine-tuning loss [28], and a study of continual pre-training of language models, focusing on improving algorithmic solutions [39]. It has also been shown that unexpected multilingual capabilities of monolingual models can be explained by language contamination: data from other languages accidentally being included in monolingual training data [6].

3 Method

3.1 Learning Scenario and Setup

We consider a scenario where a generative language model is trained in successive stages given a data input stream $D = D_1, D_2, \ldots D_n$, where each D_i contains data from a different language. After each stage, we evaluate model performance on each language seen during training by computing cross-entropy loss on held-out test sets. To study how forward and backward transfer effects depend on the order and characteristics of languages involved, we systematically vary the order of languages.

In this study, we experiment with data from four languages, namely English (en), Danish (da), Icelandic (is), and Norwegian (no). We consistently take the initial data D_1 from English, and then add Danish, Icelandic and/or Norwegian in all possible orders, which gives us 15 different training sequences of different lengths.[1] All models are randomly initialized before being trained on the first language of the sequence. For each successive continual learning stage i ($i > 1$), all model parameters are initialized from the last checkpoint obtained after stage $i - 1$ and then updated using the standard next-step prediction objective with data batch D_i as input.

The size of the training sets is constant across languages, which allows us to study the effect of language shift independently from data set size. Regarding tokenization, we assume that a multilingual tokenizer, covering all the languages studied, is available to us before training. This is often the practice case as well, given the abundance of openly available tokenizers with large language coverage [7,35], although it is an unrealistic assumption for many low-resource languages. For comparison, we also train two types of baseline models: one monolingual model for each language (en, da, is, no), trained on the corresponding training batch D_i, and one multilingual model (en+da+is+no) trained on the concatenation of the three training sets $D_1 + D_2 + D_3$.

---

[1] 3 sequences of length 2 (English + 1 language) and 6 sequences each of length 3 and 4.

3.2 Model Architecture and Training

We use the standard GPT architecture [33], as implemented in the Nvidia NeMo library.[2] To study the role of model capacity, we run experiments with three model sizes: 126M, 356M, and 1.3B parameters. Exact model hyperparameter values, following [11], are given in Table 5 in Appendix A. For tokenization, all models use the GPT-SW3 tokenizer [11,37], which is a BPE tokenizer with 64K tokens, trained on the Nordic Pile [29], a multilingual corpus including English, Danish, Icelandic, Norwegian, and Swedish. For each language pre-training stage during continual learning, the model is trained for 35K steps (approximately 36M sequences of 2048 tokens each). The monolingual baselines are also trained for 35K steps, while the multilingual baseline model is trained for 105K steps, thereby covering the training sets for all three languages. Following [11], we start by warming up the learning rate for 250 steps, increasing its value to the maximum, and then decreasing it according to the cosine decay function until reaching its minimum value, where it will stay constant for the last 4.9K steps.

Table 1. Data per language (Lang) and number of training sequences in millions (N).

English	N	Danish	N	Icelandic	N	Norwegian	N
The Pile [13]	32	Wiki da	0.36	Icelandic Gigaword [5]	24	NCC [20]	18
Wiki en	4	mc4 (da) [40]	18	Wiki is	0.23	Wiki no	0.37
		OSCAR (da) [30]	16	mc4 (is) [40]	6	mc4 (no) [40]	11
		Danish Gigaword [38]	0.36	OSCAR (is) [30]	6	OSCAR (no) [30]	7

3.3 Datasets

Models are trained on a subset of the Nordic Pile corpus [29].[3] The list of datasets used per language, together with the number of training sequences per dataset is presented in Table 1. Each of the validation and test sets per language consists of 51K samples and has the same proportions of datasets as the training set.

---

[2] https://github.com/NVIDIA/NeMo.
[3] See Appendix A for more information about the total corpus and data weighting scheme.

3.4 Language Similarity Metrics

One of our hypotheses is that the degree of shift between language corpora will have a significant effect on the model knowledge and hence we explore two types of metrics for estimating the similarity between languages and their distributions. First, the linguistic similarity between two languages is used, as estimated by the pre-computed distance[4] of their corresponding *lang2vec* language vectors which are part of the URIEL database [25]. Different types of features are considered: syntactic (SYN), phonetic (INV), and phonological (PHON).[5] Second, we introduce a data-driven metric of similarity between two language corpora, which we call *token distribution similarity* (TDS). The idea is that similarity should take into account the difference in token distributions between two datasets. Hence, we create one vector per dataset, counting the number of occurrences for each token, and then compute the cosine similarity between vectors of different datasets. This is similar to the metric of [14], which estimates the similarity between two vocabularies by the percentage of token overlap between them. The advantage of our method is that we do not take into account only the list of common tokens, but also the frequency of their occurrence. We sample 50K sequences for each language, according to the data weighting scheme used for pre-training. We tokenize each language sample using the model tokenizer and create a vector where position i corresponds to the frequency of vocabulary token i in the sample. Next, we compute the cosine similarity of the two token frequency vectors, which is used as a proxy for language similarity.

The similarity values between all language pairs are shown in Table 2. According to the TDS metric, Danish and Norwegian have by far the highest degree of similarity, followed by English-Norwegian and English-Danish, while English and Icelandic have the lowest similarity. For the URIEL features, we see very different patterns for syntactic features, where English, Danish and Norwegian are most similar while Icelandic is an outlier, and for phonetic and phonological features, where the Scandinavian languages are highly similar and different from English. Our hypothesis is that positive transfer will be observed for languages with the highest TDS value, as this metric derives from the overlap between the data inputs, but we also expect syntactic similarity to be relevant.

3.5 Language Contamination

Another potentially important factor is language contamination, meaning that a language corpus may contain (small) amounts of text in other languages. The degree of contamination has been shown to correlate strongly with the performance of English monolingual models on other languages [6]. To investigate such correlations in our scenario, we quantify the number of out-of-language raw text examples included in each language corpus, using the fastText language identification module [18].

---

[4] http://www.cs.cmu.edu/~dmortens/projects/7_project/.
[5] While phonetic and phonological features are not directly relevant for written texts, they can be seen as proxies for orthographical features.

Table 2. Language similarity values estimated by our TDS metric and by URIEL syntactic (SYN), phonetic (INV) and phonological (PHON) features [25].

	en-da	en-is	en-no	da-is	da-no	is-no
TDS	0.58	0.35	0.60	0.50	0.92	0.54
SYN	0.50	0.21	0.41	0.26	0.42	0.31
INV	0.40	0.40	0.40	1.00	1.00	1.00
PHON	0.43	0.43	0.43	0.99	0.99	0.99

Table 3. Percentage of examples per language in each language pre-training corpus, as classified by the fastText language identification model.

Language corpus	en	da	is	no	sv
English	99.79	0.06	0.0002	0.05	0.08
Danish	1.32	97.56	0.0003	0.98	0.12
Icelandic	0.48	0.07	99.32	0.11	0.01
Norwegian	3.16	1.12	0.001	95.40	0.24

The language contamination analysis results are presented in Table 3. English has the largest miscellaneous presence in the other corpora: around 3% of the Norwegian and around 1% of the Danish raw examples were classified as English. By contrast, the English training corpus contains only small amounts of data from the other languages, although Danish and Norwegian are more common than Icelandic. Icelandic, finally, is almost absent from the other training sets and also has the most homogeneous training set with less than 0.5% of English and about 0.1% each of Danish and Norwegian. We hypothesize that these asymmetries will affect the transfer effects between languages.

4 Results

Table 4 shows test loss on all languages after each training stage. Starting with the baselines, the monolingual models in most cases reach a lower test loss than the corresponding jointly pre-trained model. The main exception is English, where joint pre-training leads to a lower test loss than monolingual training for all three model sizes. This is likely due to language contamination, as both the Danish and Norwegian training sets contain non-negligible amounts of English text. This hypothesis is supported by the fact that monolingual Danish and Norwegian models attain competitive test loss on English.

4.1 Forward and Backward Transfer

To visualize the forward transfer effects, Fig. 1 shows the progression of model test loss on Danish, Icelandic and Norwegian when learned in the 2nd, 3rd or

Table 4. Test loss on English (en), Danish (da), Icelandic (is), and Norwegian (no), measured at the end of each pre-training sequence, for three model sizes.

	Model: GPT 126M				Model: GPT 356M				Model: GPT 1.3B			
	en	da	is	no	en	da	is	no	en	da	is	no
en	3.45	4.85	6.38	5.44	3.08	4.39	5.99	5.04	2.79	4.05	5.70	4.73
da	3.34	2.60	5.82	3.89	3.20	3.19	4.97	2.82	2.92	2.20	5.43	3.38
is	4.60	5.25	2.55	5.79	4.34	4.95	2.33	5.50	4.15	4.73	2.14	5.27
no	3.50	3.47	5.35	3.11	3.15	2.41	5.70	3.65	2.95	2.95	4.64	2.57
en+da+is+no	2.75	2.84	2.69	3.29	2.50	2.56	2.39	2.97	2.22	2.26	2.05	2.61
en-da	3.17	2.54	-	-	4.01	-	2.23	-	2.57	2.03	-	-
en-is	4.43	-	2.58	-	3.02	-	-	2.74	3.89	-	2.26	-
en-no	3.41	-	-	3.03	2.84	2.25	-	-	2.73	-	-	2.48
en-da-is	4.32	4.84	2.52	-	3.91	4.39	2.19	-	3.57	3.91	1.99	-
en-da-no	3.35	3.34	-	3.04	2.99	2.99	-	2.69	2.72	2.70	-	2.43
en-is-da	3.20	2.52	5.33	-	2.86	2.23	4.88	-	2.61	2.01	4.30	-
en-is-no	3.42	-	4.96	3.07	3.03	-	4.48	2.71	2.82	-	4.10	2.51
en-no-da	3.16	2.50	-	3.65	2.83	2.22	-	3.26	2.57	1.99	-	2.94
en-no-is	4.32	-	2.54	5.30	3.91	-	2.19	4.92	3.57	-	1.97	4.43
en-da-is-no	3.35	3.36	4.94	3.03	3.00	3.01	4.43	2.68	2.72	3.83	2.77	2.43
en-da-no-is	4.31	4.85	2.49	5.37	3.89	4.36	2.17	4.89	3.55	3.92	1.96	4.39
en-is-da-no	3.35	3.33	5.12	3.02	2.98	2.98	4.68	2.67	2.70	2.71	4.13	2.42
en-is-no-da	3.16	2.50	5.43	3.63	2.84	2.22	5.01	3.24	2.62	2.01	4.54	2.96
en-no-da-is	4.31	4.83	2.50	5.41	3.89	4.34	2.18	4.94	3.53	3.86	1.96	4.45
en-no-is-da	3.17	2.50	5.27	3.70	2.84	2.21	4.82	3.31	2.59	1.99	4.17	3.00

4th training stage, in comparison to the monolingual baseline (1st stage). We observe a consistent positive trend for all languages, with diminishing returns as the length of the sequence training grows. The only exceptions are found for the 126M model, where there is a slight loss increase for Icelandic in 2nd position and for Norwegian in 3rd position. This suggests that model capacity could be a limiting factor for positive forward transfer.

The research question about backward transfer concerns what happens to previously seen languages when a new language is learned. The left-hand side of Fig. 2 visualizes the evolution of model test loss on English, which is always the first training language, as other languages are added in different order. The patterns here are strikingly different from the ones for forward transfer in that they do not relate to the position of a language in the sequence, but to the

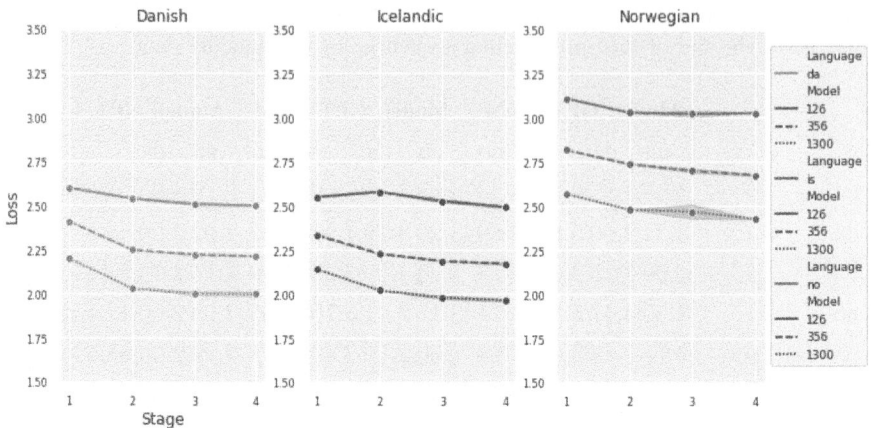

Fig. 1. Test loss on Danish, Icelandic, and Norwegian when learned at different stages. *Clear improvement in the loss is observed when the language is learned later in the sequence, except for the 126M model trained on Icelandic.*

choice of language. More specifically, training a model on Icelandic always leads to an *increase* in the English test loss, which is a sign of catastrophic forgetting or negative transfer. By contrast, training a model on Danish always leads to a *decrease* in the English test loss, regardless of where it occurs in the sequence. Training on Norwegian, finally, has an effect in between the two extremes; it mostly leads to a decrease in English test loss, although smaller than for Danish, but it leads to an increase when added directly after Danish.

The right-hand side of Fig. 2 instead visualizes the test loss for Danish, Icelandic, and Norwegian, respectively, as the other two languages are added in different order. In this case, Icelandic still hurts the other two and also suffers a degradation in performance when one of the other two is added. By contrast, Danish and Norwegian have a much weaker negative effect on each other when adjacent, and mutually help each other when added after Icelandic. The overall pattern emerging from these observations about negative transfer is that Icelandic interacts negatively with all the other languages, while Danish and Norwegian mostly have a weak positive effect on English and each other, partly dependent on their position in the training sequence.

4.2 Model Size

Our final research question concerns the effect of model size on forward and backward transfer. As shown in Fig. 3 (left), overall test loss decreases as expected with increasing model size. However, contrary to our expectations, the forgetting patterns do not change noticeably when increasing the model size. This is reflected already in the largely parallel curves in Figs. 2, but is visualized directly in Fig. 3 (right), which shows the trade-off between test loss in the current language and the cumulative loss increase on previous languages. Symbols

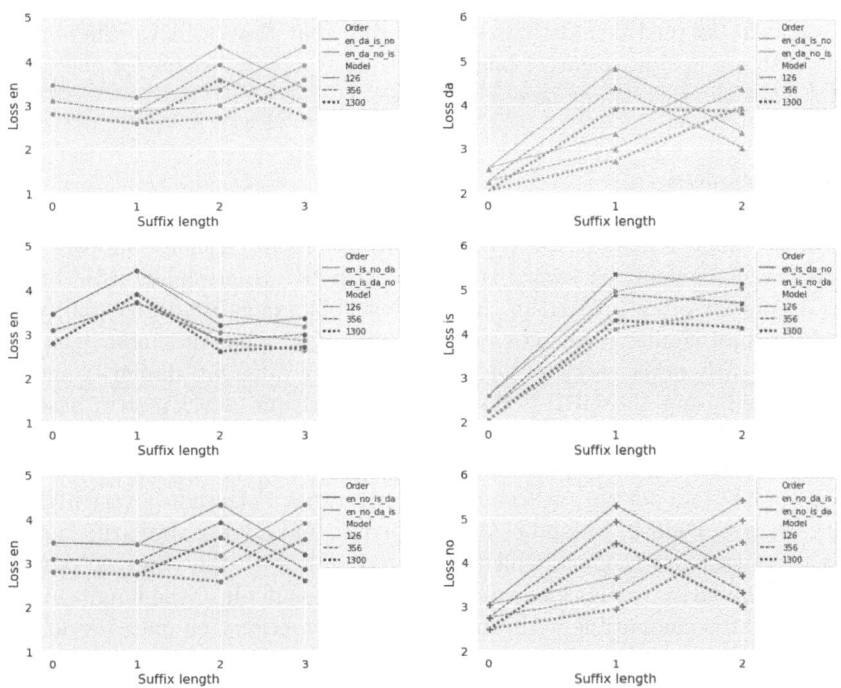

Fig. 2. Suffix length refers to the number of languages added after then one visualised. Test loss on English (**Left**) and on Danish, Icelandic, and Norwegian (**Right**) for models with varying size and language suffixes. *Overall, Icelandic always causes forgetting to the other languages, while positive (or weaker negative) transfer is observed between Danish and Norwegian, and from those two languages to English.*

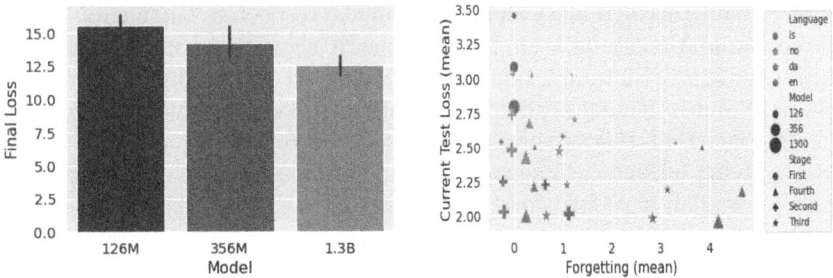

Fig. 3. Left: Cumulative loss over all language test sets, averaged per model size at the final stage. *Growing the model size from 126M to 356M and then to 1.3B leads to a drop of the model test loss on average by 8% and 11% respectively.* **Right:** Test loss in the current language vs forgetting (i.e. loss growth on previous languages). *For a given language and stage (color and shape), increasing the size of the model consistently decreases the current loss while forgetting remains mostly the same.*

representing different model sizes for the same language and sequence position are vertically aligned in most cases, indicating that they achieve different loss levels with a constant negative backward transfer effect, except for the largest model when the target language is Icelandic.

5 Discussion

One of our main results is that, while positive forward transfer appears to be language-independent, backward transfer is sensitive to cross-linguistic interaction. A concept often invoked in this context is language similarity [32], where the idea is that similar languages support each other, while dissimilar languages may harm each other. Section 3.4 reports on four different similarity metrics, three derived from the URIEL database [25] and one novel metric based on token distribution similarity (TDS).

Starting with the URIEL metrics, we note that the phonetic (INV) and phonological (PHON) do not appear to be informative; they both show a very high similarity between Danish, Icelandic and Norwegian, and a lower similarity between these languages and English, which is incompatible with the observed patterns. Syntactic similarity (SYN) looks more promising, as it ranks the language pairs in a way that is compatible with the observed interactions: en-da > da-no > en-no >is-no > da-is > en-is. The TDS metric generally shows the lowest values for pairs involving Icelandic, which is compatible with the observed negative backward transfer effects for this language. However, only the value for Icelandic and English is significantly lower than for other language pairs. Moreover, despite a very high similarity between Danish and Norwegian, we do not observe a strong positive backward transfer between these languages, so the explanatory value of the TDS metric seems limited.

For language contamination, [6] shows that the performance of (supposedly) monolingual models for English on other languages is strongly correlated with the amount of data (accidentally) included from the respective languages. The language contamination statistics in Sect. 3.5 clearly show that Icelandic is extreme both in having less contamination in its training set (especially for Danish and Norwegian) and in having negligible amounts of data in the other training sets, which suggests that this is the primary cause of the strong negative transfer effects involving Icelandic. Language contamination could also explain why Danish and Norwegian have stronger positive effects on English than on each other, given that these training sets contain the largest amount of English data among the non-English training sets. However, it does not explain why Danish has a stronger positive effect than Norwegian, since the amount of English data is more than twice as large in the Norwegian training set. The fact that Danish shows a larger syntactic similarity to English than Norwegian suggests that we need to take both contamination and linguistic similarity into account.

6 Conclusion

In this paper, we study forward and backward transfer in a language modeling task – standardly used for pre-training of large language models – in a novel multilingual continual learning scenario. Our experimental results with three model sizes and four languages indicate that forward transfer is consistently positive. However, backward transfer can be positive or negative, depending on the choice of languages and their order. In an attempt to further analyze the transfer patterns observed, we find that no single factor can explain all patterns and that a combination of language contamination and syntactic similarity best fits our empirical results. Increasing model size improves performance for all languages, but all transfer patterns remain essentially the same.

We hope that our findings will help design large-scale continual learning language models The next step would be to explore a larger and more diverse set of languages to assess the generality of the transfer patterns. As a long-term goal, this could improve resource efficiency while reducing negative transfer in the computationally extensive setup of continual language model pre-training.

Acknowledgments. The research presented in this paper was supported by the Swedish Research Council (grant no. 2022-02909). A significant part of the computations was enabled by the Berzelius resource provided by the Knut and Alice Wallenberg Foundation at the National Supercomputer Centre at Linköping University, Sweden (Berzelius-2023-178). In addition, the authors gratefully acknowledge the HPC RIVR consortium (www.hpc-rivr.si) and EuroHPC JU (eurohpc-ju.europa.eu) for funding this research by providing computing resources of the HPC system Vega at the Institute of Information Science (www.izum.si). Magnus Boman acknowledges funding from the Swedish Research Council on Scalable Federated Architectures.

A Experimental Details

Table 5 shows parameters for each model size. Table 6 shows dataset weights.

Table 5. Model and training parameters for the three model sizes.

Parameter Name	126M Model	356M Model	1.3B Model
Sequence length	2048	2048	2048
Number of layers	12	24	24
Hidden size dimension	768	1024	2048
Number of attention heads	12	16	32
Tensor parallelism	1	2	4
Micro batch size	4	4	4
Global batch size	1024	1024	1024
Min learning rate	$6e-5$	$3e-5$	$2e-5$
Max learning rate	$6e-4$	$3e-4$	$2e-4$

Table 6. Datasets with size and assigned weighted contribution to the language corpus.

Dataset	Category	Weight	Dataset size (GB)	Normalised weight
books3 (The Pile)	Books	1	89	0.43
PubMed, arXiv (The Pile)	Articles	0.9	33	0.15
Stackexchange (The Pile)	Miscellaneous	1	35	0.17
NCC	Miscellaneous	1	39	0.49
Icelandic Gigaword	Miscellaneous	1	9.6	0.69
Danish Gigaword	Miscellaneous	1	3.47	0.06
Pile Openwebtext	Web CC	0.5	58	0.14
mc4, oscar (no)	Web CC	0.5	78	0.49
mc4, oscar (is)	Web CC	0.5	8.4	0.34
mc4, oscar (da)	Web CC	0.5	93	0.93
Wiki en	Wiki	1.5	15	0.1
Wiki no	Wiki	1.5	0.5	0.01
Wiki is	Wiki	1.5	0.05	0.006
Wiki da	Wiki	1.5	0.4	0.01

Our experiments ran in two clusters: Vega[6] a GPU cluster part of the EuroHPC network consisting of 60 GPU nodes of 4 NVidia A100 each and Berzelius[7] which is located in Linköping, Sweden and consists of 94 GPU nodes of 8 NVidia A100 each. A single monolingual training phase of the 126M model required 352 GPU hours on Vega, while the same run for the 356M model took 352 GPU hours on Berzelius. All 1.3B models were trained on Berzelius and each of the pre-training stages on a single language took 2272 GPU hours.

References

1. Abnar, S., Dehghani, M., Neyshabur, B., Sedghi, H.: Exploring the limits of large scale pre-training. In: International Conference on Learning Representations (2021)
2. Aljundi, R., et al.: Online continual learning with maximal interfered retrieval. In: Advances in Neural Information Processing Systems (2019)
3. de Masson d'Autume, C., Ruder, S., Kong, L., Yogatama, D.: Episodic memory in lifelong language learning. In: Advances in Neural Information Processing Systems (2019)
4. Badola, K., Dave, S., Talukdar, P.: Parameter-efficient finetuning for robust continual multilingual learning. arXiv preprint arXiv:2209.06767 (2022)
5. Barkarson, S., Steingrímsson, S., Hafsteinsdóttir, H.: Evolving large text corpora: Four versions of the Icelandic Gigaword Corpus. In: Proceedings of the Thirteenth Language Resources and Evaluation Conference, pp. 2371–2381 (2022)
6. Blevins, T., Zettlemoyer, L.: Language contamination helps explains the cross-lingual capabilities of English pretrained models. In: Proceedings of the 2022 Conference on EMNLP (2022)

---

[6] https://doc.vega.izum.si/.
[7] https://www.nsc.liu.se/systems/berzelius/.

7. Conneau, A., et al.: Unsupervised cross-lingual representation learning at scale. In: Proceedings of the 58th Annual Meeting of the ACL, pp. 8440–8451 (2020)
8. Coria, J.M., et al.: Analyzing BERT cross-lingual transfer capabilities in continual sequence labeling. In: Proceedings of the First Workshop on Performance and Interpretability Evaluations of Multimodal, Multipurpose, Massive-Scale Models, pp. 15–25. International Conference on Computational Linguistics (2022)
9. Devlin, J., Chang, M.W., Lee, K., Toutanova, K.: BERT: pre-training of deep bidirectional transformers for language understanding. In: Proceedings of the 2019 Conference of the North American Chapter of the Association for Computational Linguistics: Human Language Technologies, Volume 1 (Long and Short Papers), pp. 4171–4186 (Jun 2019)
10. Douillard, A., Ramé, A., Couairon, G., Cord, M.: Dytox: transformers for continual learning with dynamic token expansion. arXiv preprint arXiv:2111.11326 (2021)
11. Ekgren, A., et al.: GPT-SW3: an autoregressive language model for the Nordic languages. arXiv preprint arXiv:2305.12987 (2023)
12. French, R.M.: Catastrophic forgetting in connectionist networks. Trends Cognit. Sci. 128–135 (1999)
13. Gao, L., et al.: The Pile: an 800GB dataset of diverse text for language modeling. arXiv preprint arXiv:2101.00027 (2020)
14. Garcia, X., Constant, N., Parikh, A.P., Firat, O.: Towards continual learning for multilingual machine translation via vocabulary substitution. arXiv preprint arXiv:2103.06799 (2021)
15. Han, R., Ren, X., Peng, N.: ECONET: effective continual pretraining of language models for event temporal reasoning. In: EMNLP 2021, pp. 5367–5380 (2021)
16. Jang, J., et al.: TemporalWiki: a lifelong benchmark for training and evaluating ever-evolving language models. In: Proceedings of the 2022 Conference on Empirical Methods in Natural Language Processing, pp. 6237–6250 (2022)
17. Jiang, A.Q., et al.: Mixtral of experts. arXiv preprint arXiv:2401.04088 (2024)
18. Joulin, A., Grave, E., Bojanowski, P., Mikolov, T.: Bag of tricks for efficient text classification. arXiv preprint arXiv:1607.01759 (2016)
19. Ke, Z., Shao, Y., Lin, H., Konishi, T., Kim, G., Liu, B.: Continual pre-training of language models. In: The Eleventh International Conference on Learning Representations (2023)
20. Kummervold, P., Wetjen, F., de la Rosa, J.: The Norwegian colossal corpus: a text corpus for training large Norwegian language models. In: Proceedings of the Thirteenth Language Resources and Evaluation Conference, pp. 3852–3860 (Jun 2022)
21. Lazaridou, A., et al.: Mind the gap: assessing temporal generalization in neural language models. In: Advances in Neural Information Processing Systems, pp. 29348–29363 (2021)
22. Lesort, T., et al..: Challenging common assumptions about catastrophic forgetting and knowledge accumulation. In: Chandar, S., Pascanu, R., Sedghi, H., Precup, D. (eds.) Proceedings of The 2nd Conference on Lifelong Learning Agents. Proceedings of Machine Learning Research, 22–25 Aug 2023, vol. 232, pp. 43–65. PMLR (2023). https://proceedings.mlr.press/v232/lesort23a.html
23. Lesort, T., Caccia, M., Rish, I.: Understanding continual learning settings with data distribution drift analysis. arXiv:2104.01678 (2021). https://arxiv.org/abs/2104.01678
24. Lin, X.V., et al.: Few-shot learning with multilingual generative language models. In: Proceedings of the 2022 Conference on Empirical Methods in Natural Language Processing, pp. 9019–9052 (2022)

25. Littell, P., Mortensen, D.R., Lin, K., Kairis, K., Turner, C., Levin, L.: URIEL and lang2vec: representing languages as typological, geographical, and phylogenetic vectors. In: Proceedings of the 15th Conference of the European Chapter of the Association for Computational Linguistics: Volume 2, Short Papers, pp. 8–14 (2017)
26. Loureiro, D., Barbieri, F., Neves, L., Espinosa Anke, L., Camacho-collados, J.: TimeLMs: diachronic language models from Twitter. In: Proceedings of the 60th Annual Meeting of the Association for Computational Linguistics: System Demonstrations (2022)
27. Magdalena Biesialska, Katarzyna Biesialska, M.R.C.J.: Continual lifelong learning in natural language processing: a survey. In: Proceedings of the 28th International Conference on Computational Linguistics, pp. 6523–6541 (2020)
28. Mao, Y., et al.: Less-forgetting multi-lingual fine-tuning. In: Advances in Neural Information Processing Systems, pp. 14917–14928 (2022)
29. Öhman, J., et al.: The Nordic Pile: A 1.2 TB Nordic dataset for language modeling. arXiv preprint arXiv:2303.17183 (2023)
30. Ortiz Suárez, P.J., Romary, L., Sagot, B.: A monolingual approach to contextualized word embeddings for mid-resource languages. In: Proceedings of the 58th Annual Meeting of the Association for Computational Linguistics, pp. 1703–1714 (2020)
31. Parisi, G.I., Kemker, R., Part, J.L., Kanan, C., Wermter, S.: Continual lifelong learning with neural networks: a review. Neural Networks 54 – 71 (2019)
32. Philippy, F., Guo, S., Haddadan, S.: Towards a common understanding of contributing factors for cross-lingual transfer in multilingual language models: a review. In: Proceedings of the 61st Annual Meeting of the Association for Computational Linguistics (Volume 1: Long Papers), pp. 5877–5891 (2023)
33. Radford, A., Narasimhan, K., Salimans, T., Sutskever, I., et al.: Improving language understanding by generative pre-training. OpenAI (2018)
34. Ramasesh, V.V., Lewkowycz, A., Dyer, E.: Effect of scale on catastrophic forgetting in neural networks. In: International Conference on Learning Representations (2022)
35. Scao, T.L., et al.: BLOOM: a 176B-parameter open-access multilingual language model. arXiv preprint arXiv:2211.05100 (2022)
36. Scialom, T., Chakrabarty, T., Muresan, S.: Fine-tuned language models are continual learners. In: Proceedings of the 2022 Conference on Empirical Methods in Natural Language Processing, pp. 6107–6122 (2022)
37. Stollenwerk, F.: Training and evaluation of a multilingual tokenizer for GPT-SW3. arXiv preprint arXiv:2304.14780 (2023)
38. Strømberg-Derczynski, L., et al.: The Danish Gigaword corpus. In: Proceedings of the 23rd Nordic Conference on Computational Linguistics (NoDaLiDa), pp. 413–421 (2021)
39. Winata, G.I., et al..: Overcoming catastrophic forgetting in massively multilingual continual learning. arXiv preprint arXiv:2305.16252 (2023)
40. Xue, L., et al.: mT5: a massively multilingual pre-trained text-to-text transformer. In: Proceedings of the 2021 Conference of the North American Chapter of the Association for Computational Linguistics: Human Language Technologies, pp. 483–498 (2021)

Neural Spell-Checker: Beyond Words with Synthetic Data Generation

Matej Klemen[1(✉)], Martin Božič[1], Špela Arhar Holdt[1,2], and Marko Robnik-Šikonja[1]

[1] Faculty of Computer and Information Science, University of Ljubljana, Večna pot 113, 1000 Ljubljana, Slovenia
{matej.klemen,martin.bozic,marko.robnik}@fri.uni-lj.si,
spela.arharholdt@ff.uni-lj.si
[2] Faculty of Arts, University of Ljubljana, Aškerčeva cesta 2, 1000 Ljubljana, Slovenia

Abstract. Spell-checkers are valuable tools that enhance communication by identifying misspelled words in written texts. Recent improvements in deep learning, and in particular in large language models, have opened new opportunities to improve traditional spell-checkers with new functionalities that not only assess spelling correctness but also the suitability of a word for a given context. In our work, we present and compare two new spell-checkers and evaluate them on synthetic, learner, and more general-domain Slovene datasets. The first spell-checker is a traditional, fast, word-based approach, based on a morphological lexicon with a significantly larger word list compared to existing spell-checkers. The second approach uses a language model trained on a large corpus with synthetically inserted errors. We present the training data construction strategies, which turn out to be a crucial component of neural spell-checkers. Further, the proposed neural model significantly outperforms all existing spell-checkers for Slovene in both precision and recall.

Keywords: Spell-checking · Large language models · Synthetic data construction · Morphological lexicon · Less-resourced languages

1 Introduction

Correct spelling enhances the clarity, effectiveness, comprehensibility, and consistency of written communication. Spell-checkers are tools that detect incorrectly spelled words, and, optionally, suggest their correct versions. Spelling error detection and correction are mostly considered separate problems, although they are sometimes tackled jointly. In this work, we focus on spelling error detection without suggesting corrections.

Existing work on spelling error detection can be divided into traditional non-neural and recent neural approaches. The former approaches validate the correctness of words by comparing them against a reference set of correctly spelled

---

M. Klemen and M. Božič—Equal contribution.

© The Author(s), under exclusive license to Springer Nature Switzerland AG 2024
E. Nöth et al. (Eds.): TSD 2024, LNAI 15048, pp. 85–96, 2024.
https://doi.org/10.1007/978-3-031-70563-2_7

words or their statistics. The neural approaches learn from large amounts of texts corrupted in an opaque way. The neural approaches typically perform better as they can learn complex text interactions and detect errors beyond misspelled words, e.g., words that should be split into two or merged into one. However, they are computationally more demanding due to the use of large models and may be harder to use in practical applications such as text editing software. In this work, we introduce two new spell-checkers for the Slovene language: a non-neural morphological lexicon-based approach (SloSpell) and a neural approach (SloNSpell). These tools support both computationally less restrictive and more restrictive settings; however, even SloNSpell is designed to be computationally efficient and usable for most contexts.

Most existing work on spelling error detection has focused on broadly-spoken languages such as Chinese and English. However, some coverage exists for a relatively large pool of languages, including less-resourced ones. In our work, we focus on the morphologically rich Slovene language, for which four off-the-shelf spell-checkers exist: the commercial, rule-based tool Amebis Besana, the freemium open source LanguageTool, the open source Hunspell used by many open source applications such as LibreOffice, and the Hunspell-based Loris. Besana is a rule-based grammar checker that can detect and correct various spelling and grammatical issues, including misplaced punctuation. Hunspell is a dictionary-based tool for spell-checking with support for Slovene. The word list used in this tool was released in 2006 and was compiled from a significantly smaller data source than is available today. Traditional spell-checking systems take into account limited surrounding context and are unlikely to identify instances of spelling errors that are formally the same as another valid word form (e.g., 'came' instead of 'cane'). In our work, we first present the traditional dictionary-based system SloSpell with an updated dictionary of valid word forms based on a morphological lexicon. Second, we present the SloNSpell neural system capable of handling more complex spelling errors.

We make the following contributions:

- *SloNSpell*, an efficient neural spell-checker based on the BERT transformer model, together with methodology and data generators for the synthetic construction of spelling errors.
- *SloSpell*, classical spell-checker, using by far the largest word list for Slovene, based on the Sloleks 3.0 morphological lexicon [4].
- *Evaluation* on synthetic and authentic language data showing strengths and weaknesses of the proposed approaches together with their comparison with existing baselines. The evaluation includes two scenarios: texts written by young learners, sampled in the Šolar-Eval dataset [1], and texts written by professional adult writers, sampled in the Lektor-Spelling dataset.

We publish the source code and fine-tuned model used in our experiments under permissible open source licenses and release the neural spell-checker in an online interface[1].

---

[1] The code is available at https://github.com/matejklemen/slonspell and the model at https://hf.co/cjvt/SloBERTa-slo-word-spelling-annotator.

The remainder of this paper is structured as follows. In Sect. 2, we outline the development of spell-checkers and existing work in Slovene. In Sect. 3, we describe the morphological lexicon Sloleks and how the classical spell-checker is built on top of it. In Sect. 4, we describe the neural approach based on the BERT model and synthetically generated errors. In Sect. 5, we describe the evaluation protocol and compare our systems against strong baselines. Last, in Sect. 6, we provide conclusions and possibilities for future research.

2 Related Work

We review related work that focuses on either isolated spelling error detection or spelling error correction, which encompasses the detection step. Existing work can be divided into classical and neural approaches. Non-neural approaches dominated earlier research but are nowadays rare. Nevertheless, these approaches remain in use due to their computational efficiency and practical usability. Below, we only review recent, neural approaches.

Neural approaches, by definition, use neural networks, in particular language models (LMs), to test the validity of a given word. For example, Li et al. [10] use the ELECTRA [5] discriminator model to detect incorrect Chinese characters. As spelling errors are commonly characterized by external factors such as pronunciation patterns, some approaches inject additional knowledge into LMs to improve detection accuracy. For example, Ji et al. [8] include radical and Pinyin information as additional visual and phonetic features in Chinese spelling error detection.

Although most existing research focuses on broadly spoken languages such as Chinese and English, some coverage exists for a relatively large pool of languages, including less-resourced ones such as Croatian [11] and Urdu [3]. For Croatian, the authors train an XLM-RoBERTa model on a Croatian dataset with synthetically generated errors.

3 SloSpell: a Spell-Checker Using a Morphological Lexicon

To construct a pattern-matching spell-checker, we need a high-quality, comprehensive list of valid word forms. For morphologically rich languages, the number of word forms multiple times exceeds the number of word lemmas – for Slovene, approximately nine times. We use the open-source morphological lexicon Sloleks 3.0 [4], containing all word forms for a large set of lexemes, part of which are manually validated.

In this section, we first describe the morphological lexicon and then describe how we built the spell-checker on top of it.

Sloleks 3.0 is a Slovene morphological lexicon containing 365 340 entries, identifiable by their lemmas. Version 3.0 expands version 2.0 with approximately 265 000 new lemmas that frequently occur in reference corpora. In addition to the lemma, each entry contains its part of speech category, inflected word forms, and morphosyntactic information annotated automatically following the MULTEXT-East specifications for Slovene [6]. The entire Sloleks 2.0 was manually validated. Among the new lemmas in Sloleks 3.0, the verbs, adjectives, adverbs, and common nouns were manually validated, while the accentuated forms were automatically generated. The total number of unique word forms in Sloleks 3.0 is 3 028 666.

In our **lexicon-based spell-checker**, the input text is first tokenized into words, and each word is processed in isolation without context. For each word, several checks are performed to circumvent some limitations due to the nature of the lexicon. The word is treated as correct if it is a number or a numeral, a URL, a punctuation, or an otherwise special symbol. These exceptions are not covered by the lexicon either because they are not "normal" words or because they can take on infinitely many forms. If none of the stated conditions are true, the word is looked up in the Sloleks lexicon, implemented as a hash map: if it exists, it is considered correct; otherwise, it is flagged as incorrect.

One advantage of using this simple architecture is that future versions of the morphological lexicon can replace the existing one without further adaptations.

4 SloNSpell: a Neural Spell-Checking Approach

We design an approach using the Slovene BERT model SloBERTa. We first describe the specifics of fine-tuning the SloBERTa model in Sect. 4.1, followed by synthetic datasets for different types of spelling errors in Sect. 4.2.

4.1 The BERT Model Fine-Tuning

In this section, we describe the fine-tuning of the Slovene SloBERTa [13] model for the identification of misspelled words in a text. As BERT models use only the encoder part of the transformer architecture [14], their inference is parallel and considerably faster than generative models, such as T5 or GPT, that generate one output at a time. The dataset we created for training the SloBERTa model is described next, in Sect. 4.2.

To fine-tune the SloBERTa model, we introduce a special mask token after each word in a model input vector. This ensures that the prediction always corresponds to a word, even if it is split into multiple tokens by the tokenizer. At the same time, we construct a vector with labels indicating if the word before the mask token is spelled correctly (0), incorrectly (1), needs to be combined with another word (2), or should be split into two words (3). An example of the input vector paired with its corresponding labels is provided in Table 1.

Table 1. A training example of a masked sentence (split into two rows) with labels used for the SloBERTa detector of misspelled words. The word "Vosil" ("Drving") is misspelled, indicated by "1" in the labels vector. Words "av" ("ca") and "to" ("r") should be combined, indicated by "2" in the labels vector. Using the BERT model, we focus on the output tokens that are adjacent to the <mask> tokens, ignoring all other output tokens, when calculating the loss. The number of labels corresponds to the number of <mask> tokens. Labels are tokens with indices 0, 1, and 2.

masked sentence	labels	translated masked sentence
Mečka <mask> spi <mask> na	1 0	Cot <mask> sleeps <mask> on
<mask> tip <mask> kovnici <mask>	0 2 2	<mask> key <mask> board <mask>

4.2 Synthetic Data for Misspelling Detection

To prepare the synthetic training dataset, we initially segment sentences from the Gigafida 2.0 corpus [9] and group them to form sentence groups, ensuring that their total length does not exceed 128 tokens. In each sentence group, we identified individual words and randomly modified some of them. We introduced six different word modification methods. The first component splits a given word into two words, the second concatenates two words, while the input and output of the remaining components are individual words. We show the outline of synthetic data preparation in Fig. 1.

Each word modification method is selected with a probability determined by preliminary experiments on separate synthetically generated data; we tested multiple probabilities in the range from 1% to 10%. In the process of synthetic data generation, we aim to generate realistic errors and preserve sentence structure and meaning. Therefore, if a certain modification method is selected, we apply it and then allow for its further selections with the same probability. This simulates multiple spelling errors in the same word.

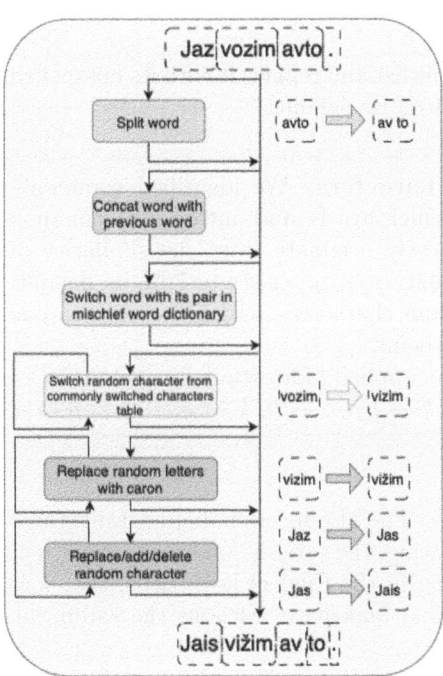

Fig. 1. The process of synthetic dataset generation used to train the SloNSpell model. The last three modification methods can be applied multiple times.

Word Split Generator. The generator selects a random position within a word as a candidate for inserting a space. In 99% of cases, we insert a space only if both parts of the split are recognized as valid words in the Sloleks lexicon (see Sect. 3). In 1% of cases, we introduce the space without such a validation.

Within the synthetic dataset for error detection, we apply the word split process with probability p_{word_split} set to 3% by default.

Generating Concatenated Words. Conceptually similar to splitting words, we concatenate them with the same default probability $p_{conc} = 3\%$. We randomly select two words. With probability $p_{conc\_exists} = 99\%$, we combine these words if the concatenated word exists in the Sloleks lexicon, and with the probability $1 - p_{conc\_exists} = 1\%$ we concatenate them unconditionally.

Introducing Commonly Misspelled Words. We collected words that commonly deviate from their conventional spelling (also called "mischief words"), in the Mischief list. We gathered these words from textbooks used in Slovene primary and secondary schools.

For each word, we check if it exists in this list and replace it with its misspelled version with probability $p_{mischief}$ set to 10% by default.

Generating Commonly Switched Characters. We identified commonly switched or replaced characters in mischief words and introduced common character replacements. The switches mostly originate from the similarity of phonemes. We select a word with probability p_{switch_chr} set to 70% by default. In the selected word, we select four random characters, and if they match one of the characters in the table, we switch them.

Common character switches in Slovene, applied bidirectionally, include: $n \leftrightarrow nj$, $l \leftrightarrow lj$, $t \leftrightarrow d$, $v \leftrightarrow u$, $u \leftrightarrow el$, $i \leftrightarrow j$, $k \leftrightarrow kj$, $k \leftrightarrow h$, $k \leftrightarrow g$, $s \leftrightarrow z$, $p \leftrightarrow b$, $š \leftrightarrow ž$, $v \leftrightarrow l$, $u \leftrightarrow l$, $t \leftrightarrow tj$, and $i \leftrightarrow ij$.

Replacement of Letters with a Caron. The Slovene alphabet contains three letters with a caron: č, š, and ž. In handwriting, the caret is sometimes forgotten, and in typing, these characters are missing in the English keyboards, therefore they are sometimes substituted with their counterparts without the caron, i.e. c, s, and z.

To generate this type of synthetic error, we select each word containing these characters with probability p_{caron} set by default to 5% and rewrite it by switching them with their non-caron counterparts.

Replacement of Random Characters. In this error generator, we modify words by switching, replacing, and inserting randomly chosen characters. We sequentially apply each of the five possible modifications to every word in the text: With probability $p_{vowel} = 5\%$, we select a random vowel and swap it with another random vowel. With probability $p_{consonant} = 5\%$, we switch a randomly selected consonant with another random consonant. With probability $p_{subst\_chr} = 2\%$, we substitute a random character with another character from the Slovene alphabet. With probability $p_{del\_chr} = 4\%$, we remove a randomly selected character from a word. With probability $p_{insert\_chr} = 3\%$, we insert a random character at a random position within a word.

Training Dataset. The final training dataset for the SloNSpell model includes 31 682 971 words distributed across 297 041 sentences. When tokenized with the SloBERTa model tokenizer, the entire training corpus contains 39 143 264 tokens.

5 Experimental Evaluation

In this section, we describe the evaluation of the proposed spell-checkers. We first describe the evaluation datasets and baseline systems, then move on to the quantitative and qualitative performance analysis.

5.1 Used Evaluation Datasets

We utilize three datasets that sample various types of texts across different evaluation settings: synthetic data and authentic texts from both young learners and adult professional writers. Their summary is displayed in Table 2.

Table 2. Summary of the evaluation datasets for spelling detection.

Dataset	text source	#words	#sentences	% errors
Synthetic-Eval	generated	53 990	959	5.29
Šolar-Eval	learners	60 271	740	1.27
Lektor-Spelling	professionals	57 336	1080	1.76

Synthetic-Eval. We construct the Synthetic-Eval dataset following the same methodology used for the SloNSpell training dataset, as outlined in Sect. 4.2. This involves randomly selecting sentences from the Gigafida 2.0 corpus [9], ensuring they are not already part of the model's training data. However, in contrast to the training corpus, where we use larger error probabilities to promote learning, in the evaluation, we reduce the likelihood of specific errors to one-eighth of their original values. This adjustment results in a lower frequency of mistakes so that the error proportion in the Synthetic-Eval dataset is halfway between the error proportion in the two authentic language datasets and the error proportion used during training. The Synthetic-Eval dataset comprises a total of 57 336 words, which are distributed across 959 sentences.

Šolar-Eval. Šolar-Eval [1] is a dataset containing 109 essays written by Slovene primary and secondary school students. The essays are sampled from the Šolar 3.0 [2] corpus which contains essays corrected by school teachers. As the focus of authentic teacher corrections is on providing educational feedback and not necessarily on annotation completeness, Šolar-Eval was manually annotated by linguist researchers [7] to ensure completeness and consistency and enable its use in the evaluation of grammatical error correction tools. The resource is annotated with a diverse range of error categories, including spelling, morphology, vocabulary, syntax, and orthography; in our evaluation, we only consider spelling errors.

Lektor-Spelling. Due to a lack of reference evaluation datasets for Slovene spell-checking, we decided to create a new one for our task. We named this dataset Lektor-Spelling and included 1080 examples, randomly sampled from the Lektor corpus [12], known for containing nonliterary texts together with corrections made by proofreaders. The annotation of Lektor-Spelling uses a similar but not fully interchangeable error category system as that used in Šolar. Following the protocol of Šolar-Eval, our evaluation dataset includes annotated spelling errors, as well as errors related to the conjoining or separating of words and word capitalization. The dataset was cleaned of irrelevant segments (e.g., non-Slovene texts or bibliographical references) and double-checked for consistency of error annotation by two independent annotators.

5.2 Baseline Systems

As currently there exists no neural spell-checker for Slovene, we utilize three baseline systems: two classical baselines (HunSpell and LanguageTool), as well as ChatGPT 4 as a proxy for neural spell-checker. We present their short description below.

HunSpell$_{SL}$. HunSpell is an open-source spell-checker with support for multiple languages used by applications such as LibreOffice and Mozilla Firefox. Support for Slovene was added in 2006[2] – its dictionary contains 246 856 entries.

LanguageTool$_{SL}$. LanguageTool is an open-source style and grammar correction tool supporting multiple languages, including Slovene. It uses the dictionary from HunSpell and additional rules covering certain error patterns such as the use of duplicate words, unclosed parentheses, etc. It has a free and a premium version – in our work, we use the free version via the API.

ChatGPT4. As a proxy for neural spell-checker, we employ ChatGPT powered by the GPT-4 model, specifically the "gpt-4-0125-preview" version.

In structuring prompts for ChatGPT, we first clearly define the problem by directing ChatGPT to detect and highlight specific spelling mistakes in a given

---

[2] Prepared by Amebis, Tomaž Erjavec, Aleš Košir, and Primož Peterlin: https://cgit.freedesktop.org/libreoffice/dictionaries/tree/sl_SI.

Table 3. The evaluation results for different systems on the three Slovene datasets. We report macro $F_{0.5}$ scores.

System	Synthetic-Eval	Šolar-Eval	Lektor-Spelling
HunSpell$_{SL}$	0.63	0.53	0.51
LanguageTool$_{SL}$	0.84	0.75	0.63
ChatGPT4$_{1S}$	0.83	0.65	0.57
SloSpell	0.88	0.79	0.62
SloNSpell	**0.97**	**0.92**	**0.65**

text. We enhance the prompt with one generic example that includes misspelled words and words incorrectly written together or apart. We instruct ChatGPT to tag these errors using three distinct labels ("<mistake>", "<split>" and "<concat>"). Later, during model output post-processing, we merge these labels into a single category. Our preliminary experimentation indicates that requesting labels for each type of mistake leads to enhanced performance.

5.3 Analysis of Results

We first analyze the quantitative results (shown in Table 3) using the $F_{0.5}$ score as the metric, as is common in grammatical error correction literature. The difference to the better known F_1 score is that $F_{0.5}$ score gives more emphasis to precision than to recall; this is a natural choice for spelling detection systems, where the importance of correct error identification is greater than capturing all errors.

The results show the superiority of the proposed neural spell-checker SloN-Spell over the tested baselines on all three benchmarks. Surprisingly, our lexicon-based SloSpell spell-checker achieves second or third best scores, showcasing the strength of a simple spell-checker using a large, high-quality, curated lexicon. Most notably, SloSpell and LanguageTool both surpass ChatGPT4 in the one-shot learning setting. The poor performance of ChatGPT4 on Šolar-Eval can potentially be attributed to the relatively specific vocabulary used in school essays (e.g., names of characters in Slovene literary works) to which ChatGPT4 was not exposed during its training. The lower score on the Synthetic-Eval dataset may be caused by the synthetic errors resulting in very unnatural words, whereas ChatGPT4 has presumably only observed a tiny portion of (more) natural Slovene text. The worst scores overall are achieved by the HunSpell$_{SL}$ spell-checker, likely due to the relatively small and outdated Slovene dictionary. Time-wise, SloNSpell is computationally much more demanding than SloSpell – on Šolar-Eval, SloNSpell processes approximately three examples per second (running on an Apple M3 Pro GPU) while SloSpell processes approximately 250 examples per second (running on an Intel i7-1165G7 CPU). Next, we qualitatively analyze the results to outline the strengths and weaknesses of the approaches.

We analyzed the predictions made by the top-performing system, SloNSpell, with a primary focus on its false positive and false negative predictions, and to a lesser extent, its true positive predictions, through a qualitative assessment across the Šolar-Eval (Š-E) and Lektor-Spelling (L-S) datasets. In Š-E, we examined 907 predictions, while in L-S, we reviewed 355 predictions. Notably, the categories below reflect the distinctions between the text types in both datasets (e.g., error types typical for learner texts in Š-E versus the presence of rare, specialized vocabulary in L-S's technical texts) and differences stemming from error annotation methods. Some identified issues with evaluation datasets will be addressed in our future development.

The occurrence of false negatives (FNs) in both datasets is comparable: 152 (16.8%) in Š-E and 64 (18%) in L-S. Within this category, misspelled *proper names* are highly typical for learner texts (32.2%) but are absent in the L-S dataset. Both datasets feature falsely accepted misspelled forms that are formally *identical to another legitimate word form* (23.7% in Š-E and 14.1% in L-S). Spell-checkers also struggle with *language variants* that are acceptable in certain contexts but not in others (5.3% FNs in Š-E and 1.6% in Lektor-Spelling). Unique to the L-S annotation are FNs related to ambiguous instances of *writing words separately or together* (20.3%) and the occurrence of *redundant, repeated words* (18.8%). These FN types are traditionally challenging for spell-checking and are anticipated by potential users of the system. However, some unexpected types were also discovered, ranging from falsely accepted *foreign words* (4.7% in L-S) and *archaic forms* (2% in Š-E and 3.1% in L-S) to completely *non-existent forms* (16.4% in Š-E and 6.3% in L-S). Similarly perplexing is the treatment of the Slovene *preposition s/z*, which is selected according to the phonetic form of the following word. Despite the relative simplicity of the selection rules, the system rather randomly errs in this regard (9.9% in Š-E).

The evaluation of false positives (FPs) also revealed both expected and unexpected categories. In Šolar-Eval, 118 (13%) and in L-S, 229 (64.5%) FPs were found. *Proper names* were often misidentified as misspelled (53.4% in Š-E and 21.4% in L-S). *Rare words or forms* presented another expected category, with 7.6% in Š-E and 23.1% in L-S (mainly technical terminology). Similarly to FNs, problems pertaining to *language variants* (5.1% in Š-E and 1.7% in L-S), words of *foreign origin* (1.7% in L-S), and *ambiguous word separation* (15.3% in Š-E and 2.6% in L-S). Unique to the L-S annotation are FNs related to *numerals, special symbols* (9.6%), *single letters* (5.7%), *abbreviations* (8.7%), and *word capitalization* (4.8%). Again, there are unexpected issues with the *preposition s/z* (0.9% in L-S) and a considerably large group of frequently occurring words that were flagged as problematic for *inexplicable reasons* (16.9% in Š-E and 10.5% in L-S).

Contrary to false positives, true positives were more frequent in Šolar-Eval, with 637 instances (70.2%), compared to Lektor-Spelling, which had 62 instances (17.5%). Most of the categories mentioned with FNs and FPs can also be found here, with correct predictions by the system (96.4% in Š-E and 88.7% in L-S).

A set of issues arose from the internal composition of Lektor-Spelling, notably including examples of transcribed spoken language, which represented 4.8% of TPs, 23.4% of FNs, and 9.2% of FPs. These instances differ significantly from the dataset's intended genre and should be removed in future revisions. A smaller proportion of TPs (3.6% in Š-E and 6.5% in L-S), FNs (2.6% in Š-E and 4.7% in L-S), and FPs (1.7% in Š-E) can be attributed to dataset issues such as extraction and annotation errors. Moreover, a significant portion of FNs (7.9% in Š-E and 3.1% in L-S) revealed that non-standard word forms were included in the Sloleks lexicon without proper labeling. We will address all these issues in our forthcoming work.

Finally, the analysis revealed that specific words, such as proper names, might be accepted in one instance and flagged in another within the same context. From the user's perspective, this undermines the system's reliability and is thus a main priority for further improvement.

6 Conclusion

In this study, we created two new spell-checkers for Slovene: SloSpell, a classical approach with a considerably larger vocabulary compared to existing ones, and SloNSpell, a novel neural approach. To train SloNSpell, we developed a novel synthetic data creation tool. We also tested two new evaluation datasets, a synthetically generated one and the other from the Lektor corpus, which we have fully annotated with spelling errors. We compared the newly created methods with three existing classical and one neural approach.

The SloNSpell model, a fine-tuned BERT model, outperformed all other methods, achieving the highest $F_{0.5}$ scores across the three evaluation datasets. This model excelled on the Synthetic-Eval and Šolar-Eval datasets, with $F_{0.5}$ scores exceeding 0.9, but with lower performance on the Lektor dataset, due to the presence of error types not covered in the training data, such as word duplication, unnecessary words, and slang.

In further work, the identified weaknesses will be used to form additional types of training data. Refining ChatGPT's prompts could boost its now disappointing performance. However, considering the cost and speed, the SloNSpell approach is likely to remain superior for the spelling correction task, while SloSpell's large vocabulary shall replace existing Slovene vocabulary in various tools.

Acknowledgments. The work was partially supported by the Slovenian Research and Innovation Agency (ARIS) core research programme P6-0411, a young researcher grant, as well as projects J7-3159, CRP V5-2297, and L2-50070.

References

1. Arhar Holdt, Š., Gantar, P., Bon, M., Gapsa, M., Lavrič, P., Klemen, M.: Dataset for evaluation of Slovene spell- and grammar-checking tools Šolar-Eval 1.0 (2023). http://hdl.handle.net/11356/1902, Slovenian language resource repository CLARIN.SI
2. Arhar Holdt, Š., et al.: Developmental corpus šolar 3.0 (2022). http://hdl.handle.net/11356/1589, Slovenian language resource repository CLARIN.SI
3. Aziz, R., Anwar, M.W., Jamal, M.H., Bajwa, U.I.: A hybrid model for spelling error detection and correction for Urdu language. Neural Comput. Appl. **33**(21), 14707–14721 (2021). https://doi.org/10.1007/s00521-021-06110-7
4. Čibej, J., et al.: Morphological lexicon Sloleks 3.0 (2022). http://hdl.handle.net/11356/1745, Slovenian language resource repository CLARIN.SI
5. Clark, K., Luong, M.T., Le, Q.V., Manning, C.D.: ELECTRA: pre-training text encoders as discriminators rather than generators. In: ICLR (2020)
6. Erjavec, T., Krstev, C., Petkevič, V., Simov, K., Tadić, M., Vitas, D.: The MULTEXT-east morphosyntactic specification for Slavic languages. In: Proceedings of the 2003 EACL Workshop on Morphological Processing of Slavic Languages, pp. 25–32 (2003). https://aclanthology.org/W03-2904
7. Gantar, P., Bon, M., Gapsa, M., Arhar Holdt, Š: Šolar-Eval: Evalvacijska množica za strojno popravljanje jezikovnih napak v slovenskih besedilih. Jezik in slovstvo **68**(4), 89–108 (2023)
8. Ji, T., Yan, H., Qiu, X.: SpellBERT: a lightweight pretrained model for Chinese spelling check. In: Moens, M.F., Huang, X., Specia, L., Yih, S.W.t. (eds.) Proceedings of EMNLP 2021, pp. 3544–3551 (2021). https://doi.org/10.18653/v1/2021.emnlp-main.287
9. Krek, S., et al.: Gigafida 2.0: the reference corpus of written standard Slovene. In: Proceedinggs of LREC 2020, pp. 3340–3345 (2020). https://aclanthology.org/2020.lrec-1.409
10. Li, J., Wu, G., Yin, D., Wang, H., Wang, Y.: DCSpell: a detector-corrector framework for Chinese spelling error correction. In: Proceedings of the 44th International ACM SIGIR Conference on Research and Development in Information Retrieval, pp. 1870-1874 (2021). https://doi.org/10.1145/3404835.3463050
11. Mitreska, M., Mishev, K., Simjanoska, M.: NLP-based typo correction model for Croatian language. In: 2022 45th Jubilee International Convention on Information, Communication and Electronic Technology (MIPRO), pp. 942–947 (2022). https://doi.org/10.23919/MIPRO55190.2022.9803646
12. Popič, D.: Revising translation revision in Slovenia. New Horizons Transl. Res. Educ. **2**, 72 (2014)
13. Ulčar, M., Robnik-Šikonja, M.: SloBERTa: slovene monolingual large pretrained masked language model. In: Proceedings of SI-KDD Within the Information Society 2021, pp. 17–20 (2021)
14. Vaswani, A., et al.: Attention is all you need. In: Advances in neural information processing systems, pp. 5998–6008 (2017). https://proceedings.neurips.cc/paper_files/paper/2017/file/3f5ee243547dee91fbd053c1c4a845aa-Paper.pdf

CoastTerm: A Corpus for Multidisciplinary Term Extraction in Coastal Scientific Literature

Julien Delaunay[1,2], Hanh Thi Hong Tran[1,3,4(✉)],
Carlos-Emiliano González-Gallardo[1(✉)], Georgeta Bordea[1(✉)],
Mathilde Ducos[1(✉)], Nicolas Sidere[1(✉)], Antoine Doucet[1(✉)],
Senja Pollak[4(✉)], and Olivier De Viron[2(✉)]

[1] University of La Rochelle, L3i, La Rochelle, France
{julien.delaunay,carlos.gonzalez_gallardo,georgeta.bordea,
mathilde.ducos,nicolas.sidere,
antoine.doucet,olivier.viron,thi.tran}@univ-lr.fr
[2] University of La Rochelle, LIENSs, La Rochelle, France
[3] Jožef Stefan International Postgraduate School, Ljubljana, Slovenia
[4] Jožef Stefan Institute, Ljubljana, Slovenia
senja.pollak@ijs.si

Abstract. The growing impact of climate change on coastal areas, particularly active but fragile regions, necessitates collaboration among diverse stakeholders and disciplines to formulate effective environmental protection policies. We introduce a novel specialized corpus comprising 2,491 sentences from 410 scientific abstracts concerning coastal areas, for the Automatic Term Extraction (ATE) and Classification (ATC) tasks. Inspired by the ARDI framework, focused on the identification of Actors, Resources, Dynamics and Interactions, we automatically extract domain terms and their distinct roles in the functioning of coastal systems by leveraging monolingual and multilingual transformer models. The evaluation demonstrates consistent results, achieving an F1 score of approximately 80% for automated term extraction and F1 of 70% for extracting terms and their labels. These findings are promising and signify an initial step towards the development of a specialized Knowledge Base dedicated to coastal areas.

Keywords: Automatic term extraction · ATE · Automatic term classification · ATC · terminology · coastal area · littoral

1 Introduction

Coastal areas, grappling with the dual impacts of global change and human interventions, constitute a complex system wherein various dynamics continuously interact (physical, chemical, biological, societal, and others). Understanding this

---

J. Delaunay and T. H. Tran—These authors contributed equally to this work.

system necessitates examining it as an inherently anthropized environment. In this context, many agents and mechanisms can only be understood by considering human actions, such as coastal development, activities affecting land and sea, resource management, and urbanization.

The resulting interdisciplinarity provides a very interesting use case for the automatic analysis of terminology and its use. Numerous nuanced terms from various domains like environmental science, geography, ecology, and sociology emerge in coastal literature, reflecting its interdisciplinary nature.

Automatic analysis of terminology offers a systematic approach to identify and categorize these domain-specific concepts, crucial to the understanding of the dynamic nature of coastal environments. By adapting to the evolving terminological landscape, this process aids in identifying key entities within coastal systems and integrating diverse disciplinary perspectives into a coherent framework.

The ARDI framework [10] enables the identification of *Actors* influencing a territory, the *Resources* they exploit, the *Dynamics* in operation, and the *Interactions* between these agents and resources. Employing ARDI enables stakeholders to collaboratively construct a conceptual framework of the system, facilitating the development of environment protection policies and enhancing scientists' comprehension of these territories. However, identifying key entities within such intricate systems poses considerable challenges. This difficulty is exacerbated by the numerous scientific disciplines addressing the subject matter, including but not limited to biology, oceanography, chemistry, and engineering, resulting in a huge amount of scholarly literature every year. In this context, automatic term extraction (ATE) and classification (ATC) play a crucial role in unlocking the knowledge embedded within the scientific literature.

Recently, we observed a push towards several ATE approaches with different methods, from rule-based to neural approaches. However, there is still a significant gap in the performance of ATE compared to other similar natural language processing (NLP) downstream tasks partially due to the following reasons. First, terms are inherently semantically defined to refer to domain-specific concepts [19]. Besides attempts to define the meaning of a *term*, such as a "language used in a subject field and characterized by the use of specific linguistic means of expression" (ISO 1087-1), in real-life settings, terms are not consistently defined and their definitions vary across the domains and use-cases, making it difficult to develop universal term extraction methods. Secondly, there is a lack of well-documented and transparent domain-specific corpora, even if recently some valuable efforts have been made [30].

Our main contributions are threefold and are summarized as follows:

- We propose two gold-annotated multidisciplinary datasets for ATE and ATC focusing on the domain of coastal areas[1].
- Inspired by the ARDI framework [10], we propose a set of labels designed to facilitate the representation of a system independently of the study domain.

---

[1] Corpus and code are available at https://github.com/jdelaunay/coastal_area_term_extraction.

– We compare a range of ATE state-of-the-art models on our datasets and identify specific challenges inherent to our data.

2 Related Work

2.1 Term Extraction Datasets

Several manually annotated monolingual and multilingual domain-specific resources have been developed for term extraction systems [33], notably for scientific domains. The ACL Reference Dataset for Terminology Extraction and Classification (ACL RD-TEC) [29] serves as a benchmark for evaluating term extraction and classification in the scientific literature related to computational linguistics. It includes 300 manually annotated abstracts from articles in the ACL Anthology Reference Corpus (1978–2006), categorized into various classes. Augenstein et al. (2017) introduce a corpus for SemEval 2017 Task 10, featuring 500 double-annotated documents from the domains of computer science, materials science and physics [7]. This corpus addresses the identification and classification of keyphrases at the word level, categorizing keyphrases into *process*, *task*, and *material*.

In relation to environment studies, SPECIES-800 [27] is a corpus of 800 manually annotated abstracts for taxon mentions recognition. Constructed by randomly selecting 100 MEDLINE abstracts from various journals, it comprises 3,708 mentions of 718 unique species, referenced by 1,503 unique names. BiodivNERE [1] offers two gold standard corpora for named entity recognition (NER) and relation extraction (RE) within biodiversity studies, created from biodiversity metadata and abstracts and manually verified by experts. The corpus comprises 2,398 statements from 150 documents, with the NER corpus identifying entities such as *organism, environment, quality, location, phenomena*, and *matter*. COPIOUS [26] stands as a gold standard corpus sourced from the Biodiversity Heritage Library. Comprising more than 26K sentences from 668 documents, it classifies its 28K entities into five categories: *taxon names, geographical locations, habitats, temporal expressions* and *person names*.

However, to our knowledge, no annotated corpora for term extraction in the interdisciplinary study of coastal regions or related domains (such as oceans or seas) exists. Thus, CoastTerm represents a pioneering effort in this regard, paving the way for an exhaustive cross-domain and multi-disciplinary Knowledge Graph construction system for studying coastal areas.

2.2 Term Extraction Methods

Classical term extractors mainly rely on either linguistic or statistical aspects [13] or combine both [17] and applied rule-based or machine-learning methods to extract the candidate terms. More recently, the introduction of representation learning and neural networks has led to the application of various text embedding techniques for term extraction, including local-global [2], non-contextual embeddings (e.g., GloVe[2], Word2Vec, skip-gram [2,3,21,41]), contextual word embeddings (e.g., Flair[3], BERT [5,21,32]), as well as their combinations (e.g., stacked Flair + BERT [5,32]).

Neural architectures are also used as end-to-end term extraction systems, and most current systems focus on tagging-based models [15,22]. In tagging-based mechanisms, the task was formulated as (1) sequence classification where or not the binary label of a term was assigned to each possible n-gram of a fixed length of a given sentence using different variants of BERT-based models (e.g., BERT, RoBERTa, and XLMR); or (2) token classification, where the label was assigned to each word in the given sentence following the IOB annotation format using different language models.

With the advent of transformers, several pre-trained language models have been applied as token classifiers. Above all, XLMR is now considered a benchmark for several languages [34,37]. Cross-domain and cross-lingual learning was also applied to these benchmarks to enhance extraction performance in the absence of available annotated data [16,22,34–37].

In addition, there have been other experiments on ATE with the adoption of span-based methods [40] or applying generative models considering Seq2Seq models such as mBART [22] to extract candidate terms more efficiently. However, while span-based methods show their potential, the performance of generative models is still under question.

In the context of our study domains, Zhao et al. (2022) explores the extraction of knowledge from operational maritime decision-making sentences [42], Andersen et al. (2022) proposes the development of a corpus to cultivate a specialized terminology relevant to Norwegian maritime discourse [4], Mouratidis et al. (2022) performs term extraction within legal documents on maritime topics in the Greek language [25]. Similarly to our work, but in the realm of karst studies, TermFrame[4] serves as a KG constructed by extracting terms and triplets from English and Slovene karst corporas [28,39]. EcoLexicon [11,12] constitutes a KB specialized on environment, encompassing six languages (English, French, German, Modern Greek, Russian, and Spanish). Moreover, for NER tasks, TaxoNERD [23] aims to recognize taxon mentions in ecological documents, while AGRONER [38] employs unsupervised NER techniques tailored to the agriculture domain, integrating an extended BERT model with Latent Dirichlet Allocation (LDA) topic modeling.

3 CoastTerm Corpus for Term Extraction

3.1 Annotation Process

We collected a corpus of 64,000 papers from Scopus[5], spanning from 1980 to 2023, containing the terms "coastal areas" or "littoral" in their abstract or title.

---

[2] https://nlp.stanford.edu/projects/glove/.
[3] https://github.com/flairNLP/flair.
[4] https://termframe.ff.uni-lj.si/.
[5] https://www.elsevier.com/fr-fr/solutions/scopus.

We initially selected randomly 600 abstracts for manual annotation, keeping a proportion of 60% regular articles and 40% surveys. Annotation efforts first enlisted the participation of two undergraduate Master's students specialized in Earth Sciences, who were compensated for their contributions. These students were supplied with example annotations and guidelines to conduct a task of document-level joint entity and relation extraction, which involves nested ATE and ATC, coreference resolution, and document-level relation extraction. The annotation guidelines were constructed by a computer science and coastal research PhD student and a domain expert, with the assistance of an ontology expert. Only the sentences that give information about the functioning of the coastal zone were annotated, we avoided sentences that described the methodology used in the article.

We adapted the ARDI framework to extract the information related to the functioning of a described system within a scientific abstract. As a result, for ATC, we designated the following labels: "Actor" (stakeholders who consume resources and/or initiate processes) and "Resource" (goods, products, facilities, and elements, including plants and animals, utilized by stakeholders). Following the Basic Formal Ontology (BFO) [6], we replaced the "Dynamic" label in ARDI with "Process", as defined by the Environment Ontology (ENVO) [8], referring to environmental, societal, or economic processes impacting the system and inducing changes. To enhance precision on the extracted information, we introduced the "Quality" label from the Phenotype And Trait Ontology (PATO) [14], which refers to height, concentration, or a specificity, and added a "Location" label.

Along with the two students specializing in Earth Sciences, the annotation process engaged the same domain expert and PhD student, all simultaneously annotating papers. All annotators were familiar with the annotation tool, INCEpTION [18]. Over two months, the campaign aimed to achieve dual-annotator coverage for 60% of the total annotated abstracts. Ultimately, 215 abstracts were annotated, with a mean Krippendorff's alpha [20] of 43%, indicating a moderate agreement, but not sufficient to use directly the dataset, falling short of the minimum threshold (66%) required for direct dataset utilization, highlighting the difficulty of the manual annotation of terms. Annotations were then curated by the PhD student according to the guidelines.

We employed three domain-relevant knowledge bases (KBs) (i.e., AGROVOC[6], GEMET (GEneral Multilingual Environmental Thesaurus)[7], AFO (Agriculture and Forestry Ontology)[8], and TAXREF-LD[9]) to pre-annotate pertinent terms in a secondary subset of 195 abstracts. This process significantly assisted in refining term boundaries. Annotations were then carried out by the PhD student annotator following the established guidelines, utilizing insights acquired from the initial annotation process. Subsequently, the initially fully manually annotated subset was homogenized with the KB-recommended one to

---

[6] https://agrovoc.fao.org/browse/agrovoc/en/.
[7] https://www.eionet.europa.eu/gemet/en/about/.
[8] https://finto.fi/afo/en/.
[9] https://github.com/frmichel/taxref-ld.

produce two datasets intended for the study of coastal areas. In the context of the present research, both datasets were adapted manually for terminology extraction by removing relations and pronouns that indicate coreferences.

3.2 Dataset Description

The KB-recommended corpus contains 1,235 annotated sentences, spanning across 61 different keywords, and includes 6,663 annotated terms. The human-recommended corpus contains 1,256 annotated sentences, spanning across 92 different keywords, and featuring 6,543 annotated terms. From Table 1 it can be seen that the distribution of labels is consistent across both datasets. In addition, the label "Actor" is observed to be less prevalent, whereas "Resource" emerges as the most frequently represented one. Together, these two datasets collectively span across 101 unique keywords, "*aquatic science*", "*oceanography*", "*ecology, evolution, behavior, and systematics*", "*ecology*", "*pollution*", "*earth and planetary sciences*", "*environmental science*", "*management, monitoring, policy, and law*", "*water science and technology*", and "*geography, planning, and development*" being the most prominent. A comparison of the top 100 frequent terms shows a significant overlap of 52% between the two corpora. In total, the two datasets share 751 common terms (730 if lemmatized). We allocated 70% of the annotated sentences within each article to the training sets, 10% to the validation sets, and 20% to the test sets. An annotation example can be visualized in Fig. 1.

Table 1. Statistics for KB and human recommended subcorporas

	KBs recommended	Human experts recommended
Vocabulary size	7,902	6,280
Sentences	1,235	1,256
Tokens	31,147	37,983
Type-to-token ratio (TTR)	0.25	0.17
Annotated Terms	6,663	6,543
Unique Terms	3,844	4,400
Unique Terms (lemmatized)	3,539	4,110
# of "Actor"	617	602
# of "Resource"	2,236	2,052
# of "Process"	1,524	1,431
# of "Quality"	1,138	1,373
# of "Location"	1,145	1,082
# of "B"	6,663	6,543
# of "I"	4,723	6,192
# of "O"	25,415	25,710

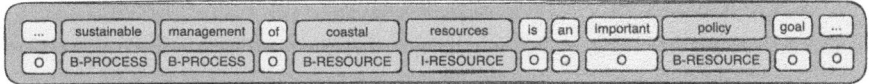

Fig. 1. A sample of the corpus annotation for term extraction and classification

4 Experiments

To evaluate the challenges of CoastTerm, we conducted extensive experiments using state-of-the-art ATE, extending our evaluation to ATC. Our in-depth analysis facilitates a discussion on potential future trajectories for interdisciplinary ATE and ATC research related to coastal areas.

4.1 Models

We considered ATE as a sequence-labeling task where the model returns a label for each token in a text sequence, using the IOB labeling mechanism [31,34,35]. We apply the same labeling scheme to ATC, with the addition of the term's class following its IOB label.

We experimented with two families of language models with both base and large versions, including:

- *Monolingual pre-trained model*: We chose RoBERTa [24], a transformer-based model pre-trained on a large corpus of English data in a self-supervised fashion.
- *Multilingual pre-trained model*: We opt for XLMR [9], a transformer-based model pre-trained on 2.5 TB of filtered CommonCrawl data containing 100 languages. This multilingual version of RoBERTa, achieves benchmark performance in ATE for rich-resourced languages (e.g. English) [30,34].

4.2 Evaluation Metrics

To assess the performance of the ATE systems on the human-recommended and KB-recommended datasets, we juxtaposed the candidate list of unique terms extracted from the entire test set against the gold standard of the test set. We performed experiments with both lemmatized and not lemmatized terms for testing. This evaluation was conducted employing strict matching criteria, with metrics including precision (P), recall (R), and micro F1-score (F1). For the ATC task, the evaluation process remained the same but each term was accompanied by its corresponding label.

4.3 Results

Table 2 presents the performances of mono- and multilingual classifiers for the ATE task on both the human- and KB-recommended datasets. Regarding the KB-recommended dataset, the results demonstrate that $XLMR_{base}$

achieves superior performance compared to RoBERTa$_{base}$ for both lemmatized and unlemmatized text. However, RoBERTa$_{large}$ outperforms XLMR$_{large}$. Conversely, on the human-recommended dataset, monolingual models exhibit better performance than multilingual ones. Additionally, it is observed that XLMR$_{large}$ performs least effectively in this context. We also performed experiments with the combined KB- and human-recommended datasets. This configuration is referred to as Fusion in Table 2 where 20% and 40% of the human-recommended dataset were allocated to the validation and test sets respectively. The KB-recommended dataset, along with the remaining 40% of the human-recommended dataset, composed the train set. Reported performances show that monolingual models outperform multilingual ones.

Table 3 displays the performance of the classifiers for the ATC task using the combined KB-recommended and human-recommended datasets. Notably, multilingual classifiers demonstrate superior performance compared to monolingual ones. Additionally, upon lemmatizing the predictions, XLMR$_{base}$ exhibits better performance than XLMR$_{large}$.

4.4 Error Analysis

To assess the impact of term length on the models' performance, we examined the proportion of correct and incorrect predictions relative to term length. Figure 2 illustrates for XLMR$_{base}$ that as the term length increases, the model encounters greater difficulty in accurately predicting it and might predict shorter terms. However, the vast majority of the terms consist of either one or two words which mitigates the negative impact on performance. This behavior is observed across all models.

We also report confusion matrices for XLMR$_{base}$ on the fusion of KB- and human-recommended datasets in Figs. 3 and 4. The results are consistent since

Table 2. Evaluation in performance of different extractors for ATE task

Models	KB			Human			Fusion		
	P	R	F1	P	R	F1	P	R	F1
Not lemmatized									
RoBERTa$_{base}$	75.51	78.12	76.79	78.65	82.07	80.32	**79.08**	80.37	79.72
XLMR$_{base}$	76.08	78.95	77.49	79.07	81.04	80.04	77.95	79.94	78.93
RoBERTa$_{large}$	**77.23**	**79.90**	**78.54**	**80.97**	**82.90**	**81.92**	79.00	80.96	**79.97**
XLMR$_{large}$	76.99	78.85	77.91	77.42	79.59	78.49	78.48	**81.45**	79.94
Lemmatized									
RoBERTa$_{base}$	75.45	78.24	76.82	78.43	82.25	80.29	**78.80**	79.97	79.38
XLMR$_{base}$	75.89	78.78	77.31	78.97	81.28	80.11	77.61	79.68	78.63
RoBERTa$_{large}$	**77.32**	**79.76**	**78.52**	**80.78**	**82.79**	**81.77**	78.74	80.70	**79.71**
XLMR$_{large}$	76.93	79.11	78.00	77.48	80.41	78.92	78.20	**81.21**	79.68

Table 3. Evaluation in performance of different extractors for joint ATE and ATC on the fusion of KB and human datasets

Models	Not lemmatized			Lemmatized		
	P	R	F1	P	R	F1
RoBERTa$_{base}$	66.36	68.93	67.62	65.96	68.74	67.32
XLMR$_{base}$	66.11	**71.69**	68.79	65.95	**71.85**	**68.77**
RoBERTa$_{large}$	66.60	71.05	68.75	66.05	70.85	68.37
XLMR$_{large}$	**66.97**	71.16	**69.00**	**66.60**	70.96	68.71

ambiguity might come from the fact that one entity might be an "Actor" at one point and a "Location" at another (Countries, for example); or a "Location" and a "Resource" or "Quality" depending on the context.

5 Conclusion and Future Works

We introduced CoastTerm, a corpus of 2,491 gold-annotated sentences for interdisciplinary automatic term extraction and classification related to the coastal area. Adapting the ARDI framework, we provided comprehensive labels applicable to various domains. Benchmarking mono- and multilingual state-of-the-art models on ATE and ATC tasks shows promising results, paving the way for interdisciplinary knowledge base construction in the coastal area domain.

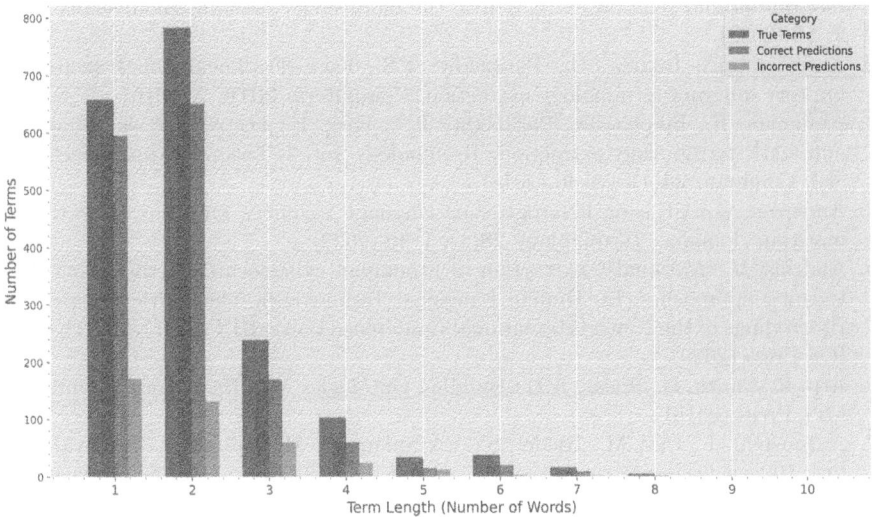

Fig. 2. Term length distribution for XLMR$_{base}$ on the fusion of KB- and human-recommended datasets

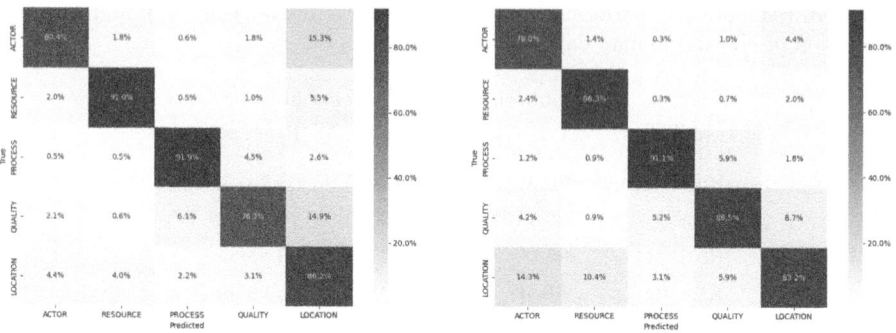

Fig. 3. Normalized on ground truth **Fig. 4.** Normalized on predictions

Acknowledgments. The work was supported by the TERMITRAD (2020-2019-8510010) project funded by the Nouvelle-Aquitaine Region, France, by the Slovenian Research and Innovation Agency core research program Knowledge Technologies (P2-0103) and the project Cross-lingual and Cross-domain Methods for Terminology Extraction and Alignment, a bilateral project funded by the program PROTEUS under the grant number BI-FR/23-24-PROTEUS006. We express our gratitude to Kenza HERMAN and Geraldine DUBOS for their invaluable assistance in the annotation process.

References

1. Abdelmageed, N., et al.: BiodivNERE: gold standard corpora for named entity recognition and relation extraction in the biodiversity domain. Biodiversity Data J. **10** (2022)
2. Amjadian, E., Inkpen, D., Paribakht, T.S., Faez, F.: Local-global vectors to improve unigram terminology extraction. Computerm **2016**, 2 (2016)
3. Amjadian, E., Inkpen, D., Paribakht, T.S., Faez, F.: Distributed specificity for automatic terminology extraction. Terminology. Int. J. Theoret. Appl. Issues Special. Commun. **24**(1), 23–40 (2018)
4. Andersen, G.: Utilising heterogeneous language resources for term extraction in maritime domains. Terminology **28**(1), 1–36 (2022)
5. Andrius, U.: Automatic extraction of lithuanian cybersecurity terms using deep learning approaches. In: Human Language Technologies–The Baltic Perspective: Proceedings of the Ninth International Conference Baltic HLT 2020. vol. 328, p. 39. IOS Press (2020)
6. Arp, R., Smith, B., Spear, A.D.: Building Ontologies with Basic Formal Ontology. MIT Press (2015)
7. Augenstein, I., Das, M., Riedel, S., Vikraman, L., McCallum, A.: Semeval 2017 task 10: scienceie-extracting keyphrases and relations from scientific publications. arXiv preprint arXiv:1704.02853 (2017)
8. Buttigieg, P.L., Morrison, N., Smith, B., Mungall, C.J., Lewis, S.E., Consortium, E.: The environment ontology: contextualising biological and biomedical entities. J. Biomed. Semant. **4**, 1–9 (2013)

9. Conneau, A., et al.: Unsupervised cross-lingual representation learning at scale. arXiv preprint arXiv:1911.02116 (2019)
10. Etienne, M., Du Toit, D.R., Pollard, S.: ARDI: a co-construction method for participatory modeling in natural resources management. Ecol. Soc. **16**(1) (2011)
11. Faber, P., León-Araúz, P., Reimerink, A.: Representing environmental knowledge in EcoLexicon. In: Bárcena, E., Read, T., Arús, J. (eds.) Languages for Specific Purposes in the Digital Era. EL, vol. 19, pp. 267–301. Springer, Cham (2014). https://doi.org/10.1007/978-3-319-02222-2_13
12. Faber, P., León-Araúz, P., Reimerink, A.: EcoLexicon: new features and challenges. GLOBALEX 73–80 (2016)
13. Frantzi, K.T., Ananiadou, S., Tsujii, J.: The *C-value/NC-value* method of automatic recognition for multi-word terms. In: Nikolaou, C., Stephanidis, C. (eds.) ECDL 1998. LNCS, vol. 1513, pp. 585–604. Springer, Heidelberg (1998). https://doi.org/10.1007/3-540-49653-X_35
14. Gkoutos, G.V., Schofield, P.N., Hoehndorf, R.: The anatomy of phenotype ontologies: principles, properties and applications. Briefings Bioinform. **19**(5), 1008–1021 (04 2017). https://doi.org/10.1093/bib/bbx035
15. Hazem, A., Bouhandi, M., Boudin, F., Daille, B.: Termeval 2020: Taln-ls2n system for automatic term extraction. In: 6th International Workshop on Computational Terminology (COMPUTERM 2020) (2020)
16. Hazem, A., Bouhandi, M., Boudin, F., Daille, B.: Cross-lingual and cross-domain transfer learning for automatic term extraction from low resource data. In: Proceedings of the Thirteenth Language Resources and Evaluation Conference, pp. 648–662 (2022)
17. Kessler, R., Béchet, N., Berio, G.: Extraction of terminology in the field of construction. In: 2019 First International Conference on Digital Data Processing (DDP), pp. 22–26. IEEE (2019)
18. Klie, J.C., Bugert, M., Boullosa, B., Eckart de Castilho, R., Gurevych, I.: The INCEpTION platform: machine-assisted and knowledge-oriented interactive annotation. In: Proceedings of the 27th International Conference on Computational Linguistics: System Demonstrations, pp. 5–9. Santa Fe, New Mexico (2018). https://www.aclweb.org/anthology/C18-2002
19. Kockaert, H.J., Steurs, F.: Handbook of terminology, vol. 1. John Benjamins Publishing Company (2015)
20. Krippendorff, K.: Computing krippendorff's alpha-reliability (2011)
21. Kucza, M., Niehues, J., Zenkel, T., Waibel, A., Stüker, S.: Term extraction via neural sequence labeling a comparative evaluation of strategies using recurrent neural networks. In: Interspeech, pp. 2072–2076 (2018)
22. Lang, C., Wachowiak, L., Heinisch, B., Gromann, D.: Transforming term extraction: transformer-based approaches to multilingual term extraction across domains. In: Findings of the Association for Computational Linguistics: ACL-IJCNLP 2021, pp. 3607–3620 (2021)
23. Le Guillarme, N., Thuiller, W.: TaxoNERD: deep neural models for the recognition of taxonomic entities in the ecological and evolutionary literature. Methods Ecol. Evol. **13**(3), 625–641 (2022)
24. Liu, Y., et al.: RoBERTa: a robustly optimized BERT pretraining approach. arXiv preprint arXiv:1907.11692 (2019)
25. Mouratidis, D., et al.: Domain-specific term extraction: a case study on Greek maritime legal texts. In: Proceedings of the 12th Hellenic Conference on Artificial Intelligence. SETN '22, New York, NY, USA. Association for Computing Machinery (2022). https://doi.org/10.1145/3549737.3549751

26. Nguyen, N.T., Gabud, R.S., Ananiadou, S.: Copious: a gold standard corpus of named entities towards extracting species occurrence from biodiversity literature. Biodiversity Data J. (7) (2019)
27. Pafilis, E., et al.: The species and organisms resources for fast and accurate identification of taxonomic names in text. PLoS ONE **8**(6), e65390 (2013)
28. Pollak, S., Repar, A., Martinc, M., Podpečan, V.: Karst exploration: extracting terms and definitions from karst domain corpus. In: Proceedings of eLex 2019, pp. 934–956 (2019)
29. QasemiZadeh, B., Schumann, A.K.: The acl rd-tec 2.0: a language resource for evaluating term extraction and entity recognition methods. In: Proceedings of the Tenth International Conference on Language Resources and Evaluation (LREC'16), pp. 1862–1868 (2016)
30. Rigouts Terryn, A., Hoste, V., Drouin, P., Lefever, E.: Termeval 2020: shared task on automatic term extraction using the annotated corpora for term extraction research (ACTER) dataset. In: 6th International Workshop on Computational Terminology (COMPUTERM 2020), pp. 85–94. European Language Resources Association (ELRA) (2020)
31. Rigouts Terryn, A., Hoste, V., Lefever, E.: HAMLET: hybrid adaptable machine learning approach to extract terminology. Terminology **27**(2), 254–293 (2021)
32. Terryn, A.R., Hoste, V., Lefever, E.: Tagging terms in text: a supervised sequential labelling approach to automatic term extraction. Terminol. Int. J. Theoret. Appl. Issues Specialized Commun. **28**(1), 157–189 (2022)
33. Tran, H.T.H., Martinc, M., Caporusso, J., Doucet, A., Pollak, S.: The recent advances in automatic term extraction: a survey. arXiv preprint arXiv:2301.06767 (2023)
34. Tran, H.T.H., Martinc, M., Doucet, A., Pollak, S.: Can cross-domain term extraction benefit from cross-lingual transfer? In: Pascal, P., Ienco, D. (eds.) DS 2022. LNCS, vol. 13601, pp. 363–378. Springer, Cham (2022). https://doi.org/10.1007/978-3-031-18840-4_26
35. Tran, H.T.H., Martinc, M., Pelicon, A., Doucet, A., Pollak, S.: Ensembling transformers for cross-domain automatic term extraction. In: Tseng, YH., Katsurai, M., Nguyen, H.N. (eds.) ICADL 2022. LNCS, vol. 13636, pp. 90–100. Springer, Cham (2022). https://doi.org/10.1007/978-3-031-21756-2_7
36. Tran, H.T.H., Martinc, M., Repar, A., Ljubešić, N., Doucet, A., Pollak, S.: Can cross-domain term extraction benefit from cross-lingual transfer and nested term labeling? Mach. Learn. 1–30 (2024)
37. Tran, T.H.H., Martinc, M., Repar, A., Doucet, A., Pollak, S.: A transformer-based sequence-labeling approach to the Slovenian cross-domain automatic term extraction (2022)
38. Veena, G., Kanjirangat, V., Gupta, D.: AGRONER: an unsupervised agriculture named entity recognition using weighted distributional semantic model. Expert Syst. Appl. **229**, 120440 (2023)
39. Vintar, Š, Martinc, M.: Framing karstology: from definitions to knowledge structures and automatic frame population. Terminology **28**(1), 129–156 (2022)
40. Wang, J., Feng, C., Liu, F., Li, X., Wang, X.: Extract then adjust: a two-stage approach for automatic term extraction. In: Liu, F., Duan, N., Xu, Q., Hong, Y. (eds.) NLPCC 2023. LNCS, vol. 14303, pp. 236–247. Springer (2023). https://doi.org/10.1007/978-3-031-44696-2_19

41. Zhang, Z., Gao, J., Ciravegna, F.: Semre-rank: improving automatic term extraction by incorporating semantic relatedness with personalised pagerank. ACM Trans. Knowl. Disc. Data (TKDD) **12**(5), 1–41 (2018)
42. Zhao, X., Lian, X., Liu, P., Gao, C.: Research on knowledge extraction of maritime operational decision-making sentences. In: C2 2022. LNEE, vol. 949, pp. 836–845. Springer, Singapore (2022). https://doi.org/10.1007/978-981-19-6052-9_75

New Human-Annotated Dataset of Czech Health Records for Training Medical Concept Recognition Models

Krištof Anetta(✉) and Aleš Horák

Natural Language Processing Centre Faculty of Informatics, Masaryk University,
Brno, Czechia
{xanetta,hales}@fi.muni.cz

Abstract. Following the widespread successes of leveraging recent large language models (LLMs) in various NLP tasks, this paper focuses on medical text content understanding. Adapting a foundational LLM to the medical domain requires a special kind of datasets where core medical concepts are accurately annotated. This paper addresses the need of better medical concept recognition in free-text electronic health records in low-resourced Slavic languages and introduces CSEHR, a new human-annotated dataset of Czech oncology health records. It describes the dataset inception, management, considerations, processing, and finally presents baseline concept recognition model results. XLM-RoBERTa models trained on the dataset using 5-fold cross-validation achieved an average weighted F1 score of 0.672 in exact and 0.777 in partial medical concept recognition ranging from 0.335 to 0.857 per different concept classes. This paper then describes future plans of bootstrapping larger annotated corpora from the CSEHR dataset and of making the dataset publicly available. This endeavor is unique in the realm of Slavic languages and already at this stage it represents a major step in the field of Slavic medical concept recognition.

Keywords: medical text analysis · electronic health records · medical concept terms · medical concept dataset · named entity recognition

1 Introduction

In recent years, the world has witnessed significant breakthroughs in computer science owing to the capabilities of transformer-based large language models (LLMs) [20], especially in the area of understanding the usage of natural language. This was made possible by the immense amount of publicly available natural language training data on the Internet and the fact that many applications, such as dialogue systems, almost do not need annotated training text – the features needed for training are inherent in the structure of natural language.

The medical domain is more demanding in terms of both training data and outcomes. Even though there are well-known public datasets of health records such as MIMIC-III and IV [8–10], and there have been attempts to employ LLMs in medicine [19], there is a crucial bottleneck: for proper machine understanding of medical text, its full density and precise meaning of the often non-linguistic elements, large corpora of annotated medical text are required. In the scale required by LLMs, this is often unachievable with human annotators only.

Furthermore, an important distinction needs to be made between work carried out in English and in other, medium- and low-resourced languages, since there is a massive difference in resource availability, both in terms of sheer amount of plain text and structured resources such as the Unified Medical Language System, abbreviated as UMLS [3,12]. Researchers often treat the work done in all non-English languages together as an area in itself [11]. A telling statistic is that while the number of English terms in the UMLS Metathesaurus is over 11 million, Spanish trails it with less than a sixth of that number, and Central European Slavic languages like Czech, Polish, and Slovak have a very limited representation of around 200,000 (less than 2% of English), 50,000 (0.5%), and 0, respectively. In Slavic languages, UMLS is often insufficient as a term database and one needs to use local resources such as national drug databases (e.g. [18] for Czech) and inconsistently maintained translations of foreign vocabularies.

This paper focuses on Czech as a representative of the low-resourced languages in case of medical text resources. For a variety of legal and institutional reasons, Czech electronic health record data are very difficult to obtain, and even when obtained, they need to be manually annotated, which requires significant personal and temporal resources. In this context, the practice of bootstrapping training data [13,15,16] becomes increasingly relevant – in other words, using a smaller set of training data to incrementally annotate a larger corpus, which can then serve as training data for LLMs.

This project follows the above logic, looking first to obtain as much human annotation as possible and then bootstrap larger annotated medical training data for Czech from the smaller human-verified data. In the following section, we introduce a new human-annotated CSEHR dataset of almost 50,000 words of medical reports with medical concept annotations. Section 3 then details experiments with the new dataset based on large pre-trained transformer models that can serve as a baseline for the medical concept recognition task.

2 The CSEHR Health Record Dataset

The medical record data used for this study were selected from a larger corpus of Czech medical records collected at the Masaryk Memorial Cancer Institute. This corpus contains a total of more than 42 million words in over 150,000 records related to more than 4,200 patients.

The subset used for human annotation contains 168 records totaling almost 50,000 words and the detailed statistics can be seen in Table 1. It represents a balanced sample of the types of text present across the whole parent corpus.

As a first step, the CSEHR dataset underwent a two-phase de-identification process. First, a standard Czech NER model was used to locate mentions of names and places which were then replaced with generic names and place names in the same grammatical form. Subsequently, to ensure complete de-identification, all text was thoroughly manually inspected by one assigned person. As a result, all names and personal data of both patients and doctors were replaced with generic names and fake data, and also other information that could be used to identify the patient, such as names of locations, institutions, and statistically rare conditions, was obfuscated.

Before handing the records to human annotators, these were pre-annotated using automated string lookup, searching for entities that are often found verbatim in texts, in this case medical abbreviations and medication names. The source of medication names was the official database of medications registered in the Czech Republic [18]. Medical abbreviations were collected from several sources [5-7] and merged into a list of 1571 abbreviations frequently used in medical texts. Regular expression search was then used to locate the relevant entities in health records. The annotators were instructed to carefully judge the pre-annotated abbreviations and medication names and adjust each annotation according to their consideration.

Table 1. CSEHR dataset statistics.

Totals	Count
Records	168
Sentences	3,566
Words	48,164
Tokens	69,699

Per record	Avg count	Min	Max
Sentences	21	1	87
Words	287	37	1,496

The annotation schema was based on the type system of Apache cTAKES [21] and its six core clinical elements, in alphabetical order: *AnatomicalSite*, *DiseaseDisorder*, *Lab*, *Medication*, *Procedure*, and *SignSymptom*. To add a degree of accuracy to this granularity of concepts, some of these elements were subdivided, and technical elements such as *DateTime*, *Negation* and *Abbreviation* were added. This produced the final set of 15 different annotation labels as seen in Table 6. One label, *Medication_instruction*, was created by post-annotation out of a subset of what the annotators marked as *Medication_dosage*.

Admittedly, including *Abbreviation* as an annotation category alongside medically meaningful categories may seem controversial, but being able to identify abbreviations and use a separate system to expand them into their unabbreviated form makes it worthwhile. In the evaluation of results, if interested in the specifically medical capabilities of the trained models, one can always choose to

Table 2. Total number of annotations in CSEHR at different stages of its creation. *Annotated tokens* include annotations from all annotators working on the same text input. *Merged overlapping annotations* present the numbers of annotated units in the final version of the dataset. The concept of *Medication_instruction* was extracted from the *Medication_dosage* concept during the annotation merging process.

Class	Annotated tokens	Merged overlapping annotations	
		tokens	entities
Abbreviation	15,026	10,009	7,415
AnatomicalSite_laterality	1,661	1,246	932
AnatomicalSite_name	3,853	2,595	2,231
DateTime	7,317	4,736	1,370
DiseaseDisorder	1,754	1,293	816
Lab_name	1,377	1,105	669
Lab_unit	1,721	1,228	781
Lab_value	2,688	1,921	1,081
Medication_dosage	1,107	358	112
Medication_instruction	0	373	120
Medication_name	1,074	693	644
Medication_strength	711	536	200
Negation	1,785	1,459	1,427
Procedure	2,542	1,990	1,492
SignSymptom	2,416	2,068	1,506
Total	**45,032**	**31,610**	**20,796**

only measure classification performance in medical categories. In this baseline study, we are interested in the general ability of the model to learn all annotated categories.

This annotation schema was made to allow any single token to be annotated with multiple annotation classes (both embedded and coextensive annotations). This is suitable for cases where, for example, a name of an anatomical site is part of a disease/disorder name, and also makes sure that the technical annotation class of *Abbreviation* does not prevent the string being annotated with its proper medical category as well.

The annotators were given a detailed manual specifying the scenarios for the identification of individual annotation classes. Shortened versions of the instruction and examples present in the manual can also be found in Table 6. The manual included the technical details of entering annotations in the BRAT annotation tool [17]. The annotations could be overlapping and/or nested and the annotators had the possibility to express their low confidence in specific annotation.

Table 3. Token counts in the training, validation and testing sets of individual models in 5-fold cross-validation.

Original				
Split	Train	Validation	Test	Total
1	55,485	4,670	9,544	69,699
2	55,856	4,520	9,323	69,699
3	56,059	4,565	9,075	69,699
4	55,575	4,610	9,514	69,699
5	55,821	4,525	9,353	69,699

No punctuation				
Split	Train	Validation	Test	Total
1	44,238	3,746	7,574	55,558
2	44,534	3,548	7,476	55,558
3	44,815	3,623	7,120	55,558
4	44,219	3,567	7,772	55,558
5	44,426	3,558	7,574	55,558

Altogether, eleven annotators were recruited among non-medical university students and there was a minimum of two annotators for every record in order to maximize both coverage (more people had the opportunity to notice a concept) and confidence (in the case of consensus). The rationale for including all annotations in the resulting gold standard dataset (not only the ones where there was consensus) was that the non-expert students' precision was likely to be much higher than recall (in other words, their choices were decent but they often missed entities) and hence removing annotations without consensus would massively diminish the newly created set of annotations, rendering it much less useful for model training. We retained the information about annotator consensus in the form of a confidence score belonging to each annotation. The annotators carried out the work over a period of three months. The numbers of annotations identified by this group of annotators can be seen in Table 2.

For each two annotators who annotated the same text, the inter-annotator agreement on token level[1] was 53.3%. In this case, annotators often noticed and annotated completely different strings, hence the Cohen's kappa measure was calculated for strings annotated by at least one annotator, considering the missing annotation of the second annotator to be class O. The Cohen's kappa value, thus defined, was 0.487 (agreement on at least one annotation) and 0.468

---

[1] calculated as the percentage of tokens annotated by at least one annotator where both annotators chose at least one identical category.

Table 4. Detailed concept classification results (entity-level based on the CoNLL evaluation methodology).

	Original			No punctuation		
Class	**P**	**R**	**F1**	**P**	**R**	**F1**
Abbreviation	0.652	0.582	**0.611**	0.804	0.721	**0.758**
AnatomicalSite_laterality	0.630	0.758	**0.647**	0.659	0.809	**0.688**
AnatomicalSite_name	0.699	0.789	**0.720**	0.706	0.777	**0.718**
DateTime	0.626	0.747	**0.678**	0.705	0.766	**0.730**
DiseaseDisorder	0.446	0.490	**0.462**	0.445	0.453	**0.444**
Lab_name	0.379	0.639	**0.411**	0.351	0.619	**0.391**
Lab_unit	0.636	0.795	**0.663**	0.623	0.777	**0.649**
Lab_value	0.621	0.791	**0.647**	0.610	0.766	**0.622**
Medication_dosage	0.600	0.517	**0.542**	0.600	0.509	**0.530**
Medication_instruction	0.379	0.249	**0.291**	0.277	0.143	**0.186**
Medication_name	0.771	0.820	**0.792**	0.777	0.820	**0.796**
Medication_strength	0.740	0.876	**0.795**	0.662	0.791	**0.714**
Negation	0.802	0.905	**0.849**	0.809	0.916	**0.857**
Procedure	0.511	0.634	**0.534**	0.490	0.615	**0.512**
SignSymptom	0.414	0.384	**0.349**	0.409	0.345	**0.335**
Weighted average	**0.619**	**0.657**	**0.631**	**0.666**	**0.693**	**0.672**

(agreement on all annotations) on token level when requiring correct B and I tags and 0.545 (agreed on at least one) and 0.477 (total agreement) when only evaluating correct token category. This indicates that although there was significant consensus, there is much to be gained from redundancy and multiple annotators are likely to present new, relevant findings.

The final annotations in the dataset have been recorded in the B-I-O format per token denoting as *B-class* the first token of an entity of type *class*, *I-class* as the second or further token in an entity and *O* as a non-entity token. All annotations have been merged and kept. Annotations entered by just one of the two or three annotators have been marked as having lower confidence. The dataset thus contains multiple possible annotations per token (with a maximum of three annotations). For the experiment evaluation any of the entered annotations was considered as a correct ground truth.

3 Experiments and Evaluation

The baseline evaluation approach was based on the Flair framework [1] fine-tuning the XLM-RoBERTa-large [4] pre-trained multi-lingual model for the token classification task. The hyper-parameters were empirically set to: the learning rate of $5 \cdot 10^{-6}$, the batch size of 8, and the training continued for 10 epochs with early stopping based on the validation subset. Concept recognition capability was evaluated using 5-fold cross-validation, training 5 separate models on 4/5 of the data and splitting the remaining 1/5 to a validation set and a testing set in the ratio of 1:2 (a train/valid/test split of 80/6.66/13.34, exact counts listed in Table 3), then averaging the results. Two main versions of the models were trained, one on the original tokenized version of the records, the other on the same records with all punctuation tokens removed which did slightly better, mostly in the *Abbreviation* category. The results were calculated on entity level utilizing the B-I-O format and the CoNLL evaluation algorithm [22] modified for multiple annotations. Results are shown in Table 4.

As the CoNLL evaluation takes only exact entity matches into account, we have also evaluated the results with the SemEval 2013 9.1 shared task measures [14] that provide detailed evaluation of partly overlapping entities and about the entity type predictions. The results in Table 5 were computed by the D. Batista's `nervaluate` tool [2] again adapted to multiple ground-truth annotations.

Categories with the lowest F1 score, ranked from the bottom, were *Medication_instruction*, *SignSymptom*, *Lab_name*, and *DiseaseDisorder*. In the cases of *DiseaseDisorder* and *SignSymptom*, this can be explained by the wide variety of different names and wordings belonging to these classes – it is difficult for a classifier to find anything in common between learned examples and new, previously unseen instances in text. The *Medication_instruction* category, on the other hand, fails because of the excessive length and low number of its representatives – it is the least "concept-like" concept in the schema, taking up an

Table 5. Detailed metrics of entity recognition accuracy with different criteria (dataset version with removed punctuation) based on the SemEval 2013 9.1 shared task metrics. The columns capture the four different evaluation schemata: *Type* checks the entity class with partial token overlap, *Partial* allows for any entity overlap, regardless of the class, *Exact* requires exact text match ignoring the class, and *Strict* checks both the text and type of the entities (this corresponds to the CoNLL evaluation). The row metrics sum to the overall numbers of the ground-truth entities in the tests (*Correct + Incorrect + Missed* = 12,312) and the predicted entities (*Correct + Incorrect + Spurious* = 12,940).

Measure	Type	Partial	Exact	Strict
Correct	9,353	9,368	9,368	8,574
Incorrect	1,022	0	1,007	1,801
Partial	0	1,007	0	0
Missed	1,937	1,937	1,937	1,937
Spurious	2,565	2,565	2,565	2,565
Precision	0.726	0.766	0.727	0.666
Recall	0.760	0.808	0.764	0.693
F1	**0.734**	**0.777**	**0.736**	**0.672**

Table 6. Annotation labels and their short descriptions.

Label	Description
Abbreviation	An abbreviation or acronym, e.g. "**RT** není **indik.**" ("***RT*** *not **indic.***")
AnatomicalSite_laterality	An expression indicating laterality or direction on the body or body part, e.g. "našla v **pravém** prsu bulku" (*"found a lump in her **right** breast"*)
AnatomicalSite_name	A name of a part or location on the human body, e.g. "našla v pravém **prsu** bulku" (*"found a lump in her right **breast**"*)
DateTime	Expressions specifying date, time, or duration, e.g. "premenopauza, dg. **9/2020**" (*"premenopause, diagnosis **9/2020**"*)
DiseaseDisorder	A word or phrase naming a disease or disorder, e.g. "**inf mononukleoza** v 15 letech" (*"**infectious mononucleosis** at the age of 15"*)
Lab_name	The name of a measured variable, laboratory or other, e.g. "**S_Estradiol**: 18 pmol/l, **S_FSH**: 3.8 IU/l"
Lab_unit	A unit of measurement, e.g. "S_Estradiol: 18 **pmol/l**, S_FSH: 3.8 **IU/l**"
Lab_value	The value of a measured variable, laboratory or other, e.g. "S_Estradiol: **18** pmol/l, S_FSH: **3.8** IU/l"
Medication_dosage	The amounts of units (tablets, drops, ...) in taking the medication, e.g. "večer: Lexaurin 1,5 mg **[0-0-1]**" (*"evening: Lexaurin 1.5 mg **[0-0-1]**"*)
Medication_instruction	Any further specification of how the medication is to be taken, e.g. "večer: Lexaurin 1,5 mg **[při bolesti** 1 tbl]" (*"evening: Lexaurin 1.5 mg **[in case of pain** 1 tbl]"*)
Medication_name	The name of a medication, e.g. "večer: **Lexaurin** 1,5 mg [0-0-1]" (*"evening: **Lexaurin** 1.5 mg [0-0-1]"*)
Medication_strength	The amount of active substance in the medication, e.g. "večer: Lexaurin **1,5 mg** [0-0-1]" (*"evening: Lexaurin **1.5 mg** [0-0-1]"*)
Negation	Markers of negation, e.g. "adjuvantní radioterapie **není** indikována" (*"adjuvant radiotherapy **not** indicated"*)
Procedure	A diagnostic or therapeutic procedure carried out by medical personnel, e.g. "benefit **adjuvantní chemoterapie** minimální" (*"minimal benefit of **adjuvant chemotherapy**"*)
SignSymptom	A word or phrase describing a sign or symptom, e.g. "při **bolestech svalů**, **teplotě**" (*"in case of **muscle pain**, **fever**"*)

average of 3.1 tokens, the longest instance being 83 characters long which is the longest annotation in the entire dataset.

In the confusion matrix (Fig. 1), it is apparent that most mistakes are related to the "*O (no annotation)*" class. Either the model misses relevant tokens (hence the vertical line above "O") or it predicts a class where there should be none (horizontal line next to "O").

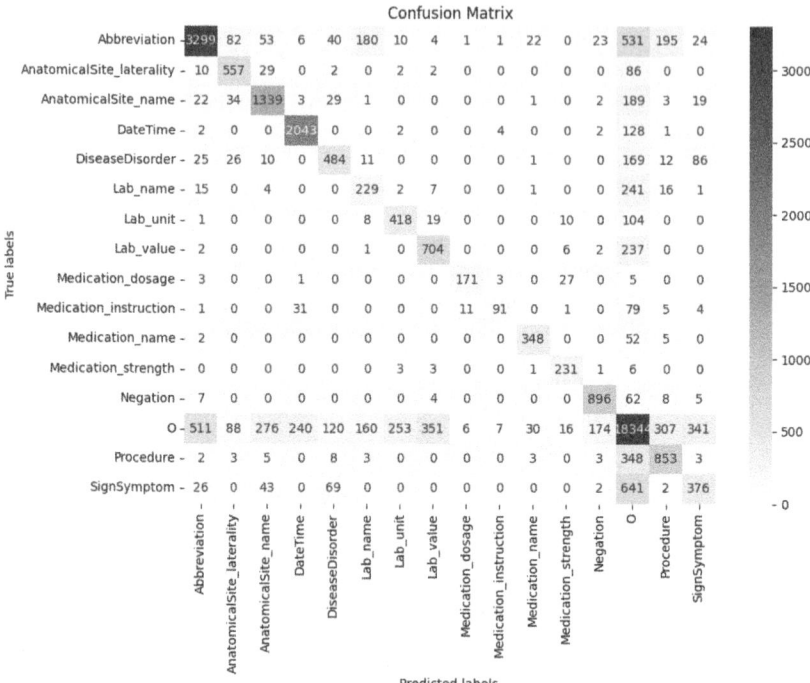

Fig. 1. Confusion matrix showing the relationship between ground truth annotation and prediction on the level of tokens.

4 Conclusion and Future Directions

This paper marks the accomplishment of having collected, processed, and human-annotated a new dataset of medical records, named CSEHR, and subsequently trained several baseline named entity recognition models, demonstrating the degree of their generalizability. However, this has only been the initial phase of a larger scheme of using small human-annotated data to bootstrap large transformer-based models for Czech medical concept recognition. In the near future, work will continue by extending human verification, and combining automated rule-based annotation and human annotation in order to build annotated corpora in the order of millions of words.

The CSEHR dataset that forms the basis of this paper is currently being prepared for public access, however, due to the sensitive nature of medical information and complex interdependencies between the institutions involved, this process is treated separately from the process of publishing results.

Acknowledgments. The analyzed Czech data was collected at the Masaryk Memorial Cancer Institute in Brno, Czech Republic within the project MUNI/G/1763/2020: *AIcope – AI Support for Clinical Oncology and Patient Empowerment.*

This work has been partly supported by the Ministry of Education, Youth and Sports of the Czech Republic within the LINDAT-CLARIAH-CZ project LM2023062.

Computational resources were provided by the e-INFRA CZ project (ID:90254), supported by the Ministry of Education, Youth and Sports of the Czech Republic.

Ethical considerations and limitations. All data underwent a two-phase de-identification process: there was both automated and manual replacement of names and personal data of both patients and doctors for generic names and fake data, and also other information that could be used to identify the patient, such as names of locations, institutions, and statistically rare conditions, was obfuscated.

References

1. Akbik, A., Bergmann, T., Blythe, D., Rasul, K., Schweter, S., Vollgraf, R.: FLAIR: an easy-to-use framework for state-of-the-art NLP. In: Proceedings of the 2019 Conference of the North American Chapter of the Association for Computational Linguistics (demonstrations), pp. 54–59 (2019)
2. Batista, D.: nervaluate Python module for evaluating Named Entity Recognition (NER) models as defined in the SemEval 2013 9.1 task (2020). https://pypi.org/project/nervaluate/
3. Bodenreider, O.: The unified medical language system (UMLS): integrating biomedical terminology. Nucleic Acids Res. **32**(suppl_1), D267–D270 (2004)
4. Conneau, A., et al.: Unsupervised cross-lingual representation learning at scale. In: Proceedings of the 58th Annual Meeting of the Association for Computational Linguistics, pp. 8440–8451. Association for Computational Linguistics (2020). https://doi.org/10.18653/v1/2020.acl-main.747, https://aclanthology.org/2020.acl-main.747
5. Institute of Endocrinology: List of abbreviations – Institute of Endocrinology — endo.cz (2009). https://www.endo.cz/files/download/seznam-zkratek.pdf. Accessed 19 Oct 2023
6. Karviná-Ráj hospital: List of abbreviations – Karviná-Ráj hospital — nspka.cz (2017). https://nspka.cz/USoubory/soubory/2017/WEB_Seznam_zkratek_29_8_2017.pdf. Accessed 19 Oct 2023
7. Jiráková, P.: List of the most common medical abbreviations – Alfabet — alfabet.cz (2014). https://www.alfabet.cz/vyvojova-vada-u-ditete/zdravotni-a-jina-pece/lekarske-zkratky/. Accessed 19 Oct 2023
8. Johnson, A., Bulgarelli, L., Pollard, T., Horng, S., Celi, L.A., Mark, R.: MIMIC-IV (2023). https://physionet.org/content/mimiciv/2.2/. Accessed 19 Oct 2023
9. Johnson, A., Pollard, T., Mark, R.: MIMIC-III (2016). https://physionet.org/content/mimiciii/1.4/. Accessed 19 Oct 2023
10. Johnson, A.E., et al.: MIMIC-III, a freely accessible critical care database. Sci. Data **3**(1), 1–9 (2016)
11. Névéol, A., Dalianis, H., Velupillai, S., Savova, G., Zweigenbaum, P.: Clinical natural language processing in languages other than English: opportunities and challenges. J. Biomed. Semant. **9**(1), 1–13 (2018)

12. NIH, National Institute of Health: UMLS Metathesaurus, National Library of Medicine (2023). https://www.nlm.nih.gov/research/umls/knowledge_sources/metathesaurus. Accessed 19 Oct 2023
13. Novotný, V., Luger, K., Štefánik, M., Vrabcová, T., Horák, A.: People and Places of historical europe: bootstrapping annotation pipeline and a new corpus of named entities in late medieval texts. arXiv preprint arXiv:2305.16718 (2023)
14. Segura-Bedmar, I., Martínez, P., Herrero-Zazo, M.: SemEval-2013 task 9: extraction of drug-drug interactions from biomedical texts (DDIExtraction 2013). In: Manandhar, S., Yuret, D. (eds.) Second Joint Conference on Lexical and Computational Semantics (*SEM), Volume 2: Proceedings of the Seventh International Workshop on Semantic Evaluation (SemEval 2013), Atlanta, Georgia, USA, pp. 341–350. Association for Computational Linguistics (2013). https://aclanthology.org/S13-2056
15. Sherborne, T., Xu, Y., Lapata, M.: Bootstrapping a crosslingual semantic parser. arXiv preprint arXiv:2004.02585 (2020)
16. Steedman, M., et al.: Bootstrapping statistical parsers from small datasets. In: 10th Conference of the European Chapter of the Association for Computational Linguistics (2003)
17. Stenetorp, P., Pyysalo, S., Topić, G., Ohta, T., Ananiadou, S., Tsujii, J.: BRAT: a web-based tool for NLP-assisted text annotation. In: Proceedings of the Demonstrations at the 13th Conference of the European Chapter of the Association for Computational Linguistics, pp. 102–107 (2012)
18. SÚKL, State Institute for Drug Control: Medicinal Products Database (in Czech Databáze léčivých přípravků DLP) (2023). https://opendata.sukl.cz/?q=katalog/databaze-lecivych-pripravku-dlp. Accessed 19 Sept 2023
19. Thirunavukarasu, A.J., Ting, D.S.J., Elangovan, K., Gutierrez, L., Tan, T.F., Ting, D.S.W.: Large language models in medicine. Nat. Med. **29**(8), 1930–1940 (2023)
20. Vaswani, A., et al.: Attention is all you need. In: Advances in Neural Information Processing Systems, vol. 30 (2017)
21. Wu, S.T., et al.: A common type system for clinical natural language processing. J. Biomed. Semant. **4**(1), 1–12 (2013)
22. Yoo, K.M.: Python implementation of the CoNLL-2000 shard task evaluation (2020). https://github.com/kaniblu/conlleval/

Analyzing Biases in Popular Answer Selection Datasets on Neural-Based QA Models

Chang Nian Chuy, Cherie Ding(✉), and Qinmin Vivian Hu(✉)

Toronto Metropolitan University, Toronto, Canada
{chang.chuy,cding,vivian}@torontomu.ca

Abstract. The amount of information available on the internet has increased exponentially over the past decade. This digitization leads to the need of automated answering system to extract useful information from different sources. Due to the high demand of automated answering systems, many large-scale QA datasets and QA models have been introduced to the field to satisfy this need. In this work, we aim to explore and shed light upon the composition of the most popular QA datasets by comparing them through statistical distribution analyses and their biases. We collect multiple open QA datasets which cover different aspects of QA features, and highlight the differences of each QA dataset and its bias by comparing its effect on multiple baseline neural QA models. Our goal is to provide a clear understanding on the relationship of QA datasets and QA models, and offer a solid foundation for future research to enhance this growing field.

Keywords: Question Answering · QA Datasets · Answer Selection model · Deep Learning model · Neural-based QA model

1 Introduction

The QA goal is to understand natural language questions and reply with the most concise and relevant answers. The state-of-the-art QA system usually takes the form of a pipeline architecture, chaining together modules that perform tasks such as document retrieval, answer candidate extraction, and answer re-ranking. Due to the complexity of the field, each task listed above has evolved into its own sub-field and is often studied and evaluated independently. Correspondingly, many QA datasets were created to target different sub-tasks of this overall objective. In general, we could classify the available QA datasets into two categories, Answer Selection (AS) and Reading Comprehension (RC). The goal of an AS field is to accurately classify whether a candidate answer is the correct answer to a given question; whereas the goal of a RC field is to extract the precise answer to a question from its corresponding context. Due to the similarity of the objective which is to identify the correct answer for the question, most QA datasets can be used for multiple sub-tasks. This creates a lot of confusion to researchers in the field, especially those who are new to it.

Large datasets have led to many breakthroughs in AI, thus, we are motivated to quantify existing QA datasets by comparing the composition through analysis of the question types and answer types, gender distribution, and sentiment analysis. According to our knowledge, there is currently no standardized baseline models for AS task to date has been published. In this paper, we established a standardized baseline models by analysing and comparing multiple QA datasets on multiple aspects to help researchers understand the structure of these datasets. The paper is designed to assist researchers to gain an overall understanding of a QA dataset by categorizing its' question and answer types using NLP classifiers. We believe our paper is a valuable contribution to the advancement of the QA field, as it could lower the barrier of entry by providing a solid starting point for new researchers and practitioners.

In summary, our contributions include:

1. We implemented multiple NLP components to quantify features of QA datasets;
2. We compared gender bias and sentiment bias which have been well-studied in the NLP field but not in the AS field;
3. We analyzed the QA datasets performance on multiple AS models.

2 Background and Related Work

Answer Selection (AS) is the fundamental component of a QA system. Traditional AS models use methods such as TF-IDF (Term Frequency - Inverse Document Frequency) and BM25 to rank candidate answers using sparse representations. Recently, deep neural models are heavily investigated to encode questions and candidate answers into latent vector space where text semantics can be measured to identify the QA pair with optimal relationship. Word embedding libraries such as GloVe[1] and word2vec[2] that convert words into vectors which are more recognizable to machines have been used with deep neural networks and have achieved tremendous success in the QA field.

In Fig. 1, we graphed the total number of AS papers published each year between 2017–2023 based on QA-type. We filtered AS papers from QA papers using keyword (answer selection) filtering. We noticed that open-domain QA is still the most dominant domain for AS task, with Community Question and Answering (CQA) gaining interest during 2018–2020 period. As shown in graph, AS task in QA is still highly interested by researchers averaging at least 10 papers per year. However, papers on Conversational Question and Answering and Multi-Lingual Question and Answering are still lacking for AS task.

2.1 Methods

Learning Approaches. In the QA field, there are three common approaches to ranking candidate answers, namely point-wise, pair-wise and list-wise [9].

---

[1] https://nlp.stanford.edu/projects/glove/.
[2] https://code.google.com/archive/p/word2vec/.

In **point-wise** approach, ranking problem is transformed to a binary classification problem and focus on optimizing the relevance score of each QA pair [3]. The training objective is to minimize the cross entropy of all labelled QA pairs in training set [7]. The **pair-wise** approach, learns the correlation among candidates in a contrastive way which distinguishes the positive and negative answers [3]. The loss function is designed to encourage the correct answer to have a higher score than an incorrect answer by a certain margin [7]. The **list-wise** approach, aims to fit the prediction of a list QA pairs with the ground-truth where the objective is to minimize the KL-divergence [7].

Neural-Network Architectures. Neural-Network (NN) models have become main stream in AS, there are three main architecture [7]: siamese, attentive, and compare-aggregate.

Siamese network, uses the same encoder to encode the vector representations of the question and candidate answers. The QA pair have no influence on the computation of each other's representation. An advantage of this architecture is that applying the same encoder to each input sentence makes the model smaller [7].

Attention mechanism focuses on the context information and interrelationship between different words in the sentences by allocating different weight to words or sub-phrases in sentences to capture interaction between input sentences during the encoding process [10]. Individual attention indicates local importance of a word while interactive attention implies the importance of a word with respect to the candidate answers [15]. While, attentive pooling allows

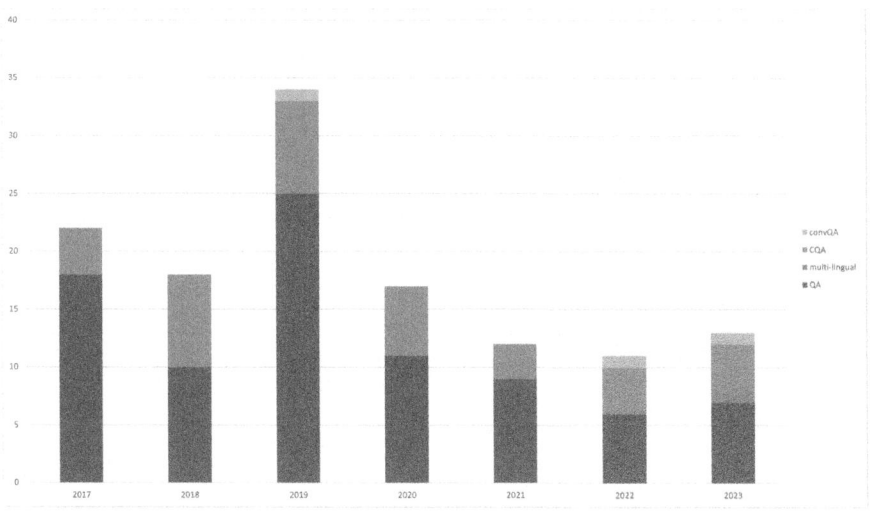

Fig. 1. Answer Selection Papers per Year

the information from the QA pair to influence the computation of each other's representation [7].

Compare-Aggregate architecture captures more interactive features between QA pair, and acquires significant improvement on performance [14]. It exploits corresponding vectors of small units with attention mechanism, then compares small units with corresponding vectors to produce comparison vectors. Finally comparison vectors are aggregated and fed into a classifier for final classification.

2.2 Bias

In [2], they have shown that NLP systems exhibit different accuracy depending on users' races caused by under-representation of certain race or stereotype bias in data. The source of bias often stems from training data, lexicons, and word embeddings that the system uses to train its prediction model. Therefore, it is likely that the systems can learn society's gender prejudices and reinforce them in their model [13]. There is currently no benchmark QA dataset for examining inappropriate biases in QA systems. Thus, in this paper we would like to partially fill this gap by investigating the gender and sentiment bias that exists in the QA datasets. The reason we particularly pick these bias over other types of biases is that gender and sentiment bias are one of the most commonly found bias in NLP dataset, and it can serve as a starting point for exploring other biases.

3 General Analyses of QA Datasets

In this section, we have chosen a few representative datasets for each QA task from different domain and compared them on various metrics using statistical analysis and NLP classification methods.

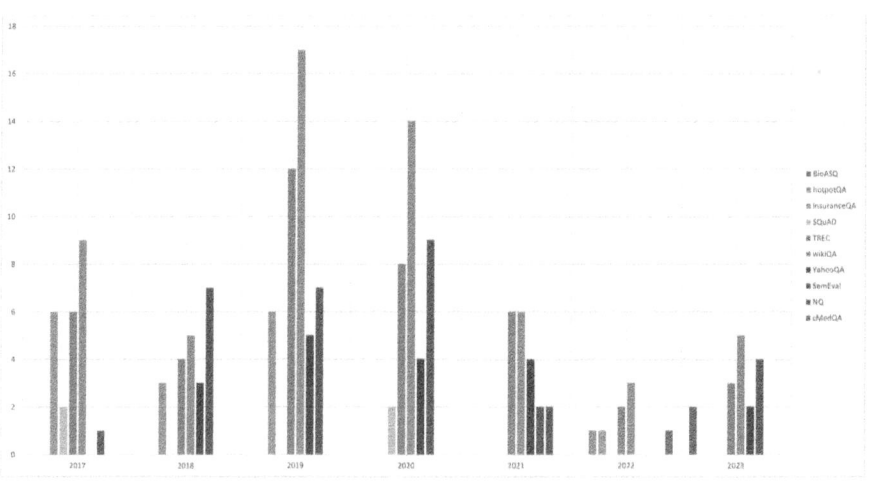

Fig. 2. Dataset used in Answer Selection Papers per Year

In Fig. 2, we graphed top 5 most used datasets for each year between 2017–2023. We noticed the most compared open-domain QA are TREC and WikiQA; and SemEval dataset is most used for CQA domain. Among these datasets, 5 were open-domain dataset and the others were CQA datasets. Among these 5 datasets, only TREC and WikiQA were originally designed for AS task, where the other 3 were for RC tasks. Thus, we hoped to provide alternative benchmark dataset for AS task by converting popular RC datasets so that researchers will be able to test their models on bigger and newer datasets.

3.1 Domain

Table 1. Overview of QA datasets

Dataset	Domain	Question Source	Answer	Year
TREC	Open	Query logs	Expert selected	2007
WikiQA	Open	Query logs	Crowd-sourced	2015
Simple Question	Freebase	Crowd-sourced	Span of words	2015
SQuAD	Wikipedia	Crowd-sourced	Span of words	2016
MS MACRO	Query logs	Expert generated	Expert generated	2016
TriviaQA	Open	Expert generated	Expert generated	2017
SearchQA	Jeopardy	Expert generated	Expert generated	2017
NewsQA	CNN	Crowd-sourced	Crowd-sourced	2017
RACE	Examinations	Expert generated	Expert generated	2017
NarrativeQA	Books and Movie scripts	Human generated	Human generated	2017
ARC	Examinations	Expert generated	Expert generated	2018
HotpotQA	Wikipedia	Crowd-sourced	Crowd-sourced	2018
DuoRC	Movie plots	Human generated	Human generated	2018
Multi RC	Multiple sources	Crowd-sourced	Crowd-sourced	2018
Natural Questions	Open	Query logs	Crowd-sourced	2019
TweetQA	Social Media	User generated	User generated	2019
CODAH	Common sense	Expert generated	Expert generated	2019
CosmosQA	Blog	Crowd-sourced	Crowd-sourced	2019
DROP	Wikipedia	Crowd-sourced	Crowd-sourced	2019
AdversarialQA	Wikipedia	Crowd-sourced	Human generated	2020
ReClor	Examinations	Expert generated	Expert generated	2020
Game of Thrones	TV Series	Expert generated	Span of words	2022

In Table 1, we ordered the datasets based on the year they were published. Most QA datasets can be categorized as open-domain where their candidate answer set are retrieved via search engine or Wikipedia site, whereas domain-specific restrict their candidate answer set to specific domains. Early QA datasets focused on open-domain where questions were queries, and QA systems are treated as an extension of IR system. Early QA model achieved high accuracy on factoid questions but suffered on non-factoid questions, thus RC datasets were introduced to examine the reasoning and logical thinking of QA models. Many QA datasets were composed from examinations designed to test the understanding of texts. Most recent QA datasets focused on conversational QA, where follow-up questions will be asked when the returned answers were undesirable.

3.2 Question and Answer Types

Table 2. Question and Answer type distribution of QA datasets

Dataset	Question Type								Answer Type					
	Who	When	Where	How	What	Why	Which	Other	Person	Date	Location	Quantity	Noun	Clause
wikiQA	10.00	11.94	10.00	12.03	56.02	0.00	0.00	0.00	13.63	12.12	17.08	15.31	22.38	19.45
NQ	36.65	16.95	10.85	4.24	16.10	0.29	3.17	11.74	31.60	16.41	11.28	7.21	12.08	23.10
trecQA	13.61	15.76	9.31	17.72	42.16	1.25	0.17	0.00	29.90	10.74	12.44	4.74	27.84	14.31
AQA	12.10	2.03	5.26	6.70	53.03	2.43	11.93	6.50	9.56	3.56	7.83	5.56	16.37	57.12
SQA	16.06	0.00	6.76	0.52	56.11	0.00	15.93	4.61	20.41	0.95	16.75	2.11	14.49	45.24
SQuAD	10.56	6.68	3.86	10.69	61.96	1.59	4.09	0.54	8.70	10.03	7.50	13.86	16.85	43.05
GoT	46.54	1.13	7.18	2.86	40.61	1.09	0.34	0.23	27.07	0.89	2.35	2.89	16.66	49.98
triviaQA	14.88	0.63	1.93	0.67	32.99	0.01	46.83	2.05	26.60	1.26	16.10	1.86	22.99	31.09
searchQA	4.60	2.70	1.07	2.26	1.25	0.12	1.77	86.19	23.51	1.37	13.88	1.46	18.14	41.59
tweetQA	32.28	1.88	5.23	7.47	47.53	1.76	2.37	1.46	19.82	4.37	5.77	5.47	8.57	55.96
ARC	0.49	6.65	1.19	4.30	17.21	2.70	51.36	15.83	1.36	1.81	2.67	4.01	2.59	87.58
CODAH	0.72	2.02	0.29	1.15	0.36	0.29	0.144	95.03	2.17	2.82	1.93	4.71	2.86	85.49
cosmosQA	1.66	0.69	1.94	5.86	55.61	31.91	0.92	1.38	4.90	2.64	2.19	1.53	3.6	84.14
DROP	8.46	1.28	0.67	56.80	9.00	0.12	21.80	1.85	15.91	8.90	6.01	42.86	12.94	13.38
duoRC	42.73	1.15	8.69	5.93	35.47	2.32	1.80	1.93	31.52	2.48	5.86	3.34	13.60	43.21
hotpotQA	24.18	2.14	2.57	2.14	35.95	0.05	25.71	7.23	32.05	7.23	10.29	8.23	21.39	20.78
MS MACRO	1.56	1.65	4.07	15.49	42.74	1.84	1.27	31.35	10.07	11.15	8.80	34.96	23.89	11.12
multiRC	17.80	3.34	3.86	11.58	44.74	7.38	5.32	5.95	13.78	6.02	7.97	7.26	13.18	51.78
narrativeQA	26.13	2.19	7.39	10.20	41.68	9.68	2.49	0.22	25.68	2.21	6.18	3.15	12.46	50.32
newsQA	22.36	4.31	7.63	7.61	49.80	0.14	2.92	5.22	16.56	7.28	11.55	12.95	14.54	37.08
RACE	2.69	4.39	1.34	4.78	20.71	4.95	13.92	47.16	5.57	4.43	4.41	6.15	7.11	72.31
ReClor	0.68	0.21	0.02	0.21	0.58	0.93	79.62	17.73	6.01	7.82	4.45	8.18	7.80	65.72

Question Types. The 'wh-type' labeling is widely used for question classification in QA field. In Table 2, the question type distribution provides an overview of the general composition of questions in the dataset. For example, the 'who, when, where' type questions are normally associated with factoid type questions, whereas the 'why, how' type ones are associated with descriptive questions. Evidently, datasets that focused on open-domain have by far the largest set of 'what' and 'who' questions because most questions are either queries or crowd-sourced factoid questions that search for an entity. Quiz-based datasets such as TriviaQA and SearchQA have more 'which' and 'how' questions as comparison and reasoning are needed to identify the entity.

Answer Types. Many works have identified how different granularity level of classification affects their QA tasks. For generality, in this paper we use a coarse-grained answer type classification that corresponds to the question type classification. A coherent dataset should have a similar answer type distribution that is in sync with their corresponding question type distribution. For example, the 'who' type question should map to a 'PERSON' type answer, and the 'why' type question should map to a 'Clause' type answer.

The answer types shown in Table 2 are more or less in sync with the question types where most datasets have a high ratio of person entity and noun phrase matching the 'who' and 'what' type questions. Interestingly, even though TriviaQA and SearchQA have different question type composition than other QA datasets, the answer type distribution is similar to other datasets which are mainly noun entities and clause, because even though the question format is different the answer to them are still mostly noun entities.

3.3 Question Answer Similarity

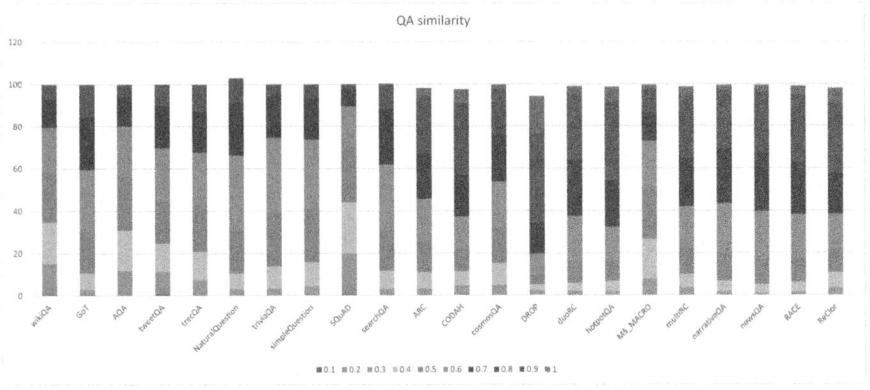

Fig. 3. QA pair similarity

Due to popular approach of employing crowd workers to come up with a question while reading a paragraph, [1] pointed out that this kind of question generation is biased, where questions are usually syntactically similar to the candidate answer. We calculated the QA pair similarity using "Sentence Transformer" library[3] to embed each word into a vector of 768 dimensions. Then, we used average-pooling to get the vector representation of question and answers. Lastly, we used the cosine similarity function to get the similarity score.

We noticed that QA datasets with longer candidate answers (sentences) have a normal QA pair similarity distribution that centered around 0.5. On the other hand, for QA datasets that have shorter answers (span of words), the similarity distribution tends to skew towards to 0.6. Our interpretation of this behavior is that dataset with long candidate answer tends to have more candidate answers mapped to a single question providing a wider coverage of topics, whereas dataset with short candidate answer tends to have less candidate answers thus giving more emphasis to QA pairs that are the correct answer. In Fig. 3, we showed the distribution of cosine similarity between QA pairs where each column represents the percentage of QA pairs. For example, the first column shows the percentage of QA pairs with cosine similarity scores between 0 and 0.1 in each dataset.

In Fig. 3, we are able to identify a trend in dataset timeline, where initial large datasets have higher QA-pair similarity largely due to the fact that most questions are from crowd workers who were asked to generate questions based on a paragraph. Because of the flaws in such datasets as shown in later research, datasets in recent years have emphasized the importance of paraphrasing to avoid using the overlapping words to generate the question. Thus, examination datasets were popular in testing the model's understanding of language.

3.4 Gender Distribution

Fig. 4. Gender distribution of QA datasets

---

[3] https://www.sbert.net/.

A popular bias that had been extensively studied in NLP dataset is gender bias such as female-to-male QA pairs ratio. Many QA datasets have imbalanced gender distribution where females are often under-represented in datasets. To mitigate gender bias, for each QA pair, we check for gender pronouns such as he, she, and her. Then, we identify people names using Spacy's Named Entity Recognition tool [5] and feed the person's name to a pre-trained name classifier to predict the gender. The name classifier is trained on traditional western names, as a result, it is prone to mislabel some foreign names. Lastly, we replace gender pronouns with gender neutral pronouns and replace names with the "name" token to create a more gender neutral dataset.

In Fig. 4, most QA pairs in dataset are neutral, even though 'who' type question and 'person' type answer are majority in these QA datasets. The remaining QA pairs are classified as neutral as they do not contain any person name or gender pronouns. Given the QA pairs with gender information, a majority of them are classified as male QA pairs. This is due to social bias and the stereotyping of job titles containing male pronoun such as policeman.

4 Experiment on QA Baseline Models

Due to the success of neural models in image field, many researchers have duplicated those models into the AS field. However, in many research papers, these models are just compared against each other, lacking some standard baselines when comparing such models. Thus, in this paper, we try to create a set of standardized baseline models for AS field. We duplicated 4 variations of LSTM models (LSTM [4] as base model, BiLSTM to utilize future context, AP-LSTM [11] for attention-mechanism and HD-LSTM [12] for mining interactions between QA pairs) and 2 CNN models (CNN [8] as base model and AI-CNN [16] for pooling interactions between QA pairs).

4.1 Experimental Settings

We conducted our experiment on Google Colab with GPU enabled. The deep learning framework and programming language used in this paper are PyTorch and Python respectively. For the above baseline AS models, we used 100-dimensional GloVe[4] word embedding which are trained based on global word co-occurrence. We pad the sentence length for all questions and answers to 100 for each QA pair that were short and truncate those that were long. The Adaptive Moment (Adam) estimation is employed for optimizing the loss in this paper. In addition, L2 regularization and the dropout methods are also used in training to avoid over-fitting. We train our models in mini-batches with vary learning rate for different datasets.

---

[4] http://nlp.stanford.edu/data/glove.6B.zip.

4.2 Result Analysis

In Table 3, we trained and tested the most compared neural-based AS models by researchers between 2017–2023 on the above selected QA datasets. Based on this result, AI-CNN model has the best performance on QA datasets with informal structure of candidate answer set. TREC candidate answer set contains mostly quotations or dialogues and TweetQA candidate answers are tweets which are mostly short and informal. The candidate set for each question in WikiQA are chosen such that most of them are similar to the entity/topic of the question. Therefore, CNN outperforms other baseline models as CNN model has shown outstanding performance in pairwise comparison. The candidate answers for SQuAD and AdversarialQA dataset are a paragraph containing the answer to the questions, even though AdversarialQA questions are mostly paraphrases, the HD-LSTM model are able to overcome it. Thus, HD-LSTM model is great at selecting corresponding sentence to the question within a paragraph. Sim-

Table 3. Baseline QA models on QA datasets

Dataset	LSTM		BiLSTM		AP-LSTM		HD-LSTM		CNN		AI-CNN	
	MAP	MRR	MAP	MRR	MAP	MRR	MAP	MRR	MAP	MRR	MAP	MRR
WikiQA	0.59	0.62	0.60	0.64	0.61	0.63	0.63	0.67	0.70	0.69	0.67	0.68
Natural Questions	0.82	0.83	0.83	0.83	0.83	0.83	0.55	0.55	0.71	0.71	0.78	0.79
TREC	0.57	0.60	0.71	0.74	0.66	0.71	0.66	0.57	0.74	0.77	0.75	0.80
AdversarialQA	0.51	0.53	0.51	0.53	0.53	0.55	0.68	0.69	0.67	0.68	0.61	0.60
Simple Question	0.63	0.61	0.66	0.64	0.65	0.64	0.60	0.59	0.65	0.62	0.59	0.60
SQuAD	0.54	0.55	0.54	0.58	0.66	0.68	0.79	0.87	0.72	0.75	0.66	0.70
Game of Thrones	0.81	0.81	0.76	0.76	0.80	0.80	0.58	0.58	0.71	0.71	0.72	0.72
TriviaQA	0.66	0.70	0.67	0.71	0.62	0.65	0.57	0.63	0.60	0.64	0.56	0.57
SearchQA	0.75	0.75	0.75	0.75	0.75	0.75	0.57	0.57	0.68	0.68	0.69	0.69
TweetQA	0.65	0.64	0.65	0.64	0.63	0.63	0.67	0.67	0.68	0.67	0.70	0.69
ARC	0.53	0.53	0.53	0.53	0.53	0.53	0.68	0.69	0.53	0.53	0.71	0.74
CODAH	0.52	0.52	0.51	0.51	0.51	0.51	0.55	0.56	0.53	0.53	0.52	0.52
cosmosQA	0.52	0.55	0.53	0.57	0.52	0.56	0.67	0.71	0.51	0.55	0.57	0.60
DROP	0.80	0.80	0.79	0.79	0.80	0.80	0.76	0.77	0.80	0.81	0.81	0.82
duoRC	0.50	0.70	0.50	0.69	0.64	0.79	0.54	0.72	0.74	0.74	0.71	0.71
hotpotQA	0.71	0.71	0.79	0.79	0.78	0.78	0.70	0.70	0.79	0.79	0.74	0.74
MS MACRO	0.53	0.58	0.56	0.61	0.49	0.54	0.77	0.82	0.33	0.34	0.66	0.7
multiRC	0.70	0.76	0.71	0.78	0.70	0.76	0.70	0.76	0.72	0.80	0.74	0.81
narrativeQA	0.75	0.78	0.76	0.79	0.76	0.79	0.68	0.72	0.69	0.69	0.68	0.71
newsQA	0.81	0.82	0.85	0.86	0.84	0.84	0.51	0.51	0.79	0.79	0.84	0.85
RACE	0.54	0.54	0.54	0.54	0.52	0.52	0.67	0.69	0.50	0.50	0.67	0.70
ReClor	0.54	0.54	0.53	0.53	0.54	0.54	0.57	0.58	0.53	0.53	0.70	0.73

ple Question, TriviaQA, Natural Questions, SearchQA and Game of Thrones datasets mostly consist of short-answer entities as their answer candidate for each question. These types of QA datasets work the best with the generic LSTM model, especially BiLSTM model that utilizes both previous and future tokens of the question. Since the candidate answers of these datasets are normally just span of words, there is minimal information that could be mined between the question and answer pair which made AP-LSTM and HD-LSTM not suitable for them. Game of Thrones dataset seems to favor LSTM more which might due to its data size, as it does not contain enough QA pairs to fully utilize the complexity of BiLSTM model.

Table 4. Baseline QA models on neutral pronoun QA datasets

Dataset	LSTM		BiLSTM		AP-LSTM		HD-LSTM		CNN		AI-CNN	
	MAP	MRR	MAP	MRR	MAP	MRR	MAP	MRR	MAP	MRR	MAP	MRR
WikiQA	0.16	0.07	0.03	-0.02	0.03	-0.02	0.05	0.02	-0.09	0.00	0.10	0.03
Natural Questions	0.02	0.01	0.01	0.01	0.01	0.01	0.00	0.00	0.06	0.07	0.00	0.00
TREC	0.16	0.11	-0.01	0.01	0.08	-0.01	0.00	0.00	-0.12	-0.08	0.05	0.07
AdversarialQA	0.00	-0.01	0.01	0.01	0.01	0.00	0.01	0.01	0.00	0.00	-0.04	-0.02
Simple Question	0.00	0.01	0.01	-0.02	0.01	-0.01	-0.03	0.02	-0.05	-0.01	0.02	-0.01
SQuAD	0.11	0.14	0.14	0.13	-0.06	-0.04	0.00	0.00	0.01	0.00	0.11	0.10
Game of Thrones	0.26	0.25	0.25	0.21	0.10	0.08	-0.21	-0.29	0.00	-0.03	0.06	0.02
TriviaQA	-0.01	-0.04	0.02	0.00	0.09	0.08	0.14	0.10	0.00	-0.04	0.00	0.04
SearchQA	0.00	0.00	-0.01	-0.01	0.00	0.00	0.00	0.00	0.04	0.04	0.01	0.01
TweetQA	-0.02	-0.02	-0.02	-0.02	0.02	0.01	0.00	0.00	0.00	0.01	-0.02	-0.01
ARC	0.00	0.00	-0.01	-0.01	0.00	0.00	-0.10	-0.11	-0.01	-0.01	0.12	0.12
CODAH	0.01	0.01	0.02	0.02	0.03	0.03	-0.01	-0.01	0.00	0.00	0.32	0.35
cosmosQA	0.00	0.01	0.00	0.00	0.01	0.01	0.00	0.00	0.01	0.01	-0.01	0.01
DROP	0.03	0.04	0.04	0.05	0.06	0.07	-0.21	-0.21	0.00	-0.01	-0.06	-0.06
duoRC	0.30	0.11	0.30	0.11	0.18	0.03	0.00	-0.19	0.02	0.03	0.06	0.05
hotpotQA	0.09	0.09	0.00	0.00	0.05	0.05	-0.12	-0.12	-0.04	-0.04	-0.01	-0.01
MS MACRO	-0.15	-0.16	-0.18	-0.18	-0.10	-0.11	-0.18	-0.24	0.00	0.00	-0.08	-0.06
multiRC	0.00	0.02	0.01	0.02	0.01	0.02	0.01	0.02	0.01	0.01	0.00	-0.02
narrativeQA	0.00	0.00	0.00	-0.01	0.00	0.00	-0.03	-0.02	0.02	0.02	0.03	0.03
newsQA	-0.06	-0.07	-0.08	-0.09	-0.08	-0.08	0.03	0.03	-0.01	-0.01	-0.18	-0.18
RACE	0.00	0.00	-0.01	-0.01	0.03	0.03	0.01	0.01	0.05	0.05	-0.05	-0.07
ReClor	0.01	0.01	0.02	0.02	0.00	0.00	0.00	0.00	0.03	0.03	-0.12	0.01

Gender Neutral Results. In [6], they have shown that gender bias exists in sentiment analysis, where gender pronouns plays a significant role in predicting sentiment polarity. To examine the gender information that exists in each QA datasets and how it affects AS models, we replaced gender pronouns such as he, she and her with neutral pronoun such as ze and zir. We then re-trained these baseline AS models on the new QA datasets and compare the MAP and MRR results. We highlighted cells in Table 4 with different colors, where green shows improvement, red shows deterioration, and white shows insignificant changes.

Based on the results, gender pronouns plays a significant role in QA datasets with long answer as candidate answers. Most QA datasets, benefit from using neutral gender pronouns where QA models have better performance on datasets that has less gender information. Gender pronouns have negative affect on MS MACRO and newsQA datasets, due to social stereotypes usually stem from news articles. Interestingly, HD-LSTM and AI-CNN models were impacted the most by removal of gender information. AI-CNN model has significant improvement on gender neutral datasets with longer candidate answers, but decrease in performance on shorter candidate answers. Most short answers in QA datasets are span of words such as name, place or numeric values, by removing gender information, we hinder the model from fully utilizing the latent features that exists in word embedding.

5 Conclusions and Future Work

With the rise of interactive information system, the number of QA datasets is growing rapidly, making it difficult for new researchers to make an informative selection of suitable QA datasets for their model. To help researchers better understand existing QA datasets, we compared and analysed different QA datasets using multiple lexical measurement as comparison. In addition, we also compared these datasets on multiple AS models that were frequently used by QA researchers during 2017–2023, these baseline models provide a good starting point for novice researchers to understand how different aspects of the QA dataset affect the CNN and RNN models. We hope that these results could be used as benchmark in AS field by future researchers.

One limitation of this paper is that we considered only English language datasets. In future work, we would like to include other language datasets and reconstruct these datasets based on different features such as domain, gender, and sentiment to provide a more well-suited QA dataset for different QA tasks.

References

1. Bartolo, M., Roberts, A., Welbl, J., Riedel, S., Stenetorp, P.: Beat the AI: investigating adversarial human annotations for reading comprehension. CoRR abs/2002.00293 (2020). https://arxiv.org/abs/2002.00293
2. Blodgett, S.L., Green, L., O'Connor, B.: Demographic dialectal variation in social media: a case study of African-American English. arXiv preprint arXiv:1608.08868 (2016)
3. Gao, H., Hu, M., Cheng, R., Gao, T.: Hierarchical ranking for answer selection (2021)

4. Hochreiter, S., Schmidhuber, J.: Long short-term memory. Neural Comput. **9**(8), 1735–1780 (1997)
5. Honnibal, M., Montani, I.: spaCy 2: natural language understanding with Bloom embeddings, convolutional neural networks and incremental parsing (2017, to appear)
6. Kiritchenko, S., Mohammad, S.: Examining gender and race bias in two hundred sentiment analysis systems. In: Proceedings of the Seventh Joint Conference on Lexical and Computational Semantics, New Orleans, Louisiana, pp. 43–53. Association for Computational Linguistics (2018). https://doi.org/10.18653/v1/S18-2005, https://aclanthology.org/S18-2005
7. Lai, T.M., Bui, T., Li, S.: A review on deep learning techniques applied to answer selection. In: Bender, E.M., Derczynski, L., Isabelle, P. (eds.) Proceedings of the 27th International Conference on Computational Linguistics, Santa Fe, New Mexico, USA, pp. 2132–2144. Association for Computational Linguistics (2018). https://aclanthology.org/C18-1181
8. LeCun, Y., Bottou, L., Bengio, Y., Haffner, P.: Gradient-based learning applied to document recognition. Proc. IEEE **86**(11), 2278–2324 (1998)
9. Liu, T.Y., et al.: Learning to rank for information retrieval. Found. Trends® Inf. Retrieval **3**(3), 225–331 (2009)
10. Liu, W., Zhang, L., Ma, L., Wang, P., Zhang, F.: Hierarchical multi-dimensional attention model for answer selection. In: 2019 International Joint Conference on Neural Networks (IJCNN), pp. 1–8 (2019https://doi.org/10.1109/IJCNN.2019.8852055
11. Santos, C.d., Tan, M., Xiang, B., Zhou, B.: Attentive pooling networks. arXiv preprint arXiv:1602.03609 (2016)
12. Tay, Y., Phan, M.C., Tuan, L.A., Hui, S.C.: Learning to rank question answer pairs with holographic dual LSTM architecture. In: Proceedings of the 40th International ACM SIGIR Conference on Research and Development in Information Retrieval, pp. 695–704 (2017)
13. Thelwall, M.: Gender bias in machine learning for sentiment analysis. Online Inf. Rev. **42**(3), 343–354 (2018). https://doi.org/10.1108/OIR-05-2017-0153, https://doi.org/10.1108/OIR-05-2017-0153
14. Wang, S., Jiang, J.: A compare-aggregate model for matching text sequences. arXiv preprint arXiv:1611.01747 (2016)
15. Wen, J., Tu, H., Cheng, X., Xie, R., Yin, W.: Joint modeling of users, questions and answers for answer selection in CQA. Expert Syst. Appl. **118**, 563–572 (2019). https://doi.org/10.1016/J.ESWA.2018.10.038
16. Zhang, X., Li, S., Sha, L., Wang, H.: Attentive interactive neural networks for answer selection in community question answering. In: Proceedings of the AAAI Conference on Artificial Intelligence (2017)

Using Neural Coherence Models to Assess Discourse Coherence

Lilia Azrou[1,3](✉), Houda Oufaida[1,3], Philippe Blache[2], and Israa Hamdine[3]

[1] Laboratoire de la Communication dans les Systèmes Informatiques (LCSI), BP, 68M Oued-Smar, 16270 Algiers, Algeria
[2] Aix Marseille Université, CNRS, LPL UMR 7309, 13604 Aix en Provence, France
[3] Ecole nationale Supérieure d'Informatique (ESI), BP, 68M Oued-Smar, 16270 Algiers, Algeria
http://lcsi.esi.dz/, https://www.esi.dz/ , https://www.lpl-aix.fr/

Abstract. Discourse coherence is an important characteristic of well-written texts and coherent speech. It is observed at several levels of discourse analysis: lexical, syntactic, semantic, and pragmatic. Recent work on discourse coherence uses deep neural network architectures to model coherence. However, most of these architectures are not linguistically explainable. In this paper, we propose a fine-tuned Large Language Model (LLM) and three interpretable approaches for modeling discourse coherence, that target different levels of discourse analysis and coherence information, capturing contextual information, semantic relatedness between adjacent sentences and paragraphs, and syntactic patterns of coherent texts. We want to determine whether these explainable approaches lead to competitive results compared to the proposed fine-tuned LLM. These architectures are evaluated on the multi-domain Grammarly Corpus of Discourse Coherence (GCDC) and compared to state-of-the-art (SOTA) and recent models. The results of our experiments show that the syntactic patterns combined with the semantic relatedness are a good indicator of the overall coherence and highlight the importance of the number of training examples for the model's ability to use the information provided by the syntactic patterns to make accurate predictions. Furthermore, the contextual information captured by the transformer-based model achieves good results that significantly outperform all other models, showing that the use of a fine-tuned LLM is now typically the best performing approach, despite being less interpretable than other methods.

Keywords: Natural language processing · Discourse coherence · Discourse analysis · Semantic level · Syntactic level · Deep neural networks

1 Introduction

Discourse coherence corresponds to elements that make a text comprehensible to the readers. It is manifested by the logical flow of ideas without interruption in the text. [18] defines coherence as a set of semantic and informative connectors that define the semantic continuity of sentences and paragraphs within text.

Coherence can be manifested by a set of lexical cohesion devices and discourse relation markers that ensure semantic continuity, such as the coreference chains and rhetoric relations. Coherence is observed at both local and global levels of discourse, i.e. sentence-to-sentence and text spans, respectively. Local coherence observes the transitions of grammatical entities between adjacent sentences using the entity grid representation of text [1]. Global coherence observes the semantic structure and the rhetorical organization of the text [11]. It considers transitions between text spans to determine the overall coherence of the discourse.

Evaluation of the discourse coherence is useful to many Natural Language Processing (NLP) tasks: machine translation, automatic summarization, etc. These systems automatically generate sentences and paragraphs that are not necessarily coherent. In fact, an incoherent summary or translation loses rapidly its informational quality. In the educational domain, automatic evaluation of essays is an inexpensive and time-saving option compared to hand-made grades. Developing sophisticated tools that automatically evaluate students' essays is extremely useful.

Recent advances in LLMs, including ChatGPT[1] demonstrate its language prowess in multiple natural language generation tasks. However, these models are very expensive, highly complex and difficult to interpret. With these models, it remains unclear which linguistic characteristic is captured and at which level of discourse analysis: lexical, syntactic, semantic, or pragmatic.

In this paper, we propose three linguistically motivated and explainable models, as well as a fine-tuned LLM, for assessing discourse coherence. We want to determine whether these explainable approaches lead to competitive results compared to the proposed fine-tuned LLM. The proposed coherence models encode information at the syntactic and semantic levels. In comparison to previous work, we aim to evaluate precisely the contribution of each level of discourse analysis to the assessment of the overall coherence of texts. This gives us a deeper insight into the behavior of our models, leading to a better explainability.

We evaluate our models using the GCDC on the task of Classification for Assessing Discourse Coherence (CADC). The results are then compared to SOTA systems. In our evaluation, our transformer-based model achieves the best results, outperforming all other models by a wide margin.

2 Related Work

Early coherence evaluation models make use of coherence features at both the lexical and discourse level. The popular Entity Grid model [1] generates for each text an entity grid representation that captures local entity transitions between adjacent sentences. The distribution of referential coherence is learned from a corpus of coherent texts and is used to evaluate how a candidate text

---

[1] https://openai.com/gpt-4.

demonstrates such a distribution. [14] uses genetic algorithms and exploits a set of semantic and syntactic features such as thematic and temporal ordering, shared entities, and a redundancy filter to generate more coherent summaries from the source texts.

Neural models for coherence modeling and assessment use several deep learning architectures to assess coherence. [10] define two neural models for discourse coherence. The first model, the sentence clique model, uses a Long-Short-Term Memory (LSTM) [6] that takes a sequence of GloVe word embeddings as input of a classifier that is trained to distinguish between coherent and incoherent discourse. The second model, a latent variable generative Markovian model, is trained to produce coherent text by capturing latent discourse sentence dependencies. [13] extends the entity grid model [1] and defines a Convolutional Neural Network (CNN) to capture long entity transitions using distributed representations. [12] use CNN and LSTM based Neural Network (NN) to capture semantic transitions between adjacent sentences. The model uses the LSTM layer to capture the semantic topic continuity between highly related words, and then the CNN layer captures the semantic patterns that mark the continuity/change of topics. [9] evaluate SOTA models (Entity grid models [1,13], lexical graphs [12] and the sentence clique model [10]) and introduce two new neural models: Sentence Averaging (SENTAVG) and Paragraph Sequence (PARSEQ) that use LSTM based NN. SENTAVG ignores the order of the sentences and contains a single LSTM to represent sentences. PARSEQ contains three stacked LSTMs to represent sentences, paragraphs, and document. [9] show that a simple neural model, like PARSEQ, which uses paragraph information, outperforms previous models on GCDC. Recent NN architectures based on transformers have shown significant improvements for many NLP tasks. [8] evaluate popular pre-trained transformers models (BERT, RoBERTa and GPT2) to the task of zero-shot shuffle test. The results show that, by simply fine-tuning a RoBERTa model, the model significantly outgrows previous work. More recently, [7] uses syntactic and semantic information along with a large-scale pre-trained language model (XLNet) that captures the focus of sentences. The motivation is that using linguistic information connections at the word level: noun phrases and proper names, captures the most usable information in the text.

3 Our Prediction Models

In what follows, we present three discourse analysis levels as well as the architectures of our deep neural coherence models grouped by discourse analysis level.

1. **Semantic level**: the semantic information to be studied for this level is **semantic relatedness** and **contextual information**. To highlight the importance of semantic information in evaluating discourse coherence, we have designed two models, presented in Sects. 3.1 and 3.2, that will allow us to answer this question: How important is the semantic information conveyed by words, sentences, and paragraphs when evaluating discourse coherence?

2. **Syntactic level**: we have considered **the grammatical categories of words** in a text as **syntactic information** to model discourse coherence. We want to answer the following questions: How does syntactic information contribute to the evaluation of discourse coherence? Is it a reliable element to judge the coherence of a text and consider it to be of good quality? The designed model **CNNPosTag**, presented in Sect. 3.3, will allow us to answer these questions through a series of experiments.
3. **Hybrid level**: Once the syntactic and semantic models are designed, we need to answer the question: How do syntactic information and semantic information contribute together to the evaluation of discourse coherence? In order to answer this question, we propose the hybrid model, **SemSyn**, which is a combination of the semantic and the syntactic levels. It represents the fusion of the two models **SemRel** and **CNNPosTag** (see Sect. 3.4).

3.1 Semantic Coherence Prediction Model: SemRel

SemRel is used to assess the impact of semantic relatedness between adjacent sentences and adjacent paragraphs in the text using cosine similarity. Its architecture, which is inspired by the two models SENTAVG and PARSEQ (developed as part of the work of [9]), is illustrated in Fig. 1:

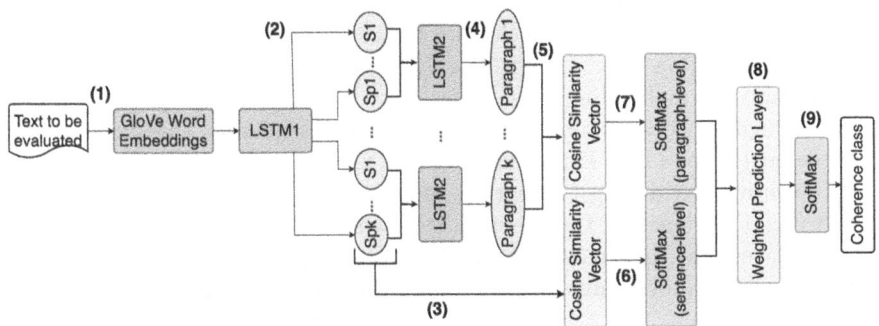

Fig. 1. Architecture of the semantic coherence prediction model SemRel.

- We generate GloVe embeddings for the words in the dataset (step **1** of Fig. 1).
- SemRel uses two LSTMs [6]: the first takes a sequence of GloVe embeddings to produce a sentence vector (step **2**), and the second takes a sequence of sentence vectors to produce a paragraph vector (step **4**).
- A cosine similarity vector is calculated between adjacent sentences (step **3**). This vector is then passed to a softmax prediction layer to predict the coherence class of the input text (for the sentence level) (step **6**).
- A cosine similarity vector is calculated between adjacent paragraphs (step **5**). This vector is then passed to a softmax prediction layer to predict the coherence class of the input text (for the paragraph level) (step **7**).

- These predictions (from sentence and paragraph levels) are then weighted by their respective weights, which are calculated by the weighted prediction layer using the Grid Search algorithm [4], and finally summed (step **8**).
- The result of the last layer is then passed to a softmax prediction layer to predict the coherence class of the input text (step **9**).

3.2 Transformer-Based Model: GPTSem

GPTSem investigates the contribution of contextual information to the evaluation of discourse coherence. It is mainly based on the pre-trained transformer model from the Hugging Face[2] platform, GPT2ForSequenceClassification[3], which represents the GPT-2 model [17] with a sequence classification head on top. This pre-trained model is widely used by the community for sequence classification tasks, while the basic GPT-2 model is designed for text generation. Fine-tuning GPT2ForSequenceClassification yields better classification results due to its specialized architecture, which includes a sequence classification head (a linear layer on top of the output) optimized for discriminating between different classes rather than generating text. The architecture of GPTSem is shown in Fig. 2:

- GPT-2 tokenization layer: it takes the text to be evaluated as input in order to distribute the words constituting it into tokens which will correspond to their respective ids in the transformer's vocabulary.
- GPT-2 embedding layer: the ids generated in the previous step will be introduced into the embedding layer in order to be transformed into a vector of embeddings (a vector representation of the text sequence) that will be used in the next step. These embeddings will be updated according to the context of the entire text during the learning phase.
- The vector obtained in the previous step will be introduced into the GPT-2 model layer and then passes through a linear layer to end up in a softmax layer in order to determine the corresponding coherence class.

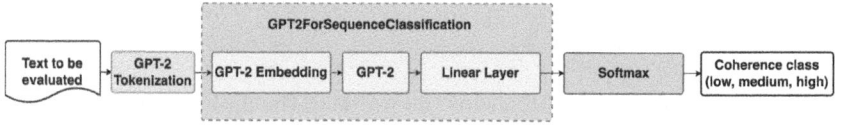

Fig. 2. Architecture of the transformer-based model GPTSem.

---

[2] https://huggingface.co/.
[3] https://huggingface.co/docs/transformers/v4.29.1/en/model_doc/gpt2# transformers.TFGPT2ForSequenceClassification.

3.3 Syntactic Coherence Prediction Model: CNNPosTag

CNNPosTag studies the contribution of syntactic information to the evaluation of discourse coherence. Its architecture is illustrated in Fig. 3:

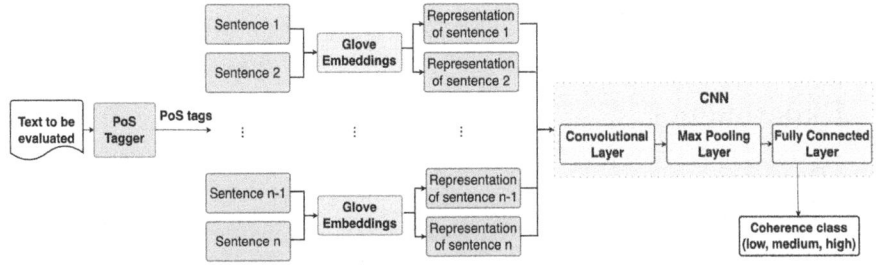

Fig. 3. Architecture of the syntactic coherence prediction model CNNPosTag.

- We use the Natural Language Toolkit (NLTK) tagger [3], to assign Part-of-Speech (POS) tags to each word. This process breaks down the text into sentences, so that each sentence is represented by a sequence of POS tags.
- A vector representation of each sentence is generated using the concatenation of GloVe embeddings for the POS tags in that sentence, and is provided as input to a CNN network.
- The CNN network takes as input the vector representations of the sentences in the document, and introduces them into a **convolutional layer**.
- The output of the convolutional layer, called the feature map, is then fed into a **max pooling layer**. The latter selects the maximum value for each feature map, which is considered to be the most representative and significant information in the sentence carried by that map.
- The output of the max pooling layer is introduced into a **fully connected layer**, in order to determine the appropriate coherence class.

3.4 Hybrid Coherence Prediction Model: SemSyn

We present here **SemSyn**, our hybrid coherence prediction model of both semantic and syntactic levels, which represents the fusion of the two models SemRel and CNNPosTag. We decided to merge these two models in order to assess the contribution of both syntactic and semantic information to the evaluation of discourse coherence. To facilitate a possible expansion to new models, we chose a linear combination using the grid search algorithm [4]: weights are assigned to the predictions of the two models and vary until the optimal combination is reached, giving the best accuracy to the final model. The architecture of the resulting model is illustrated in Fig. 4.

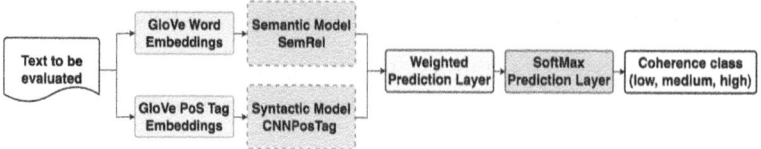

Fig. 4. Architecture of SemSyn: the hybrid coherence prediction model.

4 Experiments

4.1 Dataset

We use GCDC, a real-world discourse coherence corpus developed as part of the work of [9], to evaluate coherence models on informal texts written by non-professional writers in everyday contexts, such as emails or online reviews. It encompasses four distinct domains: emails for the Clinton and Enron domains, responses from an online forum for the Yahoo domain, and online reviews of businesses for the Yelp domain. Each text in the corpus has a consensus label that expresses how coherent it is: {low, medium, high}. For our experiments, we use the consensus label of the expert ratings as computed by [9].

4.2 Dealing with Imbalanced Data

Having observed an imbalanced distribution of expert ratings across the three coherence classes in GCDC (see Table 1) and a relatively limited number of documents in this dataset, we emphasize one of the most widely used methods for estimating model reliability and improving predictive performance by making deeper use of training data, namely Cross Validation (CV) [2]. We used the Stratified K-Fold CV, which is the most suitable CV method for our case due to the imbalanced nature of the dataset [2].

Table 1. Distribution of the expert ratings in GCDC. #Docs represents the number of texts by domain, LI = Low Instances, MI = Medium Instances and HI = High Instances.

	#Docs	LI (%)	MI (%)	HI (%)
Yahoo	1200	44.8	17.9	37.3
Clinton	1200	27.8	20.3	51.8
Enron	1200	30.1	20.3	49.6
Yelp	1200	26.8	21.7	51.6

4.3 Implementation Details

We implement our models using the PyTorch library. We employ GPT-2 for the pretrained language model [17] and use GloVe for word embeddings, trained on Google News [15], with 300-dimensional for GCDC. For all baseline models, we report results from their original papers, except for [10]'s model, for which we report performance from [9] (since it was reimplemented to evaluate its effectiveness on GCDC). The reported results for all our models are the mean of 10 runs with different random seeds. For all experiments, we perform stratified 10-fold CV and use accuracy and balanced accuracy as evaluation measures on the 3-class classification. The cross-entropy loss function is deployed for training and the ADAM optimizer is used with a learning rate of 2e-5. For GPTSem, we evaluate performance for 3 epochs on the validation set (as recommended by [5]) and 10 epochs for the other models. We use a mini-batch size of 32 with random-shuffle and fix max sequence length to 512 for GPTSem.

4.4 Results

To see the impact of the coherence information, studied for each discourse analysis level, on the evaluation of discourse coherence, we compare our models with SOTA and recent neural coherence models for the CADC task: the sentence clique model [10], SENTAVG [9], PARSEQ [9] and [7]'s model. Tables 2, 3 and 4 report the performance on GCDC.

Table 2. CADC: Mean accuracy (balanced accuracy) performance on the test sets in GCDC (∗: reported performance in [9] and ∗∗: reported performance in [7]).

Model	Yahoo	Clinton	Enron	Yelp	Average
∗ [10]'s model	53.5	61.0	54.4	49.1	54.5
∗SENTAVG	52.6	58.4	53.2	54.3	54.6
∗PARSEQ	54.9	60.2	53.2	54.4	55.7
∗∗ [7]'s model	58.4	**64.2**	55.3	57.3	58.9
SemRel	47.3 (41.7)	38.2 (41.6)	40.8 (39.4)	38.7 (38.4)	41.3 (40.3)
CNNPosTag	37.7 (37.2)	44.2 (41.6)	41.2 (38.9)	39.5 (39.3)	40.7 (39.3)
SemSyn	48.8 (46.8)	41.3 (42.2)	44.8 (44.2)	45.5 (45.0)	45.1 (44.6)
GPTSem	**62.2 (60.7)**	62.7 (61.4)	**64.4 (65.2)**	**61.1 (61.4)**	**62.6 (62.2)**

In-Domain Results. We train each model in the training set of a given domain and evaluate it in the test set of the same domain (see Table 2):

- Broadly, we observe that our transformer-based model **GPTSem** significantly outperforms the baselines, establishing a new record (an acc of 62.6%).

- The model proposed in [7] outperforms other baselines. The ablation study verifies that the latter gains improvements not only from the encoding strategy but also from the novel local coherence module.
- **SemSyn**, which represents the fusion of the two models SemRel and CNNPosTag, outperforms **SemRel**, which in turn outperforms **CNNPosTag**: these results show that the linear combination of SemRel and CNNPosTag, leads to better results than those obtained by each model separately.
- **SemRel** and **SemSyn** perform best on **Yahoo** domain. This shows that SemRel and SemSyn are better at capturing **semantic relatedness** and **the grammatical categories of words** for this domain in particular, which, according to the expert ratings, contains more documents of low coherence than the other domains (see Table 1). We can conclude that incoherence is mostly captured at the semantic and syntactic levels of discourse analysis.
- All the **baselines** outperform: SemRel, CNNPosTag and SemSyn. As for **SemRel** (and consequently **SemSyn**, since it represents the fusion of SemRel and CNNPosTag), we suppose that this is due to the type of information being encoded. Specifically, the two baselines, **SENTAVG** and **PARSEQ** (since SemRel is inspired by these two models), use LSTMs to build a hierarchical representation of a document. Unlike SemRel, the information modeled is **semantic relatedness**, based on the cosine similarity between adjacent sentences and paragraphs. We think that **semantic relatedness** is better modeled by LSTMs than by cosine similarity. Our findings are in accord with a previous study [16], which revealed that the average cosine similarity between adjacent sentences was not a significant variable.

Table 3. CADC: Mean accuracy (balanced accuracy) performance on the test sets when trained on data from all four domains in GCDC (∗: reported performance in [9]).

Model	Yahoo	Clinton	Enron	Yelp	Average
∗PARSEQ	58.5	61.0	53.9	56.5	57.5
SemRel	45.8 (41.2)	43.0 (45.3)	35.7 (38.9)	40.2 (40.9)	41.2 (41.6)
CNNPosTag	50.7 (48.8)	55.5 (53.3)	60.3 (59.0)	53.5 (51.0)	55.0 (53.0)
SemSyn	48.3 (47.3)	48.8 (51.6)	43.7 (45.8)	46.0 (48.6)	46.7 (48.3)
GPTSem	63.8 (63.2)	66.8 (67.3)	65.0 (64.5)	66.3 (66.1)	65.5 (65.3)

Merging-Domains Results. We expect to improve the predictive power of our models by providing them with more training examples. Thus, we first use the entire GCDC for training and then evaluate the proposed models in each of the four domains separately (see Table 3). Second, we use the entire GCDC for training and testing (see Table 4). Our main observations are as follows:

Table 4. CADC: Mean accuracy (balanced accuracy) performance when trained and tested on data from all four domains in GCDC.

Model	Merged GCDC
SemRel	38.3 (39.7)
CNNPosTag	**54.4** (**52.3**)
SemSyn	45.7 (46.8)
GPTSem	**65.0** (**66.4**)

- Models trained on data from all four domains perform better (except for SemRel) than models trained on individual domain data.
- The decline in SemRel's performance is likely due to domain-specific characteristics, such as unique styles in each domain (e.g., emails vs. reviews), that challenge the LSTM-based model's ability to generalize, leading to increased noise and difficulty in capturing coherent patterns across varied contexts.
- There is a clear improvement in the results for **CNNPosTag** after merging the 4 training domains and even after merging the 4 training and testing domains: this confirms the importance of the number of training examples for the model's ability to use the information provided by **the grammatical categories of words**, to make accurate predictions. We think that training **CNNPosTag** on all four domains leads to a better generalization of the resulting model. The same observation applies to **GPTSem**, **contextual information** is well captured when the model is trained on all four domains and even when the model is trained and tested on all four domains.
- CNNPosTag outperforms SemSyn likely due to suboptimal weight assignment during ensembling, as the models do not complement each other as expected. CNNPosTag performs well on its own, but combining it with the semantic model results in a performance drop.

4.5 Synthesis

The experiments we conducted with models targeting different levels of discourse analysis and coherence information allowed us to address the questions raised during the elaboration of our models. The responses are summarized below:

- While [7]'s **model** outperforms other baselines, our proposed transformer-based model **GPTSem** shows enhanced performance.
- Our analysis shows that a simple model relying on the pretrained language model GPT-2 outperforms a more sophisticated architecture ([7]'s model) which uses the Tree-Transformer architecture [19] and the pretrained language model XLNet [21] to encode sentences.
- The semantic information for modeling discourse coherence, encoded by GPT-Sem, which is **contextual information**, provided the model with enough information to properly classify documents, thanks to its conceptual features, in particular the contextualization of word embeddings (embeddings

are updated according to the context of the entire text during the learning phase). We can also explain the performance of GPTSem by noting that it is based on the pre-trained language model GPT-2 [17], which is a large transformer-based language model, pretrained using language modeling on a very large and diverse corpus of 40GB of text data.

- Our findings suggest that **semantic relatedness** between adjacent sentences and between adjacent paragraphs in the document did not provide the model with enough information to properly classify documents.
- The combination of two categories of coherence information (semantic and syntactic information) shows improved performance.
- We found that **the amount of training** data can significantly affect the predictive efficiency of the models.
- Our results suggest that modeling discourse coherence at the **syntactic level** with a large amount of training data and analyzing the **grammatical categories of words**, according to their type and the order in which they appear in the text, does indeed have a positive impact on model performance.

5 Conclusion

We have proposed a fine-tuned LLM and three interpretable neural coherence models designed to address different levels of discourse analysis and coherence information, encompassing contextual information, semantic relatedness between adjacent sentences and paragraphs, and syntactic patterns of coherent texts. This makes our models better explainable.

The results show that syntactic patterns combined with semantic relatedness are a good indicator of the overall coherence of the text. This also allows us to reveal that modeling discourse coherence at the syntactic level with a large amount of training data and analyzing the syntactic patterns does indeed have a positive impact on model performance. Our analysis shows that a simple model based on the pre-trained language model GPT-2, which captures the contextual information, achieves the best results. The results obtained with our transformer-based model GPTSem justify the tendency to use transformers for assessing discourse coherence, given the performance they offer.

Our findings suggest interesting directions for future research. Our work could be extended to a multilingual setup. Our model GPTSem is not tied to a specific pre-trained language model, but can use other LLMs such as the BERT model [5] and the XLNET model [21]. Thus, it can use a multilingual model [20], to assess the coherence of texts in other languages.

Acknowledgements. The authors would like to thank Alice Lai and Joel Tetreault for making their dataset and code available, and Sungho Jeon and Michael Strube for sharing their code. Thanks also to the anonymous reviewers for their comments.

Disclosure of Interests. The authors declare that they have no known competing financial interests or personal relationships that could have potentially influenced the results presented in this paper.

References

1. Barzilay, R., Lapata, M.: Modeling local coherence: an entity-based approach. Comput. Linguist. **34**(1), 1–34 (2008)
2. Berrar, D.: Cross-validation. In: Encyclopedia of Bioinformatics and Computational Biology, pp. 542–545. Academic Press, Oxford (2019)
3. Bird, S., Klein, E., Loper, E.: Natural language processing with Python: analyzing text with the natural language toolkit, chap. 5, pp. 179–219. O'Reilly Media (2009)
4. Brownlee, J.: Hyperparameter optimization with random search and grid search. Machine Learning Mastery (2020)
5. Devlin, J., Chang, M.W., Lee, K., Toutanova, K.: BERT: pre-training of deep bidirectional transformers for language understanding. In: Proceedings of the 2019 Conference of the North American Chapter of the Association for Computational Linguistics: Human Language Technologies, pp. 4171–4186. ACL (2019)
6. Hochreiter, S., Schmidhuber, J.: Long short-term memory. Neural Comput. **9**, 1735–1780 (1997)
7. Jeon, S., Strube, M.: Entity-based neural local coherence modeling. In: Proceedings of the 60th Annual Meeting of the Association for Computational Linguistics (Volume 1: Long Papers), pp. 7787–7805 (2022)
8. Laban, P., Dai, L., Bandarkar, L., Hearst, M.A.: Can transformer models measure coherence in text: re-thinking the shuffle test. In: Proceedings of the 59th Annual Meeting of the Association for Computational Linguistics and the 11th International Joint Conference on NLP (V2: Short Papers), pp. 1058–1064 (2021)
9. Lai, A., Tetreault, J.: Discourse coherence in the wild: a dataset, evaluation and methods. In: Proceedings of the 19th Annual SIGdial Meeting on Discourse and Dialogue, pp. 214–223 (2018)
10. Li, J., Jurafsky, D.: Neural net models of open-domain discourse coherence. In: Proceedings of the 2017 Conference on Empirical Methods in Natural Language Processing, pp. 198–209 (2017)
11. Mann, W.C., Thompson, S.A.: Rhetorical structure theory: a theory of text organization. University of Southern California, Information Sciences Institute (1987)
12. Mesgar, M., Strube, M.: A neural local coherence model for text quality assessment. In: Proceedings of the 2018 Conference on Empirical Methods in Natural Language Processing, pp. 4328–4339 (2018)
13. Nguyen, D.T., Joty, S.: A neural local coherence model. In: Proceedings of the 55th Annual Meeting of the Association for Computational Linguistics (Volume 1: Long Papers), pp. 1320–1330 (2017)
14. Oufaida, H., Blache, P., Nouali, O.: A coherence model for sentence ordering. In: Métais, E., Meziane, F., Vadera, S., Sugumaran, V., Saraee, M. (eds.) NLDB 2019. LNCS, vol. 11608, pp. 261–273. Springer, Cham (2019). https://doi.org/10.1007/978-3-030-23281-8_21
15. Pennington, J., Socher, R., Manning, C.: GloVe: global vectors for word representation. In: Proceedings of the 2014 Conference on Empirical Methods in Natural Language Processing (EMNLP), pp. 1532–1543. ACL (2014)
16. Pitler, E., Nenkova, A.: Revisiting readability: a unified framework for predicting text quality. In: Proceedings of the 2008 Conference on Empirical Methods in Natural Language Processing, pp. 186–195 (2008)
17. Radford, A., Wu, J., Child, R., Luan, D., Amodei, D., Sutskever, I.: Language models are unsupervised multitask learners. OpenAI Blog **1**(8), 9 (2019)

18. Telenyk, S., Pogorilyy, S., Kramov, A.: Evaluation of the coherence of polish texts using neural network models. Appl. Sci. **11**(7), 3210 (2021)
19. Wang, Y., Lee, H.Y., Chen, Y.N.: Tree transformer: integrating tree structures into self-attention. In: Proceedings of the 2019 Conference on Empirical Methods in Natural Language Processing and the 9th International Joint Conference on Natural Language Processing (EMNLP-IJCNLP), pp. 1061–1070. ACL (2019)
20. Xue, L., et al.: mT5: a massively multilingual pre-trained text-to-text transformer. In: Proceedings of the 2021 Conference of the North American Chapter of the Association for Computational Linguistics, pp. 483–498. ACL (2021)
21. Yang, Z., Dai, Z., Yang, Y., Carbonell, J.G., Salakhutdinov, R., Le, Q.V.: XLNet: generalized autoregressive pretraining for language understanding. In: Neural Information Processing Systems (2019)

Named Entity Linking in English-Czech Parallel Corpus

Zuzana Nevěřilová[1](✉) and Hana Žižková[2]

[1] Faculty of Informatics, Masaryk University, Brno, Czechia
xpopelk@fi.muni.cz
[2] Faculty of Arts, Masaryk University, Brno, Czechia

Abstract. We present a procedure to build relatively quickly new resources with annotated named entities and their linking to Wikidata. First, we applied state-of-the-art models for named entity recognition on a sentence-aligned parallel English-Czech corpus. We selected the most common entity classes: person, location, organization, and miscellaneous. Second, we manually checked the corpus in a suitably set annotation application. Third, we used a state-of-the-art tool for named entity linking and enhanced the ranking using sentence embeddings obtained by sentence transformers. We then checked manually whether the linking to knowledge bases was correct.

As a result, we added two annotation layers to an existing parallel corpus: one with the named entities and one with links to Wikidata. The corpus contains 14,881 parallel Czech-English sentences and 3,769 links to Wikidata. The corpus can be used for training more robust named entity recognition and named entity linking models and for linguistic research of parallel news texts.

Keywords: named entity recognition · named entity linking · parallel corpus · sentence similarity

1 Introduction

Named entity recognition (NER) aims to recognize names of persons, organizations, locations, products, artworks, events, etc., in texts. Named entity linking (NEL) is the next step in bridging the gap between the so-called unstructured (or not understandable by computers) and structured data. It aims to disambiguate named entities (NEs) and link them to knowledge base items. NEs enriched with knowledge from knowledge bases can support other tasks such as question answering, information integration, or explainability.

NEL challenges are: synonymy (variants of the word or expression that still mean the same thing and reference the same object), transliterations (e.g., *Yangtze* or *Yangzi*), abbreviations (e.g., *Charles Philip Arthur George* and *Charles III*), acronyms (*United States dollar*, *US dollar*, *USD*), and grammar variants (e.g., Czech inflection of *New York* can be *New Yorku*). At the same time, NEs are homonymous. For example, *Queen* can be a musical band, a film, or a company, or it can reference a particular person.

According to [23], NEL usually consists of the following steps: candidate generation, candidate ranking, and unlinkable mention prediction. In the candidate generation, probable items are extracted from a knowledge base. In the general domain, knowledge bases such as Wikipedia[1], Wikidata[2], YAGO[3], and others are used. In the next phase, the candidates are ranked so that the highest probable ones are most likely what the author of the text intended. The last step, unlinkable mention prediction, selects not to predict the top-ranked candidate because the entity does not have a record in the knowledge base.

The paper describes a new annotation for the existing parallel corpus of news text. The parallel texts were annotated using existing models for NER in Czech and English. The NEs mentioned in new articles are linked in Wikipedia. The NEs were manually checked before the NEL. The NEL was applied to English and then transferred to Czech data [method]. The resulting resource contains NER annotations for both languages and entity-linking annotations to English Wikipedia for both languages.

To find NEL candidates, we used the Opentapioca tool[4]. The tool provides scores with the probability of assignment of a particular entity depending on the context. In addition to the Opentapioca scoring, we used transformers to measure the semantic similarity between the candidate definitions and the source sentence.

1.1 Paper Outline

The paper is organized as follows: Sect. 2 overviews NER methods, NER-annotated corpora, and NEL. In Sect. 3, we describe the used data NER annotation process and issues in parallel data. We also describe how the manual annotations were performed. The results are described in Sect. 4. Section 5 summarizes the work and describes our plans for the future.

2 Related Work

We briefly overview NER, datasets and annotation schemas, and annotated parallel corpora. Next, current systems for named entity linking are presented.

2.1 Named Entity Recognition

Named entity recognition (NER) is a well-established NLP task used mainly in information extraction. After the rise of neural networks, the best NER models

---

[1] https://www.wikipedia.org/.
[2] https://www.wikidata.org/.
[3] https://yago-knowledge.org/.
[4] https://github.com/UB-Mannheim/spacyopentapioca.

are based on transformers since they can capture long sequences of tokens and handle properly repeating occurrences of NEs in documents.

Annotated datasets exist for training NER systems; they differ in the number and type of entities and the nature of the text. One of the best-known datasets is CoNLL-2003 [27] with annotation of four classes: person (PER), location (LOC), organization (ORG), and miscellaneous (MISC). The texts in CoNLL-2003 are in German and English, with data from the Reuters corpus, i.e., the standard language. Since the dataset is over 20 years old, it does not capture some phenomena currently common in the newspaper genre (e.g., social media aliases or platform names).

Another dataset, the OntoNotes 5.0 corpus [28], was tagged with 18 named entity types. The entity types cannot be easily converted from one format to another, e.g., geopolitical unit (GPE) in OntoNotes can be both location and organization in ConLL-2003, depending on the context. The OntoNotes corpus contains texts from the Wall Street Journal, transcriptions from broadcast shows, telephone conversations, and web data. It contains texts in English, Chinese, and Arabic; part of the corpus is parallel English-Chinese.

WikiAnn [19] is another dataset for NER, constructed automatically using the linked entities in Wikipedia pages for 282 different languages. The entity types in WikiAnn are only person (PER), location (LOC), and organization (ORG). The advantage of this dataset is its massive multilingualism and up-to-dateness. The annotation quality of a silver standard is slightly beyond the manually annotated resources. WikiAnn contains by nature the data for named entity linking, which was used, e.g., in [17]. Silver standards were further developed by [26].

Apart from WikiAnn and other silver standards, a corpus of named entities in Czech has been developed by [21]. It contains a nested annotation with 46 entity types. The corpus exists in two variants – apart from the original, a ConLL-2003 compatible version was published by [11]. All recent models for Czech NER are trained on CNEC data. One of the current multi-purpose models for Czech is the models of the Czert family, which achieved state-of-the-art results in NER [24]. We use the CZERT-B-ner-CNEC model in our work.

The multilingual nature of some resources also leads to cross-lingual NER that is held in parallel texts, e.g., by [14].

2.2 Parallel Corpora with Annotated Named Entities

Parallel corpora are traditionally used in tasks related to machine translation. The alignment of a parallel corpus allows automatic NE alignment or NE annotation in the language where NER tools cannot be used. In [7], the authors automatically aligned NEs in English, French, Spanish, German, and Czech. In the medical domain, [2] used NER tools for English and transferred annotations into French and Spanish, benefiting from existing Wikipedia links for some NEs. In domain-specific corpora, NEs often do not refer to person/location/organization names but to the domain terminology. A similar work is a semi-automatically extracted terminology in the land surveying domain [20].

2.3 Named Entity Linking and Disambiguation

Named entity linking (NEL) follows or is performed together with named entity disambiguation (NED). The result of both tasks is an increased understanding of a coherent text – NED detects that multiple NE mentions refer to the same or distinct objects. NEL adds reference, whenever possible, to an external knowledge base. Assuming a coherent text, a common approach to NED is to consider the context of multiple mentions and assume semantic relatedness of the NEs within the text. One of the earliest works, [4], measures the maximum co-occurrence of mentions in Wikipedia articles to disambiguate entities such as "Columbia". Within the evaluation, the author created the MSNBC dataset. In [10], the authors combined knowledge base popularity priors, similarity measures, and coherence into the system AIDA (Accurate Online Disambiguation of Named Entities in Text and Tables). Based on the mention-mention graph, it calculates the most probable link to the YAGO2 ontology [9]. For evaluation, they annotated 1,400 newswire articles and published the AIDA-YAGO dataset. In [1], an iterative algorithm achieved 89% F1 on the AIDA dataset. One of the largest contributions of this work is the cross-dataset evaluation and dataset consolidation into WNED [8].

Current NEL methods combine neural network approaches (transformers), knowledge graphs from data sources such as Wikipedia, YAGO, Wikidata, and reasoning approaches. An example project is Qurator [12], focusing on historical NEL in German, English, and French. SpEL [22] achieved 88.6% and 92.9% micro-F1 on the AIDA test (testa and testb, respectively), with a fast inference. Bootleg [16] developed reasoning patterns for disambiguation and reached 96.8% F1 on the AIDA dataset. Finally, we present OpenTapioca [5], the tool we decided to use. Although it achieves 87% micro-F1 on the AIDA dataset, it is easy to deploy and can be connected to up-to-date Wikidata snapshots.

3 Used Datasets and Methods

The task was to apply a cascade of existing methods with possible enhancements in individual steps to obtain a Czech dataset with linked entities. We selected the Parallel Global Voices corpus (PGV) for NER annotation in parallel data because of its suitable license, reasonable document lengths, and style.

For NER, we used one English model and two Czech models. The evaluations show that the training data for the model are more important than the model's training parameters. Next, we used an existing tool for NEL with a slight enhancement.

3.1 Parallel Global Voices

PGV [18] is a massively parallel (756 language pairs), automatically aligned corpus of citizen media stories translated by volunteers. The Global Voices community blog contains several guides, including the Translators' guide[5]. It contains

---

[5] https://community.globalvoices.org/guide/lingua-guides/lingua-translators-guide/.

recommendations to "localize" whenever possible. Also, it mentions English as the largest source language. However, according to authors of the PGV [18], the source language for the translation cannot always be reliably identified.

PGV contains texts crawled in 2015, reporting "on trending issues and stories published on social media and independent blogs in 167 countries" [18]. The corpus contains the Global Voices (GV) topics about politics and elections; civil, sexual, and socio-economic rights; disasters and the environment; demonstrations and police reaction; labor; and specific geographic regions. The sentence-level alignment has been done automatically. The Czech-English pairs (450 documents) are in aligned 1:1 in 86% of cases, the rest are 1:2, 2:1, 1:0, and 0:1 alignments.

The corpus is highly suitable for our task because it contains news texts, i.e., coherent texts in formal correct language. The average document length is 33, and the longest is 138 sentences. As with many other news texts, the articles mention a lot of NEs (including repeated mentions), many of them referring to general knowledge bases such as Wikipedia. The translation is of reasonable quality, and the topics are diverse, as mentioned above. The license for the PGV corpus is Creative Commons.

3.2 NER in Czech and English

For NER annotation in English, we used the dslim/bert-large-NER model from HuggingFace [6,27]. We tested two models for Czech: Czert-B (a BERT-like Model for language representation, evaluated on semantic text similarity, morphological tagging, semantic role labeling, sentiment classification, and multi-label document classification) and our model trained on top of the RobeCzech model [25]. We used CNEC for training.

We simplified the Czert-B annotation by discarding CNEC labels for time, numbers, bibliographic items, and by merging artwork and media names to the MISC category.

We used the automatic pre-annotations by Czert-B and checked the data manually. We prepared annotation guidelines that compile knowledge of previous guidelines, summarized in Universal NER [15]; however, we added much more details about questionable entities[6]. Although we used one of the best models for both languages (Czert-B achieved F1 81.63% on CNEC, which is slightly below the best models for Czech), many NEs were missed. Although information on inter-annotator agreement (IAA) would be interesting, we did not address it at this research stage.

For example, the annotators often have to create new annotations for persons in Twitter aliases (starting with @). Also, the media were often not annotated or misclassified (we distinguish *Facebook* as a company and *Facebook* as a publication platform; websites such as *tumblr* were typically not annotated).

Although both PGV and CoNLL-2003 contain news text, the writing style and topics have changed in just one decade.

---

[6] https://gitlab.fi.muni.cz/nlp/named_entity_linking.

The evaluation results of the three models are shown in Table 1. The models' F1 performance was considerably lower than the published results, the main cause being the low recall. We performed a more detailed evaluation based on the MUC schema [3].

Table 1. Evaluation of the English and Czech data based on MUC. The categories are CORrect, INCorrect, PARtial, MISsed, SPUrious, POSsible, and ACTual.

	robeczech-ner				czertb				bert-large-ner			
	type	partial	strict	exact	type	partial	strict	exact	type	partial	strict	exact
COR	4292	3232	2683	3232	3242	3021	2788	3021	9233	9013	8216	9013
INC	918	0	2527	1978	422	0	876	643	1170	0.00	2187	1390
PAR	0	1978	0	0	0	643	0	0	0	1390	0	0
MIS	10180	10180	10180	10180	11726	11726	11726	11726	9448	9448	9448	9448
SPU	406	406	406	406	158	158	158	158	1370	1370	1370	1370
POS	15390	15390	15390	15390	15390	15390	15390	15390	19851	19851	19851	19851
ACT	5616	5616	5616	5616	3822	3822	3822	3822	11773	11773	11773	11773
P	0.76	0.75	0.48	0.58	0.85	0.87	0.73	0.79	0.78	0.82	0.70	0.77
R	0.28	0.27	0.17	0.21	0.21	0.22	0.18	0.20	0.47	0.49	0.41	0.45
F1	0.41	0.40	0.26	0.31	0.34	0.35	0.29	0.31	0.58	0.61	0.52	0.57

The total number of NEs produced by the model is *ACTual*, and the number of gold-standard annotations is *POSsible*. Table 1 shows the number of NEs in Czech is smaller than the English ones. There are several reasons for this:

- In English, adjectives meaning nations (e.g., *Chinese*), nouns referring to languages (e.g., *English*) are marked as NEs. In Czech, such words are lowercase and usually not considered NEs (compare with [15] and CNEC annotation manual [29]). A similar situation is in the names of locations: e.g., *Hong Kong* protesters are translated as *hongkongští aktivisté*, using a non-NE adjective.
- In the GV news texts, the names are often left out in the Czech translation, especially person names in the stories.
- Sometimes, English texts are more informative. For example, in the pair *Last week, a student law clinic in Zagreb, Croatia filed a petition to the UN Working Group on Arbitrary Detention (UNWGAD) on behalf of jailed photo journalist.* and *Minulý týden podala studentská „právní klinika" z chorvatského Záhřebu jménem uvězněné fotožurnalistky petici pracovní skupině OSN pro svévolné zadržování.* it can be seen that both *Zagreb* and *Croatia* are mentioned, while in the Czech translation, only *Záhřeb* (*Zagreb*) is mentioned.

3.3 Multilingual Annotation Considerations

Our NE categories contain proper names, so we followed the typology of proper names according to [13]. The parallel annotation aimed to assign the same type of NE whenever possible. In some cases, this was not possible for the following reasons:

- Change of POS in the translation
  Proper names are capitalized in both English and Czech. English considers adjectives derived from proper names to be proper names, too, such as *Italy–Italian*. In contrast, Czech derived adjectives were not marked as NEs.
- Omission in the translation
  The sentence *Trained in international human rights doctrine by leading professors from Oxford and Zagreb Universities, the students researched, drafted, and filed petitions on behalf of a number of clients.* mentions *Oxford University* and *Zagreb University*. In the translation *Tito studenti, které školí v oboru lidských práv přední profesoři z univerzit v Oxfordu a Záhřebu, sepsali a podali petice jménem několika klientů.*, (some) universities from Oxford and Zagreb are vaguely mentioned, therefore, *Oxford* and *Zagreb* are place names.
- Differences in appellativization[7]
  Appellativization, sometimes reduced to brand genericization, affects both languages and causes variations in writing. For example, we annotated *Internet* as a non-entity since we consider it an appellative. On the other hand, entities like Muslim, Christian, and Jew are marked as NEs in both languages, even though not capitalized in Czech.

Annotation Choices. In the annotation process, we followed the annotation guidelines from previous NER tasks. Unfortunately, many of the guidelines are not available or not detailed enough. We used [15] as a source and considered other guidelines as well (namely, ConLL NER[8] and OntoNotes 5[9]). In minor cases, we did not follow the annotation guidelines (e.g., law names or currencies). Since we linked entities to Wikidata in the next step, the *linkability* to Wikidata was another decision criterion. For this reason, entities such as *Black Lives Matter* or *Pakistan Floods* were marked as events.

Our annotation guidelines focus on difficult decisions about the labels: a company providing an eponymous service (such as *Facebook*), metonymy (such as *Kremlin*), and hashtags (such as *#FreeSavchenko*).

Sometimes, the annotators have to search external knowledge bases for linkable NEs. Entities such as *Human Rights Watch Latin America* were annotated as two entities: *Human Rights Watch* as ORG and *Latin America* as LOC since there is no official entity with the compound name.

3.4 Entity Linking to Wikidata

Some of the NEL systems provide linking to more resources (typically, YAGO, Wikipedia, and Wikidata). We only selected Wikidata since it is a standard in

---

[7] The process of a proper name becoming a common noun.
[8] https://www.cnts.ua.ac.be/conll2003/ner/annotation.txt.
[9] https://catalog.ldc.upenn.edu/docs/LDC2013T19/OntoNotes-Release-5.0.pdf.

OpenTapioca. Although OpenTapioca allows training custom models, we tried NEL with the standard setting. Next, we experimented with NEL candidate reranking using sentence similarity. We benefitted from Wikidata definitions that were found for almost all candidates. We compared the embeddings of these definitions with the embedding of the sentence containing the entity.

For example, in the target sentence is *Vladimir Markin, the spokesman for Russia's Investigative Committee (another police branch of the federal government and a bureaucratic rival to the General Prosecutor), refused to comment [ru] on Zviagintsev's revelation.*, *Vladimir Markin* refers to Wikidata QName (entity identifier) Q4281950[10] with the definition "Russian journalist and politician".

Other possible QNames in Wikidata have the following definitions:

- Soviet composer
- Russian philosopher and logician
- Soviet association football player (1940-1981)
- no definition, a person not mentioned in Wikidata

The definition "Russian journalist and politician" has the most similar sentence embedding with the target sentence. OpenTapioca outputs a score based on a combination of local features and long-distance spread calculated from Markov chains. We multiply this score by the sentence similarity calculated by sentence transformers[11] which improves the results.

Manual NEL Annotation. In the second step, the annotators have to disambiguate NEs. The whole document was presented, with all candidate Wikidata entities and pre-annotation by our method (see Fig. 1). To ease the decision, we provided the Wikidata definition; however, in some cases, the annotators had to search the web to be able to decide. The median time per document was 36 s.

NEL Evaluation. The annotators were asked to remove linkings that were not valid. For example, if a person's name is mentioned in the text, the annotators have to check whether it is that person who is mentioned.

We calculated unique linked entities in texts. It means we count in once if an entity occurs in text multiple times. We evaluated 207 documents, with 1,831 linked entities in total. The default candidate ranking by OpenTapioca was correct for 1,504 NEs (82.1% accuracy); the re-ranking with sentence transformers increased the number of correct NEL to 1,619 (88.4% accuracy).

The NEL evaluation only focuses on precision, not recall. Annotators were not asked to find missing entities in Wikidata and complete the linking. In such a setting, the annotation process takes a relatively short time but can vary depending on the annotators' general knowledge.

---

[10] https://www.wikidata.org/wiki/Q4281950.
[11] https://huggingface.co/sentence-transformers/all-MiniLM-L6-v2.

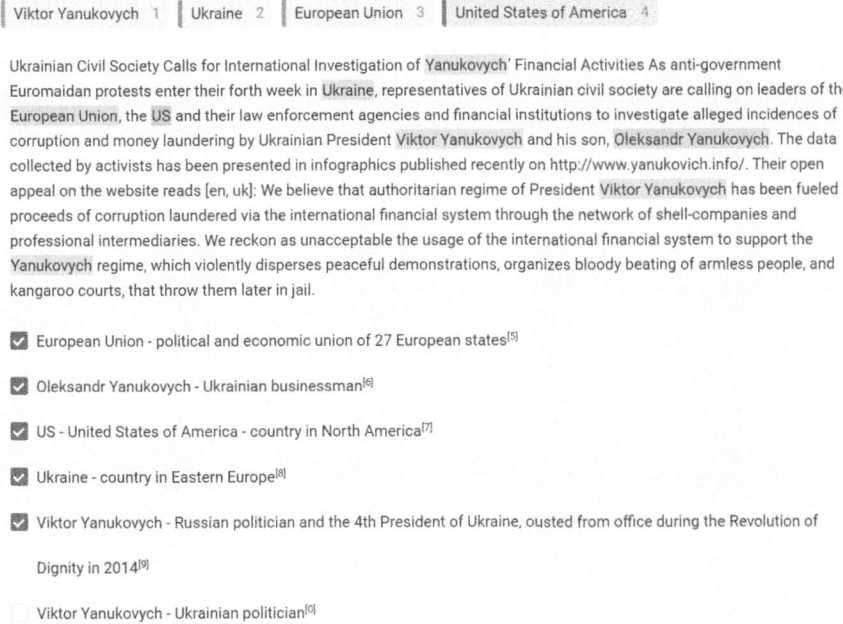

Fig. 1. NEL annotation task. The annotator has to check/uncheck the entities. Entities in text are marked to simplify the process. In this case, the annotator has to decide whether *Viktor Yanukovych* refers to the former Ukrainian president or his son Viktor Yanukovych.

4 Results

The work generated several results. We created a new dataset of NE annotations in the parallel data (the NER ground truth), published in the LINDAT/CLARIAH-CZ repository[12], together with links to Wikidata (the NEL ground truth) for English NEs. We published a state-of-the-art NER model for Czech based on RobeCzech and trained on CNEC[13]. We successfully experimented with a re-ranking procedure for NEL candidates based on embedding similarity of sentence and Wikidata definitions. All tools for data conversion between various formats to fully reproduce the evaluation are available in the project repository[14].

All data were published in the standardized formats – TSV and IOB. The NE annotated dataset can be used for NER model training. It can contribute to data heterogeneity, which is desirable for a better model generalization. The linking of parallel entities can contribute to further investigation about how NEs are

---

[12] http://hdl.handle.net/11234/1-5533.
[13] https://huggingface.co/popelucha/robeczech-NER.
[14] https://gitlab.fi.muni.cz/nlp/named_entity_linking.

used in news texts across languages, how frozen or variable NEs are in different languages, or for training cross-lingual models.

Apart from the results, the paper contributes to the methods for developing similar data resources. The manual annotation of NEs was done relatively quickly, even though the pre-annotation by existing NE models was not perfect (generally, with low recall).

The annotation tool LabelStudio[15] provides annotation time measurement. The user interface measures time spent with annotation opened (lead time). Even though it allows the user to notify the program they are interrupting their work, the annotators did not use the feature. For this reason, the time measurements are biased by outliers (when the annotator left the application window open for several hours), and therefore, the median is the most informative measure.

The lead time median for the first task in seconds, as calculated by LabelStudio, was 4.723 s and 4.482 s for annotators 1 and 2, respectively. The second task was annotated by Annotator 1 with a median time of 36 s. From the median times of the two annotations, we estimate the process could take less than 30 h of work ($A_1 = 5\,\text{s} \times 15000 \approx 21\,\text{h}$ and $A_2 = 40\,\text{s} \times 450 \approx 5\,\text{h}$).

5 Conclusion

The presented work established a possible workflow for NER/NEL with existing NER models, and NEL. The result is a new annotation for existing parallel data. We evaluated the existing NER models on newer data than are usually used for evaluation. We proposed an annotation pipeline that is reasonably efficient.

5.1 Future Work

The pipeline will be extended to Czech NEL. The workflow can be used for other language pairs as well. Finally, we plan to automate a cross-lingual annotation as much as possible.

References

1. Alani, H., Guo, Z., Barbosa, D.: Robust named entity disambiguation with random walks. Semant. Web **9**(4), 459–479 (2018). https://doi.org/10.3233/SW-170273
2. Bodnari, A., Névéol, A., Uzuner, Ö., Zweigenbaum, P., Szolovits, P.: Multilingual named-entity recognition from parallel corpora. In: Forner, P., Navigli, R., Tufis, D., Ferro, N. (eds.) Working Notes for CLEF 2013 Conference , Valencia, Spain, 23–26 September 2013. CEUR Workshop Proceedings, vol. 1179. CEUR-WS.org (2013). https://ceur-ws.org/Vol-1179/CLEF2013wn-CLEFER-BodnariEt2013.pdf
3. Chinchor, N., Sundheim, B.: MUC-5 Evaluation Metrics. In: Fifth Message Understanding Conference (MUC-5) (1993). https://aclanthology.org/M93-1007

---

[15] https://labelstud.io/.

4. Cucerzan, S.: Large-scale named entity disambiguation based on wikipedia data. In: Eisner, J. (ed.) Proceedings of the 2007 Joint Conference on Empirical Methods in Natural Language Processing and Computational Natural Language Learning (EMNLP-CoNLL), Prague, Czech Republic, pp. 708–716. ACL (2007). https://aclanthology.org/D07-1074
5. Delpeuch, A.: OpenTapioca: Lightweight Entity Linking for Wikidata (2020)
6. Devlin, J., Chang, M., Lee, K., Toutanova, K.: BERT: Pre-training of Deep Bidirectional Transformers for Language Understanding. CoRR abs/1810.04805 (2018). http://arxiv.org/abs/1810.04805
7. Ehrmann, M., Turchi, M., Steinberger, R.: Building a multilingual named entity-annotated corpus using annotation projection. In: Mitkov, R., Angelova, G. (eds.) Proceedings of the International Conference Recent Advances in Natural Language Processing 2011, Hissar, Bulgaria, pp. 118–124. Association for Computational Linguistics (2011). https://aclanthology.org/R11-1017
8. Guo, Z., Barbosa, D.: WNED datasets and results (2017). https://doi.org/10.7939/DVN/10968
9. Hoffart, J., Suchanek, F.M., Berberich, K., Lewis-Kelham, E., de Melo, G., Weikum, G.: YAGO2: exploring and querying world knowledge in time, space, context, and many languages. In: Proceedings of the 20th International Conference Companion on World Wide Web, WWW 2011, pp. 229–232. Association for Computing Machinery, New York (2011). https://doi.org/10.1145/1963192.1963296
10. Hoffart, J., et al.: Robust disambiguation of named entities in text. In: Barzilay, R., Johnson, M. (eds.) Proceedings of the 2011 Conference on Empirical Methods in Natural Language Processing, Edinburgh, Scotland, UK, pp. 782–792. Association for Computational Linguistics (2011). https://aclanthology.org/D11-1072
11. Konkol, M., Konopík, M.: CRF-based Czech named entity recognizer and consolidation of Czech NER research. In: Habernal, I., Matoušek, V. (eds.) TSD 2013. LNCS (LNAI), vol. 8082, pp. 153–160. Springer, Heidelberg (2013). https://doi.org/10.1007/978-3-642-40585-3_20
12. Labusch, K., Neudecker, C.: Named entity disambiguation and linking historic newspaper OCR with BERT. In: Conference and Labs of the Evaluation Forum (2020). https://api.semanticscholar.org/CorpusID:225072200
13. Langendonck, W.V.: Theory and Typology of Proper Names. De Gruyter Mouton, Berlin, New York (2007). https://doi.org/10.1515/9783110197853
14. Li, B., He, Y., Xu, W.: Cross-Lingual Named Entity Recognition Using Parallel Corpus: A New Approach Using XLM-RoBERTa Alignment (2021)
15. Mayhew, S., et al.: Universal NER: A Gold-Standard Multilingual Named Entity Recognition Benchmark (2024)
16. Orr, L., et al.: Bootleg: Chasing the Tail with Self-Supervised Named Entity Disambiguation (2020)
17. Papantoniou, K., Efthymiou, V., Plexousakis, D.: Automating benchmark generation for named entity recognition and entity linking. In: Pesquita, C., et al. (eds.) ESWC 2023. LNCS, vol. 13998, pp. 143–148. Springer, Cham (2023). https://doi.org/10.1007/978-3-031-43458-7_27
18. Prokopidis, P., Papavassiliou, V., Piperidis, S.: Parallel global voices: a collection of multilingual corpora with citizen media stories. In: Calzolari, N., et al. (eds.) Proceedings of the Tenth International Conference on Language Resources and Evaluation (LREC 2016), Portorož, Slovenia, pp. 900–905. European Language Resources Association (ELRA) (2016). https://aclanthology.org/L16-1144

19. Rahimi, A., Li, Y., Cohn, T.: Massively multilingual transfer for NER. In: Proceedings of the 57th Annual Meeting of the Association for Computational Linguistics, Florence, Italy, pp. 151–164. Association for Computational Linguistics (2019). https://www.aclweb.org/anthology/P19-1015
20. Rambousek, A., Horák, A., Suchomel, V., Kocincová, L.: Semiautomatic building and extension of terminological thesaurus for land surveying domain. In: Horák, A., Rychlý, P. (eds.) The 8th Workshop on Recent Advances in Slavonic Natural Languages Processing, RASLAN 2014, Karlova Studanka, Czech Republic, 5–7 December 2014, pp. 129–137. Tribun EU (2014). http://nlp.fi.muni.cz/raslan/2014/2.pdf
21. Ševčíková, M., Žabokrtský, Z., Krůza, O.: Named entities in Czech: annotating data and developing NE tagger. In: Matoušek, V., Mautner, P. (eds.) TSD 2007. LNCS (LNAI), vol. 4629, pp. 188–195. Springer, Heidelberg (2007). https://doi.org/10.1007/978-3-540-74628-7_26
22. Shavarani, H., Sarkar, A.: SpEL: structured prediction for entity linking. In: Proceedings of the 2023 Conference on Empirical Methods in Natural Language Processing, Singapore, pp. 11123–11137. Association for Computational Linguistics (2023). https://aclanthology.org/2023.emnlp-main.686
23. Shen, W., Wang, J., Han, J.: Entity linking with a knowledge base: issues, techniques, and solutions. IEEE Trans. Knowl. Data Eng. **27**(2), 443–460 (2015). https://doi.org/10.1109/TKDE.2014.2327028
24. Sido, J., Pražák, O., Přibáň, P., Pašek, J., Seják, M., Konopík, M.: Czert – Czech BERT-like model for language representation. In: Mitkov, R., Angelova, G. (eds.) Proceedings of the International Conference on Recent Advances in Natural Language Processing (RANLP 2021), pp. 1326–1338. INCOMA Ltd., Held Online (2021). https://aclanthology.org/2021.ranlp-1.149
25. Straka, M., Náplava, J., Straková, J., Samuel, D.: RobeCzech: Czech RoBERTa, a monolingual contextualized language representation model. In: Ekštein, K., Pártl, F., Konopík, M. (eds.) TSD 2021. LNCS (LNAI), vol. 12848, pp. 197–209. Springer, Cham (2021). https://doi.org/10.1007/978-3-030-83527-9_17
26. Tedeschi, S., Maiorca, V., Campolungo, N., Cecconi, F., Navigli, R.: WikiNEuRal: combined neural and knowledge-based silver data creation for multilingual NER. In: Moens, M.F., Huang, X., Specia, L., Yih, S.W.t. (eds.) Findings of the Association for Computational Linguistics: EMNLP 2021, Punta Cana, Dominican Republic, pp. 2521–2533. Association for Computational Linguistics (2021). https://doi.org/10.18653/v1/2021.findings-emnlp.215. https://aclanthology.org/2021.findings-emnlp.215
27. Tjong Kim Sang, E.F., De Meulder, F.: Introduction to the CoNLL-2003 shared task: language-independent named entity recognition. In: Proceedings of the Seventh Conference on Natural Language Learning at HLT-NAACL 2003, pp. 142–147 (2003). https://www.aclweb.org/anthology/W03-0419
28. Weischedel, R., et al.: OntoNotes Release 5.0 LDC2013T19 (2013). https://doi.org/10.35111/xmhb-2b84
29. Ševčíková, M., Žabokrtský, Z., Krůza, O.: Zpracování pojmenovaných entit v českých textech. Technical report 2007/TR-2007-36, ÚFAL MFF UK (2007). https://ufal.mff.cuni.cz/~zabokrtsky/reports/techrep-ne-2007.pdf

TamSiPara: A Tamil – Sinhala Parallel Corpus

Randil Pushpananda[1]( ), Chamila Liyanage[1], Ashmari Pramodya[2], and Ruvan Weerasinghe[1]

[1] Language Technology Research Laboratory, University of Colombo School of Computing, Colombo, Sri Lanka
{rpn,cml,arw}@ucsc.cmb.ac.lk
[2] Nara Institute of Science and Technology, Ikoma, Japan
pussewala.ashmari.ow4@naist.ac.jp

Abstract. This paper presents the development of a Sinhala-Tamil bilingual parallel corpus with sentence-level alignment. The corpus comprises source language text from contemporary writings, with all sentences translated manually. Active learning methods were employed to select sentences, ensuring the representation of effective language structures in both languages. The corpus is divided into two parts: one with translations from Sinhala to Tamil direction, consisting of 25k parallel sentences, while the other consists of translations from Tamil to Sinhala direction, comprising 22k parallel sentences. Manual translations were conducted by two teams of professional translators. The resulting final version of `TamSiPara`, the Tamil-Sinhala bilingual parallel corpus consists of a total of 47k parallel sentences.

Keywords: Parallel corpus · Bitexts · Sinhala - Tamil · Low-resource languages

1 Introduction

Parallel corpora, also known as `bitexts`, consist of large texts containing parallel translations and serve as valuable resources for researchers in a wide range of disciplines, including machine translation, computer-assisted translation, terminology extraction, computer-assisted language learning, contrastive linguistics, and translation studies [1]. Parallel corpora can be constructed through various approaches. While the automatic extraction of parallel text, such as Morishita et al. [2], can be effective for generating a large volume of parallel text, the utility of such automatic or semi-automatic corpora is limited due to the prevalence of sentences with identical syntactic structures and vocabulary, resulting in a narrow vocabulary range in both the source language (L1) and the target language (L2). Moreover, the manual creation of corpora requires substantial linguistic expertise, making it a challenging task. Nevertheless, such resources remain valuable, especially for open-domain NLP research. Thus, this research paper discusses an initiative developing a domain-free text of Bilingual Parallel Corpus (BPC) with one-to-one alignment for the Sinhala and Tamil language pair.

Among these two languages, Tamil is used by a global community of 87 million (m) speakers, comprising 79 m people as their first language, while 8 m as a second language. In terms of total speakers, Tamil ranks as the 19th most widely spoken language in the world.[1] The Tamil linguistic community spans various countries, including India, Sri Lanka, Singapore, Malaysia, Mauritius, and others. Sri Lanka, in particular, is home to 4 m Tamil speakers and is one of the two official languages in the country.[2] Conversely, Sinhala serves as the other official language in the country, with around 15 m people speaking it as their first language. Sinhala is predominantly spoken by Sri Lankans and by Sinhala diaspora communities around the world. Among the shared features of the two languages, both are identified as agglutinative languages [3]. Additionally, in terms of writing systems, both utilize abugida scripts [4].

2 Related Work

Despite the linguistic features and significance of the two languages, both are reported as low-resource languages [5,6]. However, Tamil language benefits from relatively more research and available tools for Tamil NLP works, attributed to the larger populations of Tamil speakers worldwide [5]. While Sinhala also has some language resources available for several domains in NLP, effective parallel corpora for these languages are yet to be developed, although a couple of initiatives have been reported.

Among the reported works, Sripirakas [7] is noted as the first-ever Sinhala-Tamil parallel dataset. This domain-specific dataset utilized Sri Lankan parliamentary order papers which has been developed by the Language Technology Research Laboratory of the University of Colombo School of Computing (LTRL-UCSC). The resource has been utilized for research on a Sinhala-Tamil MT system. Jayakaran [8] presented a follow-up to Sripirakas [7] and reported a parallel corpus consisting of 5,697 sentences.

Hameed et al. [9] reported on an automatic process for developing a sentence-aligned Sinhala-Tamil parallel corpus. However, neither the paper reveals the size of the data nor is the resource publicly available. Furthermore, Farhath et al. [10] conducted work on manually collecting parallel sentences, representing another initiative to collect such sentences using order papers from the government of Sri Lanka and reported that the corpus consisted of 15k parallel sentences. Vasantharajan et al. [11] also presents an automatic method for extracting parallel sentences. However, the study utilizes the *Tesseract* Optical Character Recognition Engine to recognize Tamil and Sinhala characters, resulting in the creation of a large-scale parallel corpus for Tamil-Sinhala-English languages. The reported

---

[1] https://www.ethnologue.com/insights/ethnologue200/.
[2] https://www.worlddata.info/languages/tamil.php.

corpus consists of 185.4k, 168.9k, and 181.04k sentences, and 2.11 m, 2.22 m, and 2.33 m words in Tamil, Sinhala, and English, respectively.

Although several attempts have been reported in developing parallel corpora for the Sinhala-Tamil language pair, most of these resources have not contributed to improving MT models for these languages. Accordingly, this initiative was undertaken to develop an effective parallel text corpus for the Sinhala-Tamil language pair, aimed at facilitating research in the field of MT.

3 Creating the Parallel Corpus

As mentioned in Sect. 2, the development of the parallel corpus was succeeded by an examination of the MT system for Sinhala and Tamil languages. Accordingly, the resource was developed across multiple stages in accordance with the findings of the aforementioned research. This section provides a detailed account of the approach employed in constructing the corpus and its subsequent analysis. The parallel corpus development process primarily involved four stages, as outlined below.

i *Manually collecting parallel translations from existing documents*
    The initial step in developing the parallel corpus involved manually collecting parallel translations of the Tamil-Sinhala language pair from existing documents. While this method may not have been the most efficient means of creating the specific resource, we provide a detailed discussion of it in Sect. 3.1, highlighting the challenges and drawbacks associated with its implementation.
ii *Manually translating text from Sinhala to Tamil*
    The second step of this initiative involved the manual translation of text from Sinhala to Tamil. The development of this version of the BPC and subsequent evaluation by utilizing the dataset for developing MT systems for the particular languages and their outcomes are described in Sect. 3.2.
iii *Manually translating selected text from Tamil to Sinhala*
    After evaluating the results of the MT systems, we proceeded to the third step: translating the text in the opposite direction, from Tamil to Sinhala. Despite the scarcity of translators for this direction, our aim was to develop a comprehensive parallel corpus for MT research in both directions. The details of the third step are described in Sect. 3.3 of the paper.
iv *Translation and text verification*
    To ensure the accuracy of the parallel text, a translation and text verification process was followed, as discussed in Sect. 3.4.

3.1 Manually Collecting Parallel Translations

Since the motivation was to support machine translation research, this study aimed to develop an effective Sinhala-Tamil BPC. Thus, the initial step involved manually collecting parallel translations from existing parallel documents, considering it was straightforward and cost-effective. However, the outcomes of subsequent MT research indicated that the collected corpus of parallel text was not significantly effective in achieving improved results. The process proved to be challenging for several reasons, as outlined below.

- **Shortage of parallel text for Sinhala and Tamil languages:**
  While more parallel resources exist for certain language pairs, the availability of resources for the target language pair is relatively limited.
- **Most of the available text is printed text:**
  A significant portion of the existing parallel text is not in electronic format and is predominantly found in printed media.
- **Electronic text is in PDF format and non-Unicode fonts:**
  Parallel text was extracted from various resources such as Annual Reports[3], Parliamentary Order Papers[4], Magazines[5], Books[6], and articles. However, the process presented challenges because a significant portion of electronically available documents was in PDF format and encoded in proprietary fonts. To utilize them effectively, it was necessary to convert the text in both languages to Unicode encoding. This conversion was facilitated by a Real-Time Font Encoding Converter[7] developed by the LTRL-UCSC. Moreover, some documents contained text in rarely used non-Unicode fonts, and the existing Real-Time Font Encoding Converter did not support their conversion.
- **Removal of graphics:**
  Graphics and formatting were extensively used in the documents, requiring careful treatment and subsequent removal of both graphics and formatting.
- **Disarrangement of sentences:**
  Aligning sentences posed a challenge as the Tamil sentences were not in order with respect to the sentences in the Sinhala documents. Since automatic alignment proved unworkable, the task was performed manually.

Due to the aforementioned challenges and two additional limitations: (i) the domain-specific nature of the documents and (ii) their limited vocabulary, manual text collection proved ineffective. Consequently, we decided to terminate this process.

3.2 Sinhala to Tamil (SiTa) Version

The second approach involved developing parallel text by manually translating Sinhala sentences into Tamil with the assistance of professional translators. The translation direction for this version was chosen from Sinhala to Tamil due to the scarcity of translators in the opposite direction. Initially, sentences were selected

---

[3] https://www.treasury.gov.lk/web/annual-reports-financial-statements-of-key-soes.
[4] https://www.parliament.lk/business-of-parliament/order-papers.
[5] http://www.cpalanka.org/.
[6] http://www.edupub.gov.lk/.
[7] http://www.ucsc.cmb.ac.lk/ltrl/services/feconverter/.

from the 10M Word Sinhala Text Corpus from the LTRL-UCSC [12]. Its use for sentence selection was motivated by several factors: the corpus contains written texts, the texts are in contemporary Sinhala, they include works by renowned Sinhala authors, and thus, it offers rich linguistic diversity.

The translation team for the SiTa version comprised a senior translator as lead, supported by several junior translators. All translators were native Tamil speakers fluent in Sinhala as a second language.

During this stage of corpus development, we successfully created around 26k Sinhala-to-Tamil parallel sentences. As indicated by the corpus statistics in Table 2, this version comprises a total of 252k Sinhala words, of which 37k are unique. Additionally, there are 219k Tamil words in the corpus, with 53k being unique. While the average number of words in the source sentences is 10, the target text has an average of 9 words. A sample of parallel sentences are depicted in the Fig. 1.

Evaluation of the Dataset

To evaluate the initial parallel corpus containing 26k parallel sentences, we developed multiple MT systems using both statistical and neural MT techniques with default settings. We conducted experiments for translation directions from Sinhala to Tamil and Tamil to Sinhala. The quality of the translation systems was evaluated using the BLEU evaluation technique, and the results from these approaches are presented in Table 1.

Table 1. BLEU scores for SMT and NMT Baseline systems

Model	BLEU Score	
	Ta to Si	Si to Ta
Statistical MT	13.61	10.11
Neural MT (Transformer)	7.35	2.98

As indicated in Table 1, it is evident that the quality of Tamil to Sinhala translations surpasses that of Sinhala to Tamil translations for both Statistical Machine Translation (SMT) and Transformer models. An error analysis was conducted to identify the factors contributing to lower results in the translation process, and it was performed as follows.

- Count the number of total words (TotW) and unique words (UniW) in each training (Tr) and testing (Te) datasets.
- Count the number of out-of-vocabulary (OOV) words in the test dataset (as a percentage of test dataset).
- Count the number of untranslated words (UntransW) (as a percentage of test dataset).
- Count the number of translated words which are not in the reference dataset (TargetOOV) (as a percentage of test dataset).

Table 2. The results of the error analysis

Description	Sinhala-Tamil		Tamil-Sinhala	
	UniW	TotW	UniW	TotW
Training Source Dataset (Average)	35k	225k	49k	195k
Testing Source Dataset (Average)	9k	25k	10k	21k
Source OOV (%)	24.54	9.10	34.43	17.10
UntransW (%)	43.77	18.33	49.87	26.47
UntransW - OOV (%)	19.23	9.23	15.44	9.37
Reference Output Dataset	10k	21k	9k	25k
Target OOV (%)	72.88	69.34	64.76	60.71

The results of the error analysis of MT systems for both directions, namely Sinhala to Tamil and Tamil to Sinhala, are presented in Table 2. Since the dataset was originally translated from Sinhala to Tamil, the same Sinhala words or phrases have been translated into several different Tamil words especially for synonyms or phrases, indicating an out-of-vocabulary issue with Sinhala. Based on this analysis, we made the decision to develop a parallel text for the Tamil to Sinhala direction.

Sinhala	එක අතකින් අද කම්කරු ව්‍යාපාරය තුළ නිර්ධන පන්ති දේශපාලනයක් නැත.
SiTrans	eka atakin ada kamkaru vyāpāraya tuḷa nirdhana panti dēśapālanayak næta
Tamil	ஒருவழியில் இன்று தொழிலாளர்களின் நடவடிக்கைகளில் பாட்டாளி வர்க்கத்தினர் ; அரசியலில் இல்லை.
TaTrans	Oruvaḻiyil iṉṟu toḻilāḷarkaḷiṉ naṭavaṭikkaikaḷil pāṭṭāḷi varkkattiṉar; araciyalil illai
English	On the one hand, there is no proletariat politics in today's labor movement.
Sinhala	කල්වේලා ප්‍රමාද නොකොට අපි රෝගියා ද සමග ඔරුවට නැගුණෙමු.
SiTrans	kalvēlā pramāda nokoṭa api rōgiyā da samaga oruvaṭa næguṇemu
Tamil	காலதாமதமின்றி நாம் நோயாளியோடு படகில் ஏறினோம்.
TaTrans	Kālatāmatamiṉṟi nām nōyāḷiyōṭu paṭakil ēṟiṉōm
English	Without delay, we boarded the boat with the patient.
Sinhala	ගෙදර එළිය කරපු පහන තිවෙයි කියල හීනෙකින්වත් හිතුවේ නැහැ.
SiTrans	gedara eḷiya karapu pahana niveyi kiyala hīnekinvat hituve næhæ
Tamil	வீட்டுக்கு ஒளியேற்றிய விளக்கு அனைந்துவிடும் என கனவிலும் நினைக்கவில்லை.
TaTrans	Vīṭṭukku oḷiyēṟṟiya viḷakku aṉaintuviṭum eṉa kaṉavilum niṉaikkavillai
English	I never dreamed that the lamp that lit the house would go out.

Fig. 1. A sample of parallel sentences from SiTa version

3.3 Tamil to Sinhala (TaSi) Version

This version of the corpus involved the manual translation of sentences in the opposite direction, i.e., from Tamil to Sinhala. This effort aimed to construct a robust parallel corpus encompassing a wider range of vocabulary and syntactic structures within a concise set of sentences for several reasons. Firstly, since the translation process is conducted manually, working with a minimal set of sentences proves to be time-consuming and cost-effective. Secondly, a concise yet comprehensive set of sentences will also be significant in decreasing the computational resources required during the MT training process.

Accordingly, we employed the Active Learning technique [13] to extract the most concise sentences for translation. This approach iteratively selects the most informative sample of sentences, minimizing the amount of data needed to train a high-performing translation model. For this version of the corpus, we utilized the 4M word Tamil text corpus [14] as the primary source for selecting sentences to translate. Given that the ultimate goal of the translation model is to produce general translations, we selected the UCSC 4M words Tamil Corpus, a news corpus containing exclusively Tamil documents and online newspapers published in Sri Lanka. This approach allowed us to extract more concise sentences, resulting in the translations of 22k sentences. Despite the limited number of sentences, the corpus demonstrates a notably strong coverage in terms of syntactic structures and vocabulary.

Given the statistics presented in Table 3, the parallel sentences in the TaSi version consisted of 259k source language (Tamil) words and 292k target language (Sinhala) words. Further, the dataset comprised 72k Tamil unique words and 36k Sinhala unique words. Compared to the list of unique words occurring in the SiTa version of the parallel text, TaSi version includes 52k new unique words for Tamil and 21k new unique words for Sinhala.

Table 3. Statistics of the both SiTa and TaSi versions of the parallel text

Description	SiTa version		TaSi version	
	Sinhala	Tamil	Tamil	Sinhala
Total number of words	252k	219k	259k	292k
Unique words	37k	53k	72k	36k
V/T	14.6%	24.2%	NA	NA
Average Number of words per sentence	10	9	10	13
New words occur in TaSi version	NA	NA	52k	21k

3.4 Translation and Text Verification Process:

Finally, a translation verification process was conducted to ensure the accuracy of the translations. This task was assisted by a computational linguist and a senior translator specializing in translations for particular languages. The text verification process was performed by the computational linguist using the unique word lists. Despite linguistic reviews conducted prior to the translation process, some text errors or typing mistakes were present in the dataset. Therefore, verification was necessary for both textual errors and translations.

During the text verification process, the frequency-1 words from both Tamil and Sinhala unique word lists were initially examined. Subsequently, the unique words in both languages were reviewed, and words with errors were identified and marked.

Having classified the errors and mistakes in the text, the following several types of errors and mistakes were identified in the dataset.

Forms with errors	Correct form	Meaning
1. දරුවන්ගෙං'	දරුවන්ගේ	(children's)
2. සංඛියාවක්	සංඛ්‍යාවක්	(a number)
3. පරිණතභාවයක්	පරිණත භාවයක්	(a maturity)
4. දෙවියන්වහත්සේට	දෙවියන් වහත්සේට	(for god)
5. කාලපරිච්ඡේදයෙහි	කාල පරිච්ඡේදයෙහි	(in the period)
6. ස්වයංපැහැදිලි	ස්වයං පැහැදිලි	(self, clear)
7. සත්ත්වයෙතිඋරගයෙති	සත්ත්වයෙති උරගයෙති	(animals, reptiles)

Fig. 2. A sample of text errors occurred and corresponding corrected forms

- **Text input errors**: Since both Sinhala and Tamil are complex scripts, the input method must function properly on the translators' computers to ensure correct typing. When the input system malfunctions, it can affect the translated text and lead to typing errors. Examples of text input errors can be seen in numbers 1 and 2 of Fig. 2.
- **Missing white-space**: There are instances of words with missing white space, which are mistakes made by the translators. Examples of this error can be found in numbers 6 and 7 of Fig. 2.
- **Word segmentation issues**: One challenge in Sinhala writing is word segmentation. Although several institutes have provided guidelines for Sinhala word segmenpation, We referred the new edition of Sinhala lēkhana ritiya [15], the Sinhala writing guide published by the National Institute of Education, Sri Lanka. However, there were words that did not adhere to the NIE-Book guidelines, as exemplified in numbers 3 and 4 of Fig. 2.
- **Visually similar character issue**: Another issue arises from translators using visually similar but distinct characters. While this occurred infrequently, a few instances were noted in the dataset, as depicted in number 5 of Fig. 2.

These specific errors were manually corrected one by one across the entire corpus.

In the translation verification process, the senior translator randomly selected sentences and reviewed them. If any mistranslations were identified, the corresponding translations were reviewed accordingly. This process facilitated the creation of an error-free and accurate parallel corpus.

4 Discussion

Since the sentences selected for translation in both versions of the corpus were drawn from text corpora of contemporary writing, they encompass diverse syntactic structures. For instance, Fig. 1 illustrates four parallel sentences from the SiTa version of the BPC. Among these, the last one represent spoken language utterances, while the first two are written language texts. This demonstrates that the corpus covers a wide range of syntactic structures.

According to the corpus statistics depicted in Table 2, there are discrepancies in the data between the two versions of the BPC. Specifically, the SiTa version displays 37k unique Sinhala words, while the Tamil text exhibits 53k, indicating a significant discrepancy of 16k unique words. In contrast, the TaSi version exhibits 72k unique Tamil words and 36k unique Sinhala words, indicating a difference of 36k, which is half the number of unique words in the Tamil list.

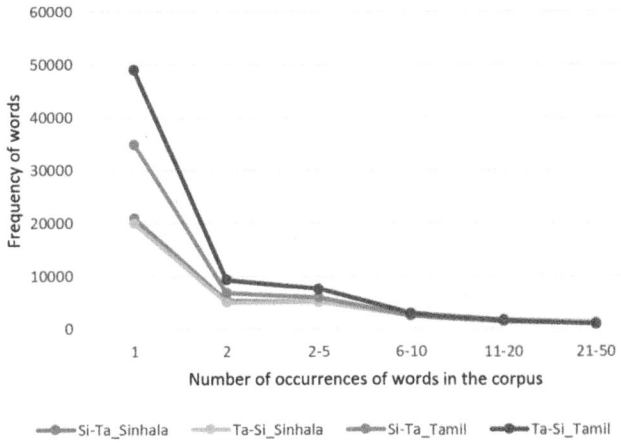

Fig. 3. Frequency of words occur in both SiTa and TaSi versions

Furthermore, the total number of words from both languages in the two versions (SiTa and TaSi) reveals that the Sinhala text has the highest numbers in both versions, with 252k and 292k words in Sinhala texts, compared to 219k and 259k in Tamil texts, which are relatively lower in both SiTa and TaSi versions, respectively. However, conversely, in both versions, the Tamil text exhibits the highest number of unique words, with 37k and 36k unique words occurring in the Sinhala text, while 53k and 72k unique words are present in the Tamil text. Since an active learning technique was employed for selecting sentences to translate in the TaSi version, a significant number of new Tamil words (51k) emerge in the Tamil text. However, there are relatively fewer new words in the Sinhala text, with only 21k new words occurring in the same version of the BPC.

Table 4. Statistics of the TamSiPara BPC (Total of around 47k sentence pairs)

	Tamil	Sinhala
Total number of words in the BPC	448k	544k
Unique words in the BPC	124k	72k
Total number of characters	4,184k	3,450k
word-to-character ratio	9.3	6.3

This is further illustrated in Fig. 3, which indicates the frequency of words occurring in both the SiTa and TaSi versions of the BPC, respectively. According to Fig. 3, the frequency of occurrence of words in Tamil unique words is higher than that in both versions of Sinhala unique words. In terms of the frequency of unique Sinhala words, there are approximately 21k occurrences in SiTa, while around 20k occur in TaSi. However, the Fig. 3 depicts a significant differentiation from frequency 1 Tamil words to frequency 2 or more occurring words in the lists. As indicated in the Fig. 3, 35k in SiTa and 49k in TaSi versions, respectively, are relatively higher than the frequency 1 Sinhala unique words in both versions. Furthermore, in the entire corpus, while the Tamil text reports a total of 105k unique words, the Sinhala text includes only 58k total unique words.

Moreover, according to the character count in the BPC (in both versions), as depicted in Table 4, the Tamil text comprises a total of 4,184k characters across 448k words, resulting in a word-to-character ratio of 9.3. Conversely, the Sinhala text contains 3,450k characters for 544k words, resulting in a word-to-character ratio of 6.3. For instance, consider the example of lengthy words from the Tamil unique words of the BPC depicted in Fig. 4. While some words may exhibit word segmentation issues, many are accurate but notably lengthy, largely owing to the use of word compounding and Sandhi in Tamil language.

Based on the statistics and subsequent analysis of both versions of the BPC, the following three reasons are identified to illustrate the data discrepancy between Tamil and Sinhala language texts, as outlined below.

- **Morphological richness of the two languages**
  Both Tamil and Sinhala languages are reported to be morphologically rich, resulting in a higher number of unique words in both languages.
- **Frequent use of Sandhi words**
  Sandhi refers to the process of joining two words or characters, where morphophonemic changes occur at the point of joining [16]. For instance, the first

couple of words in Fig. 4 are sandhi words, which are formed by joining two (or more) words to create longer ones.
- **Utilization of word compounds**
  Whitespace removal for word compounding occurs in both Tamil and Sinhala for two reasons: i. intentionally to create word compounds, and ii. unintentionally as a result of whitespace removal. In both cases, two or more lexical entries merge into a single lexical entry, as shown in numbers 5 and 6 in Fig. 4.

1 தடுமாறிக்கொண்டிருக்கிறார்கள்   (They are stumbling)
2 முன்வைக்கப்படவில்லை   (Not presented)
3 பிரகடனப்படுத்துவார்   (will declare)
4 பயிர்செய்வதைமேம்படுத்தல்இத்துறையில் (Cultivation improvement in this sector)
5 போராட்டத்துக்குஷிழிணிவி   (Help for the struggle)
6 சேதுராமன்இலக்கியங்கள்   (Chethuraman Literature)

Fig. 4. A sample of Tamil unique words from `TaSi` version

Although both Tamil and Sinhala are reported to be morphologically rich languages, the statistics of the `TamSiPara` BPC indicate that Tamil text exhibits greater complexity in its morphology compared to Sinhala. Furthermore, the statistics demonstrate that Tamil more frequently employs white-space removal between words to showcase *Sandhi* and word compounding, in contrast to its usage in Sinhala writing.

4.1 Availability and Access

This resource is available under a `CC BY-NC` license, and interested parties can access it through the `ltrl.ucsc.lk/tools-and-resourses` website.

5 Conclusion and Future Work

Developing language resources for low-resource languages is crucial for NLP research on specific languages. This paper explores the process of creating an efficient parallel corpus for the Sinhala and Tamil language pair, presenting detailed statistics on the corpus. The parallel text comprises approximately 47k manually translated sentences, which were divided into two versions based on the translation direction. Thus, the SiTa version includes 25k parallel sentences translated from Sinhala to Tamil, while the TaSi version comprises 22k parallel sentences translated from Tamil to Sinhala. Since the motivation behind this development was to support MT research for these languages, this corpus will be valuable for both machine translation and machine-assisted translation initiatives in the respective languages.

Our future goal is to incorporate annotations for the entire dataset. This addition would significantly benefit numerous NLP research endeavors for the specified languages.

Acknowledgements. The first phase of this research was funded by the ICTA of Sri Lanka, and we appreciate their support. Furthermore, we acknowledge the partial funding received from the University of Colombo School of Computing through the Research Allocation for Research and Development. We also thank all the translators and the members of the LTRL of UCSC for their various contributions to making this work successful.

References

1. Paulussen, H., Macken, L., Vandeweghe, W., Desmet, P.: Dutch parallel corpus: a balanced parallel corpus for Dutch-English and Dutch-French. In: Essential Speech and Language Technology for Dutch: Results by the STEVIN Programme, pp. 185–199 (2013)
2. Morishita, M., Suzuki, J., Nagata, M.: Jparacrawl: a large scale web-based English-Japanese parallel corpus. arXiv preprint arXiv:1911.10668 (2019)
3. Thampoe, H.D.: Sinhala and Tamil: a case of contact-induced restructuring. Ph.D. thesis, Newcastle University (2017)
4. Daniels, P.T.: Writing systems. In: The Handbook of Linguistics, pp. 75–94 (2017)
5. De Silva, N.: Survey on publicly available sinhala natural language processing tools and research. arXiv preprint arXiv:1906.02358 (2019)
6. Sarveswaran, K., Dias, G., Butt, M.: Thamizhi morph: a morphological parser for the Tamil language. Mach. Transl. **35**(1), 37–70 (2021)
7. Sripirakas, S.: Statistical Machine Translation for Sinhala and Tamil, unpublished BSc thesis, University of Colombo (2010)
8. Jeyakaran, M.: A novel kernel regression based machine translation system for Sinhala-Tamil translation, unpublished BSc thesis, University of Colombo (2013)
9. Hameed, R.A., et al.: Automatic creation of a sentence aligned Sinhala-Tamil parallel corpus. In: Proceedings of the 6th Workshop on South and Southeast Asian Natural Language Processing (WSSANLP 2016), pp. 124–132 (2016)
10. Farhath, F., Theivendiram, P., Ranathunga, S., Jayasena, S., Dias, G.: Improving domain-specific SMT for low-resourced languages using data from different domains. In: Proceedings of the Eleventh International Conference on Language Resources and Evaluation (LREC 2018) (2018)
11. Vasantharajan, C., Tharmalingam, L., Thayasivam, U.: Adapting the tesseract open-source OCR engine for Tamil and Sinhala legacy fonts and creating a parallel corpus for Tamil-Sinhala-English. In: 2022 International Conference on Asian Language Processing (IALP), pp. 143–149. IEEE (2022)
12. Language Resources of LTRL-UCSC: UCSC 10M Word Sinhala Text Corpus. Language Technology Research Laboratory, University of Colombo School of Computing, Sri Lanka. LTRL resources, 1.0 (2007)
13. Cohn, D.A., Ghahramani, Z., Jordan, M.I.: Active learning with statistical models. J. Artif. Intell. Res. **4**, 129–145 (1996)
14. Language Resources of LTRL-UCSC: 4M Word Sri Lanka Tamil Text Corpus. Language Technology Research Laboratory, University of Colombo School of Computing, Sri Lanka. LTRL resources, 1.0 (2013)
15. Sinhala lēkhana rītiya - New Edition: NIE. National Institute of Education, Sri Lanka (2015)
16. Devadath, V., Kurisinkel, L.J., Sharma, D.M., Varma, V.: A sandhi splitter for Malayalam. In: Proceedings of the 11th International Conference on Natural Language Processing, pp. 156–161 (2014)

Automatic Ellipsis Reconstruction in Coordinated German Sentences Based on Text-to-Text Transfer Transformers

Marisa Schmidt(✉)[ID], Karin Harbusch[ID], and Denis Memmesheimer[ID]

Computer Science Faculty, University of Koblenz, Universitätsstr. 1,
56070 Koblenz, Germany
{marisaschmidt,harbusch,denismemmesheimer}@uni-koblenz.de

Abstract. Ellipsis reconstruction, i.e., revealing omitted syntactically obligatory words in a sentence, is still a challenging task in Natural Language Processing (NLP) technologies, even though this information is essential for advanced human-computer dialogues. Corpora for the training of omitted word reconstruction are increasingly appearing in the literature. Here, we focus on ellipsis phenomena in coordinated sentences—also called Clausal Coordinate Ellipsis (CCE)—in German. We report results with a Unified Text-To-Text Transfer Transformer (T5) model (Transfer Learning). A pre-trained model of written German is fine-tuned with a parallel corpus of pairs containing a reduced sentence and its canonical form, in which all omitted elements are explicitly listed. We compare the results for two parallel CCE corpora here, both of which are extracted from existing treebanks of German newspaper articles. We achieve a BLEU score of .8196 for testing the parallel TüBa-D/Z CCE corpus, and .6093 for testing a pre-release of the new parallel TIGER CCE corpus, respectively. The results can be improved to .8349 and .7543, respectively, for the training with the two CCE corpora together.

Keywords: Clausal Coordinate Ellipsis (CCE) · parallel Ellipsis reconstruction resources · Large Language Model · T5 · Transfer Learning

1 Introduction

Ellipsis, i.e., the linguistic phenomenon that leads to the omission of syntactically obligatory words in a sentence (for a good overview, see, e.g., [5] and for a focus on German, see, e.g., [10]), occurs in spoken and written text. Its automatic reconstruction is required for various understanding tasks, such as dialog act prediction and semantic role labeling (cf. [30] or [2] for an argument that ellipsis (and coreferences) is a means to make human-machine conversations more fluent and natural, with illustrative suboptimal/failing examples). As Cavar and colleagues show [4], ellipsis constructions are challenging for State-of-the-art (SotA) Natural Language Processing (NLP) technologies (see Sect. 3). For the reconstruction of elided words in English, the evaluation of GPT-4 using the zero-shot strategy achieved an accuracy of .25 [18].

Here, we focus on the automatic reconstruction of ellipsis in coordinated sentences—hereafter called *Clausal Coordinate Ellipsis (CCE)*—in German. The individual phenomena have been linguistically well studied (cf. Sect. 2). According to Harbusch [7], the frequency of coordinated sentences is 14% in the TIGER treebank of written German newspaper text (50,474 trees; [3]) and 10% in the TüBa-D/S treebank of spoken German (also called VERBMOBIL with dialogues about appointment making and planning a trip together; 38,282 tree structures [24]). Within the set of coordinated sentences, CCE occurs in 56% of the written cases and in 35% of the spontaneous utterances. All kinds of phenomena occur in both modalities, text and speech. Therefore, attention to improved CCE analysis/reconstruction is important in advanced NLP systems—for both written and spoken language applications.

In the area of rule-based approaches, systems explicitly designed for ellipsis reconstruction have been proposed (see, e.g., the Gapping detection system in Finnish [15]). Recently, a BLEU score [19] of .928 was reported for an evaluation with the parallel TüBa-D/Z-based CCE-corpus [13] using a probabilistic parsing model with explicit heuristics for the CCE reconstruction in German [14]. In Sect. 3, we review existing Machine Learning (ML) approaches—hereafter called *Large Language Model (LLM)*—for ellipsis recognition/reconstruction as starting point for our research question: *How far can we get with a Large Language Modeling approach using the existing parallel TüBa-D/Z-based CCE-corpus?*

In terms of *Unified Text-To-Text Transfer Transformer (T5)* models (cf. [26] for its vanilla shape and [22] where the limits of *Transfer Learning* with a Unified Text-To-Text Transformer are explored), our task is a typical *downstreaming* task [6]. Accordingly, we fine-tune a large pre-trained model of German obtained from unlabeled data with labeled data—in our case, two parallel corpora that pair reduced sentences and the reconstructed sentences, where omitted elements are explicitly listed. We employ the parallel TüBa-D/Z-based CCE-corpus [13]. In addition, a new pre-release based on the TIGER treebank [9] is used for comparisons. We achieve a BLEU score of .8196 for testing the parallel TüBa-D/Z CCE corpus, and .6093 for testing a pre-release of the new parallel TIGER CCE corpus as supervised dataset, respectively.

The paper is organized as follows. Section 2 describes the clausal coordinate ellipsis phenomena in German we explore here. Section 3 gives a brief overview of the state of the art in the area of Large Language Modeling with a focus on ellipsis reconstruction approaches. In Sect. 4, the datasets used in our system are delineated. In Sect. 5, we outline the system architecture, and present the evaluation results. In Sect. 6, we draw conclusions and address future work.

2 Clausal Coordinate Ellipsis Phenomena in German

Clausal Coordinate Ellipsis (CCE) refers to the omission of words in either the first or the second conjunct of a coordinated sentence. The phenomena have been well studied for many different languages. Here, we focus on German (for a general linguistic description, see, e.g., [10], for a psycho-linguistic investigation of Dutch and German, see [11], and for a specific linguistic formalism, see, e.g., [20] for Lexical Functional Grammar; focusing on a relaxed approach to word order analyses of ellipsis and coordination—as is the case in German—in Head-Driven Phrase Structure Grammar, see [1,17], and references therein).

Here, we follow the definitions originally presented by Harbusch and Kempen [8], since the parallel corpora we use (cf. Sect. 4) are encoded according to this terminology. The following four CCE phenomena are encoded: *Gapping, Forward Conjunction Reduction (FCR), Backward Conjunction Reduction (BCR)*, and *Subject Gap in Clauses with finite or fronted Verb (SGF)*. For all four types of CCE, the basic requirement for an omission is *lemma-identity*—the term goes back to [12]—of at least one element in both conjuncts. The so-called *remnants*, i.e., the counterpart of an omission, are underlined in the following examples. The omitted elements are struck through. The subscripts indicate the ellipsis mechanism at work: "g"=all kinds of Gapping, "f"=FCR, "b"=BCR, "s"=SGF. By *canonical form*, we refer to the reconstructed sentence. The examples here fold up the pairs in the parallel CCE corpora used in our system, assuming the crossed out words are present or absent, respectively.

In *Gapping*, at least the finite verb in the second conjunct is omitted (cf. Example (1); all examples in this section are taken from [13] and extended by the German translations to be tested with our system; cf. Table 3 in the evaluation section). However, the remnant and the reconstructed omitted finite verbform do not have to match in their morphological features, but the subject in the second conjunct determines the morphological shape of the finite verbform—an additional obstacle in any reconstruction process. Gapping can also cover one or more lemma-identical major, i.e., complete, constituents that are omitted in the second conjuncts. In the case of *Stripping*, there is only one remnant left over in the second conjunct. This constituent has to be accompanied by a so-called *Stripping particle* (cf. Example (2) with *auch* 'too'). Two more Gapping variants involve more than one verbform. In the case of *Long Distance Gapping (LDG)*, all verbforms of a coordinated sentence are omitted (cf. Example (3)). *Subgapping* allows to omit only the highest verb(s) in the dominance structure, rather than all of them (cf. Example (4)).

(1) *Henry lebt in Boston und Peter lebt$_g$ in Chicago.*
 Henry lives in Boston and Peter lives$_g$ in Chicago

(2) *Henry lebt in Boston und alle seine Kinder leben$_g$ auch in Boston$_g$.*
 Henry lives in Boston and all his children live$_g$ also in Boston$_g$
 'Henry lives in Boston and all his children too.'

(3) *Meine Frau möchte ein Auto kaufen, mein Sohn möchte$_g$ ein Motorrad kaufen$_g$.*
 My wife wants a car to_buy my son wants$_g$ a motorcycle to_buy$_g$
 'My wife wants to buy a car, my son wants to buy a motorcycle.'

(4) *Der Fahrer wurde getötet und die Mitfahrer wurden$_g$ schwer verletzt.*
 The driver was killed and the passengers were$_g$ severely wounded

In *Forward Conjunction Reduction (FCR)*, wordform-identical constituents with the same grammatical function in both conjuncts can be elided in the left periphery of the second conjunct. See Examples (5) and (6), which illustrate the local periphery interpretation using square brackets.

(5) [^{S} *Meine Schwester* lebt in Berlin und ~~*meine Schwester*~~$_f$ arbeitet in Frankfurt.]
    My   sister   lives in Berlin and ~~my   sister~~$_f$ works in Frankfurt

(6) *Amsterdam ist die Stadt,* [^{S} *in der* Jan *lebt und* ~~*in der*~~$_f$ *Piet arbeitet.*]
    Amsterdam is the city   where Jan lives and ~~where~~$_f$ Piet works

In *Backward Conjunction Reduction (BCR)*, lemma-identical words in the right periphery can be elided in the first conjunct. Thus, BCR is almost a mirror image of FCR. However, it works word by word. So, it can cut into major constituents. See Example (7), where we mark the right-peripheral PP/NP with square brackets.

(7) *Anne kam   [vor [1 ~~Uhr~~$_b$]$_{NP}$]$_{PP}$ und Susi ging [nach [3 Uhr]$_{NP}$]$_{PP}$.*
    Anne arrived before one ~~o'clock~~$_b$ and Susi left after three o'clock

Subject Gap in Clauses with Finite/Fronted Verb (SGF) is triggered by subject-verb-inversion in the first conjunct. If the sentence-initial constituent is not an argument, the subject in the second conjunct can be omitted (cf. Example (8)).

(8) *Warum bist   du gegangen und* ~~*du*~~$_s$ *hast mich nicht gewarnt?*
    Why   are$_{aux}$ you left   but ~~you~~$_s$ have me   not   warned
    'Why did you leave but didn't warn me?'

3 State of the Art in Large Language Models with Focus on Ellipsis Recognition/Reconstruction

Here, we focus on the state of the art in automatic ellipsis reconstruction based on Machine Learning approaches. In addition, we briefly address Transfer Learning here.

Zhang and colleagues introduced a dataset for the completion of ellipsis with an end-to-end pointer network. They found that the results of such completion models can be incorrect or incomplete. Their hybrid approach, which takes both the elliptical utterance and the completion into consideration, was evaluated for English by comparing two sequence-to-sequence models that were trained on their dataset. One model has an additional copying mechanism. These models achieve a BLEU score of .42 (without copying) and .72 (with copying) [30].

ELLie, a dataset, consisting only of elliptical utterances, was proposed by Testa and colleagues. It has been tested on several tasks with two Large Language Models: BERT [6] and GPT-2 [21]. In one of these tasks, the models were asked to return the missing verb in the elliptical sentence. For GPT-2, this was a text generation task; for BERT, it was a fill-mask task. In 55.8% of the examples, the correct word was the one with the highest probability, and in 32.5%, the correct verb was not one of the top five suggested words. The authors interpreted their results as evidence that LLMs have a general problem with the reconstruction of ellipsis [25].

Brabant and colleagues proposed a model for ellipsis and coreference detection based on DistilBERT [23]. They obtain an F1 score of 77 for ellipsis. They surmise that distinguishing between the specific kinds of ellipsis would improve the results [2].

Cavar and colleagues defined a large multi-lingual source of ellipsis data consisting of sentence pairs. Each pair consists of the elliptical utterance and the reconstructed sentence. All examples exhibit ellipsis. To test their dataset, a GPT-4 model [18] was trained on the ellipsis data and tested on multiple tasks. For ellipsis reconstruction, they report an accuracy of .25 in a zero-shot scenario. The authors conclude that the regular training of LLM does not enable them to deal with ellipsis [4].

With respect to our research question, we focus on the *Unified Text-To-Text Transfer Transformer (T5;* cf. [16] for a recent survey), which is suitable for all kinds of applications where the input and output are always text strings (*Transfer Learning*). This framework provides a consistent training objective for both pre-training and fine-tuning. Specifically, the model is trained with a maximum likelihood objective regardless of the task. To specify which task the model should perform, a task-specific text prefix is added to the original input sequence before it is fed to the model. In Sect. 5, we apply this model to existing CCE downstream data—as a proof of concept.

4 Data Collection and Analysis

We employ three datasets in our approach: (1) the dataset used to train a German T5 model, (2) the parallel ellipsis corpus of TüBa-D/Z [13], and (3) the new pre-release of another German treebank of newspaper text [9].

The GC4 corpus[1] is a subset of the Common Crawl[2] corpus containing German sentences. In the next section, we use thirty files of its 'head' part to train *GSmall* and *GBase*, since this part consists of high quality texts such as newspapers (number of sentences used for training in our system: about 8M).

The parallel TüBa-D/Z CCE corpus contains all sentences of the evaluation part of TüBa-D/Z [13]. Out of these 5,000 exemplars, 1,803 sentences contain at least one CCE occurrence (total data size used in for training: 1,442 after ignoring local CCE and further data cleaning). According to the encoding format, different gold standard options are provided in the corpus, so that the number of cases in Table 1 does not add up to 1,803. The authors argue with cases where a CCE phenomenon is not the only encoding option. For instance, local NP coordination can also be assumed. Example (14) from [13] (cf. Example (9) with two gold standards) illustrates such a case.

---

[1] https://German-nlp-group.github.io/projects/gc4-corpus.html.
[2] https://commoncrawl.org.

(9) *Sie   verloren* Arbeit, [~~sie   verloren~~]$_{fg}$ Wohnung und [~~sie   verloren~~]$_{fg}$ Familie.
They lost   work, [~~they lost~~]$_{fg}$ housing and [~~they lost~~]$_{fg}$ family.
Sie verloren [Arbeit, Wohnung, und Familie]$_{NP}$.

The "fg" encoding condenses two alternative CCE interpretations: (1) FCR, where the constituents that have been omitted in the left periphery cover the verbform, or (2) Gapping. In Gapping, however, the word order in the second conjunct can vary. But the corpus spells out only one possible order. This underspecification may interfere with the generalized learning of Gapping.

Additionally, we evaluated a pre-release of the parallel TIGER CCE corpus [9]. In fact, the TIGER treebank is merely half the size of TüBa-D/Z (about 105,000 sentences). However, explicit *secondary edges* in the tree structure facilitate a near-automatic reconstruction of the canonical forms with high accuracy. In addition, the TIGER treebank uses VP coordination with crossing edges supported by the encoding in XML. All these cases were (semi-)automatically expanded to the same encoding format as used in the parallel TüBa-D/Z CCE corpus. See Table 1 for the number of CCE occurrences in both parallel corpora to be used for training and testing.

Table 1. Number of sentences with individual CCE phenomena in the CCE recources.

Underlying treebank	Size	Local CCE	Gapping variants	FCR	BCR	SGF	Training set	Test set
TüBa-D/Z eval. part	5,000	988	286	507	71	91	1,442	361
TIGER	50,474	13,237	1,793	3,575	821	401	2,950	736

5 Transfer Learning of Ellipsis Reconstruction

In this section, we first present the implementation for our research question. In the evaluation section, we compare the results of the two CCE resources.

5.1 Training

After reviewing and briefly experimenting with state-of-the-art Large Language Models, we chose Google's T5 model (see, e.g., [22] or [16]). To the best of our knowledge, there is no off-the-shelf translation model that has been trained for German-to-German translation. Therefore, we first train two models *GSmall* with about 60M parameters and *GBase* with 200M parameters—according to the usual model checkpoints—by adding the prefix "translate German to German:" in front of every sentence in the dataset from GC4 (cf. the left half of the flowchart in Fig. 1). In turn, these models are fine-tuned for CCE reconstruction with the individual parallel CCE corpora[3]. In the following, we refer to the CCE corpora simply by "TüBa" and "TIGER", respectively.

During the pre-processing phase, we download the files corresponding to GC4 and then convert them into the required format for T5[4]. According to the released model checkpoints, we set up the two pre-trained models *GSmall* and *GBase* (in the figure, inherently addressed by the paths "T5-Small"/"T5-Base" pre-trained with CG4).

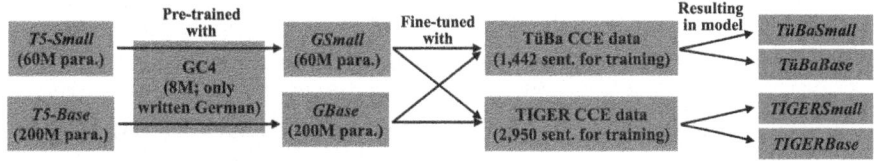

Fig. 1. Flowchart for obtaining our models for automatic CCE reconstruction.

In the next step, specific Huggingface datasets of the TüBa and TIGER resources are created[5]. For each corpus with randomized numbers of sentence pair, the top 80% of the data is used for training purposes, while the remaining 20% is reserved for testing. The same split ratio is applied to both CCE datasets, resulting in 1,442 sentences for training with the TüBa dataset, and 2,950 for TIGER, respectively.

The two parallel CCE corpora are used to fine-tune the German-to-German translation models *GSmall* and *GBase*, respectively, with the prefix: "reconstruct the ellipsis in this sentence:". Details of the training procedure can be found in the code. There, the parameters are set depending on the selected dataset, i.e., the model to be trained.

---

[3] In general, we use two libraries from Huggingface (cf. https://huggingface.co): (1) datasets (https://huggingface.co/docs/datasets), and (2) transformers (https://huggingface.co/docs/transformers). For the full implementation, see https://github.com/marisaschmidt/CCE-reconstruction on GitHub.

[4] We filter the data so that only sentences that are at least 98% German are considered. NLTK (cf. https://www.nltk.org) is used to tokenize the texts into sentences. For our purposes, the final sentence pairs are written to a jsonl file. This file consists of a column for sentence ids, a column for the sentences and a column for the gold standards, which is the same as the sentence in our case.

[5] First, a Pandas (cf. https://pandas.pydata.org) data frame is created, replacing NaN values with empty strings and converting all data to a string datatype. Next, each row in the data frame is converted to a string in json format.

5.2 Evaluation

Here, we discuss the results obtained for the two pre-trained models *GSmall* and *GBase*, fine-tuned with the CCE corpora Tüba and TIGER, resulting in *TüBaSmall* (TB_S), *TIGERSmall* (T_S), *TüBaBase* (TB_B) and *TIGERBase* (T_B), respectively (cf. Table 2[6]). For comparison purposes, we have kept the two resources separate here (both CCE corpora in combination as a larger training set improve the results to a BLEU score=.8349 for the Tüba test sentences and .7543 for TIGER, respectively).

Table 2. Results from training and testing the two corpora.

Mode	Model name	Training time (h)	BLEU train	BLEU test
Pre-trained (60M para.)	*GSmall*	86.5	.9995	not applicable
+ CCE from TüBa	TB_S	.5	.8557	**.8196**
+ CCE from TIGER	T_S	2.0	.7245	**.6093**
Pre-trained (200M para.)	*GBase*	172.5	.9929	not applicable
+ CCE from TüBa	TB_B	.75	**.8711**	.7986
+ CCE from TIGER	T_B	4.0	**.8487**	.5933

Here, one specific fact is particularly surprising. The *GSmall*-based models perform better—contrary to the expectation according to [22], where is stated that the larger models generally perform better. To get a better insight into how the learning performance changes over the number of epochs to help us diagnose problems with learning that may lead to an underfit/overfit model (cf. the differences between training and validation results), see the loss curves during training and evaluation for all four models in Fig. 2. In essence, the slopes look promising. Certainly, it is possible that additional epochs could potentially lead to improved results—this topic is on our list for future work.

Since we have kept the fine-tuning resources disjunct, we can test the two TIGER models with the full TüBa CCE dataset, and *vice versa*, the two TüBa models with TIGER. Interestingly, *TIGERSmall* achieves a BLEU score of .8169, and *TIGERBase* achieves a score of .7749 for testing with TüBa. Conversely, the TüBa models achieve lower BLEU scores on the Tiger dataset (*TüBaSmall*: .5218; *TüBaBase*: .5059). So, we plan to analyze the system behavior with a wider range of model checkpoints.

---

[6] We ran our system on a virtual machine with 700GB storage, 32GB RAM and a NVIDIA A100 graphics card with 40GB memory, which allowed us to take advantage of the optimized CUDA functionality offered by Huggingface.

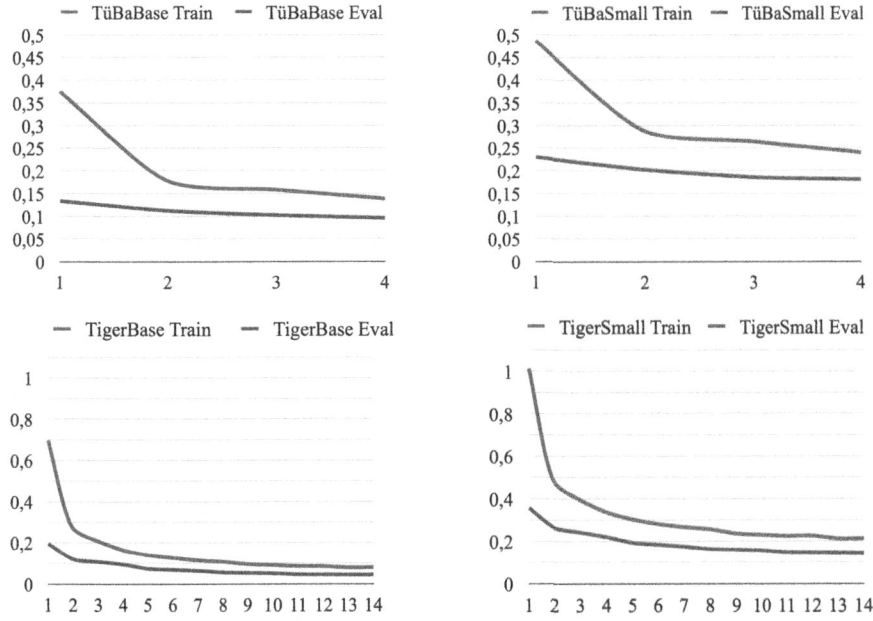

Fig. 2. Loss Curves for training and testing our four models.

For a more in-depth investigation, inspired by the suggestion in [2] to differentiate between the different kinds of ellipsis (cf. Sect. 3), we examine the BLEU scores per phenomenon. See Fig. 3, where we mark the four models with (M), and the two test sets with (T); additionally, we depict the BLEU scores for the examples from Sect. 2—which do not belong to any training or test set. Panel (a) shows at first glance that all four models cover the test set of Tüba much better than TIGER and the examples from Sect. 2 (in the data table, the best model is framed). Panel (b) highlights the same observation, however, grouped for each individual CCE phenomenon—in this bar chart, the x and y axes of the data table provided in Panel (a) are exchanged. Obviously, the ability of all four models to generalize is the highest for Tüba. FCR and SGF tested with TIGER perform considerably better than Gapping and BCR. Surprisingly, for the unseen prototypical CCE example sentences, the overall performance is the lowest. For an impression of the reconstruction quality, we delineate the examples from Sect. 2 in Table 3. FCR in the main clause—cf. Example (5)—worked in all models but *TüBaBase*. *TIGERSmall* produces reconstruction attempts for all examples. Perfectly reconstructed canonical forms (cf. *exact match (EM)* in the evaluation metrics of huggingface, where all labels must match exactly, including upper/lower case sensitivity, for the sample to be correctly classified in the test set) are highlighted in green. For the examples, $EM(T_S) = .35$, $EM(T_B) = .25$, and .0 for $TB_{B/S}$.

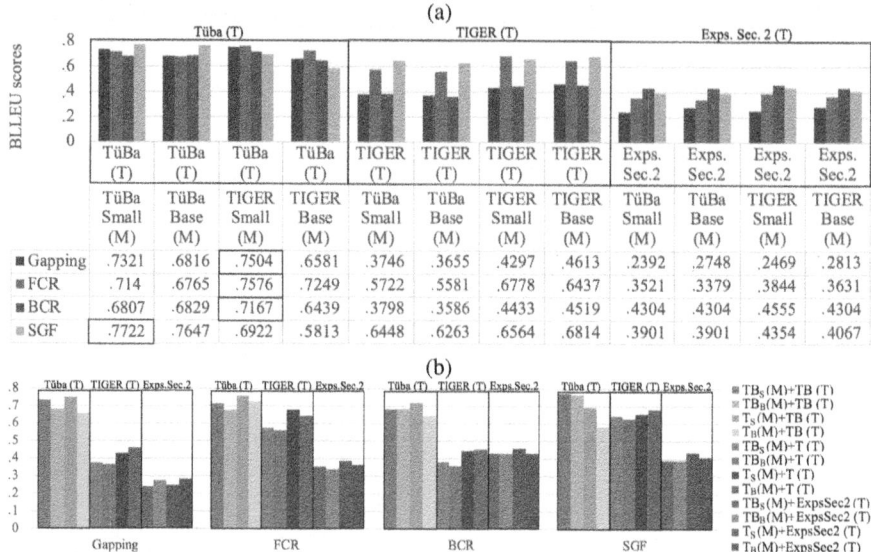

Fig. 3. Evaluations: Panel (a): per test corpus; Panel (b): per CCE phenomenon.

Table 3. Reconstructions for the eight example sentences in Sect. 2. Here, we only enumerate cases, where a reconstruction is attempted. So, *TüBaBase* (TB_B) is skipped here. Reconstructions that result in a perfect canonical form are highlighted in green.

ExNo	Prediction by the model	Error characteristics
(2_{TB_S})	Henry lebt in Boston und **Henry** alle seine Kinder auch.	Wrong FCR/Gapping of **Henry**, missing Gapping of **lebt** in the 2nd conjunct
(4_{TB_S})	Der Fahrer wurde getötet und **der Mitfahrer** schwer verletzt.	Forbidden change of number for the subject, although Subgapping of **wurde** is missing
(5_{TB_S})	Meine Schwester lebt in Berlin und **mein Schwester** arbeitet in Frankfurt.	Correct FCR including adaptation to lower case however, wrong morphological form **mein**
(1_{T_S})	Henry lebt in Boston und Peter lebt in Chicago.	Correct Gapping
(2_{T_S})	Henry lebt in Boston und **Henry lebt** alle seine Kinder auch.	Wrong FCR/Gapping of **Henry**, correct Gapping of **lebt**, however, in singular
(3_{T_S})	Meine Frau möchte ein Auto kaufen, mein Sohn **möchte** ein Motorrad.	Partially correct LDG of **möchte**
(4_{T_S})	Der Fahrer wurde getötet und **der Mitfahrer wurde** schwer verletzt.	Correct Subgapping, however, wrong morphological form (3rd Person, singular)
(5_{T_S})	Meine Schwester lebt in Berlin und **Meine Schwester** arbeitet in Frankfurt.	Correct FCR, however, preserving the capitalisation in the reconstructed possessive pronoun
(6_{T_S})	Amsterdam ist die Stadt, in der Jan lebt und in der Piet arbeitet.	Correct FCR of in der in the relative clause
(7_{T_S})	Anne kam vor 1 Uhr und Susi ging nach 3 Uhr.	Correct BCR of Uhr in the 1st conjunct
(8_{T_S})	Warum bist du gegangen und **warum** hast mich nicht gewarnt?	FCR of **warum** instead of SGF; thus, no subject in the 2nd conjunct
(1_{T_B})	Henry lebt in Boston und Peter lebt in Chicago.	Correct Gapping
(2_{T_B})	Henry lebt in Boston und **Henry lebt** alle seine Kinder auch.	Wrong FCR/Gapping of **Henry**, correct Gapping of **lebt**, however, wrong morphological form
(3_{T_B})	Meine Frau möchte ein Auto kaufen, mein Sohn **möchte** ein Motorrad.	Partially correct LDG of **möchte**
(4_{T_B})	Der Fahrer wurde getötet und die Mitfahrer **wurde** schwer verletzt.	Correct Subgapping, however, in singular
(5_{T_B})	Meine Schwester lebt in Berlin und meine Schwester arbeitet in Frankfurt.	Correct FCR including adaptation of meine to lower case
(8_{T_B})	Warum bist du gegangen und **warum** hast mich nicht gewarnt?	FCR of **warum** instead of SGF of **du**

6 Conclusions and Future Work

We have presented a proof of concept that CCE reconstruction in German can be realized as a so-called downstreaming task of a T5 model with small fine-tuning corpora. For a first evaluation, we pre-trained a German-to-German T5 model with 60M (Small) and 200M (Base) parameters. For our four models, we used two parallel corpora (TüBa with about 1,500 and TIGER with 3,000 sentence pairs). The evaluation looks promising. Specific next steps have been mentioned in the previous section. However, individual parallel CCE-specifically optimized models have to pay attention to the fact that several phenomena can be active at the same time (cf. *Meine Frau <u>möchte</u> ein Auto ~~kaufen~~$_b$, mein Sohn ~~möchte~~$_g$ ein Motorrad <u>kaufen</u>.*—the BCR+Subgapping variant of the canonical form of Example (3), prevailing in German). Probing ChatGPT with this sentence works well. However, *Kauf bitte ein Brot und wenn es Eier gibt, Acht.* 'Please buy a loaf of bread and if there are eggs, eight.' fails: *Kauf ein Brot. Wenn es frische Eier gibt, achte darauf.* (ill-formed imperative of 'to take care' vs. One-Anaphora).

In general, many suggestions have been made to optimize LLMs (see, e.g., [16]). For reasons of space, we only discuss *Masking* here (see, e.g., [28]). In more detail, Zeng and colleagues [29] or Wettig and colleagues [27] evaluate the effects for a wide range of linguistic phenomena, including ellipsis. As for future work, we want apply these findings to our system. Moreover, similar models for spoken German (including a new CCE data set for TüBa-D/S [24]) are under development.

References

1. Abeillé, A., Chaves, R.P.: Coordination. In: Müller, S., Abeillé, A., Borsley, R.D., Koenig, J.P. (eds.) Head-Driven Phrase Structure Grammar: The Handbook (Empirically Oriented Theoretical Morphology and Syntax), Berlin, Germany, pp. 725–776. Language Science Press (2021). https://doi.org/10.5281/zenodo.5599848
2. Brabant, Q., Rojas-Barahona, L.M., Gardent, C.: Active learning and multi-label classification for ellipsis and coreference detection in conversational question-answering. In: Procceedings of the 12th International Workshop on Spoken Dialog System Technology (IWSDS 2021) (2021). https://hal.science/hal-03533906/document
3. Brants, S., et al.: TIGER: linguistic interpretation of a German corpus. Res. Lang. Comput. **2**(4), 597–620 (2004)
4. Cavar, D., Mompelat, L., Abdo, M.: The typology of ellipsis: a corpus for linguistic analysis and machine learning applications. In: Proceedings of the 6th Workshop on Research in Computational Linguistic Typology and Multilingual NLP, St. Julian's, Malta, pp. 46–54 (2024). https://aclanthology.org/2024.sigtyp-1.6
5. van Craenenbroeck, J., Temmerman, T.: The Oxford Handbook of Ellipsis. Oxford University Press, Oxford (2018)
6. Devlin, J., Chang, M.W., Lee, K., Toutanova, K.: BERT: pre-training of deep bidirectional transformers for language understanding. In: Proceedings of the 2019 Conference of the North American Chapter of the Association for Computational Linguistics (ACL): Human Language Technologies (NAACL), pp. 4171–4186.

Minneapolis, Minnesota (2019). https://doi.org/10.18653/v1/N19-1423. https://aclanthology.org/N19-1423

7. Harbusch, K.: Incremental sentence production inhibits clausal coordinate ellipsis: a treebank study into Dutch and German. Dialogue Discourse **2**(1), 313–332 (2011). https://doi.org/10.5087/dad.2011.112

8. Harbusch, K., Kempen, G.: ELLEIPO: a module that computes coordinative ellipsis for language generators that don't. In: Proceedings of the 11th of Conference of the European Chapter of the ACL (EACL): Posters & Demonstrations, Trento, Italy, pp. 115–118 (2006). https://aclanthology.org/E06-2008/

9. Harbusch, K., Memmesheimer, D.: A parallel corpus for the TIGER treebank of written German with reconstructed omitted elements due to ellipsis in coordinated sentences. In: Proceedings of Conference on Form and Meaning of Coordination (FMC), Göttingen, Germany (2024)

10. Haspelmath, M.: Coordination. In: Shopen, T. (ed.) Language Typology and Linguistic Description, vol. 2, pp. 1–51. Cambridge University Press, Cambridge (2007)

11. Kempen, G.: Clausal coordination and coordinate ellipsis in a model of the speaker. Linguistics **47**(3), 653–696 (2009). https://doi.org/10.1515/LING.2009.022

12. Kempen, G., Huijbers, P.: The lexicalization process in sentence production and naming: indirect election of words. Cognition **14**(2), 185–209 (1983) https://doi.org/10.1016/0010-0277(83)90029-X. https://www.sciencedirect.com/science/article/pii/001002778390029X

13. Memmesheimer, D., Harbusch, K.: A German parallel clausal coordinate ellipsis corpus that aligns sentences from the TüBa-D/Z treebank with reconstructed canonical forms. In: Proceedings of the 26th International Conference on Text, Speech, and Dialogue (TSD), Pilsen, Czech Republic, pp. 116–128 (2023). https://www.springerprofessional.de/a-german-parallel-clausal-coordinate-ellipsis-corpus-that-aligns/25949856

14. Memmesheimer, D., Harbusch, K.: Exploring the feasibility of accurate reconstruction of clausal coordinate ellipsis in German. In: Proceedings of the 7th International Conference on Statistical Language and Speech Processing (EACB 2023), Amherst, MA, USA, July (2023)

15. Muhonen, K., Purtonen, T.: Rule-based detection of clausal coordinate ellipsis. In: Proceedings of the 8th International Conference on Language Resources and Evaluation (LREC 2012), Istanbul, Turkey, pp. 1955–1959 (2012). https://aclanthology.org/L12-1146/

16. Naveed, H., et al.: A comprehensive overview of large language models (2024). https://arxiv.org/abs/2307.06435

17. Nykiel, J., Kim, J.B.: Ellipsis. In: Müller, S., Abeillé, A., Borsley, R.D., Koenig, J.P. (eds.) Head-Driven Phrase Structure Grammar: The Handbook (Empirically Oriented Theoretical Morphology and Syntax), Berlin, Germany, pp. 847–888. Language Science Press (2021). https://doi.org/10.5281/zenodo.5599856

18. OpenAI, Achiam, J., et al.: (279 additional authors not shown), S.A.: GPT-4 technical report (2024). https://cdn.openai.com/papers/gpt-4.pdf

19. Papineni, K., Roukos, S., Ward, T., Zhu, W.J.: BLEU: a method for automatic evaluation of machine translation. In: Proceedings of the 40th Annual Meeting of the ACL, Philadelphia, Pennsylvania, USA, pp. 311–318 (2002). https://doi.org/10.3115/1073083.1073135. https://aclanthology.org/P02-1040

20. Patejuk, A.: Coordination. In: Dalrymple, M. (ed.) Handbook of Lexical Functional Grammar, Berlin, Germany, pp. 309–374. Language Science Press (2023). https://langsci-press.org/catalog/book/312

21. Radford, A., Wu, J., Child, R., Luan, D., Amodei, D., Sutskever, I.: Language models are unsupervised multitask learners (2019). https://api.semanticscholar.org/CorpusID:160025533
22. Raffel, C., et al.: Exploring the limits of transfer learning with a unified text-to-text transformer. J. Mach. Learn. Res. **21(140)**, 1–67 (2020). https://api.semanticscholar.org/CorpusID:204838007
23. Sanh, V., Debut, L., Chaumond, J., Wolf, T.: DistilBERT, a distilled version of BERT: smaller, faster, cheaper and lighter. arXiv: 1910.01108 (2020)
24. Stegmann, R., Telljohann, H., Hinrichs, E.W.: Stylebook for the German treebank in VERBMOBIL. Technical report, 239, DFKI, Saarbrücken, Germany (2000)
25. Testa, D., Chersoni, E., Lenci, A.: We understand elliptical sentences, and language models should too: a new dataset for studying ellipsis and its interaction with thematic fit. In: Proceedings of the 61st Annual Meeting of the ACL, Toronto, Canada, pp. 3340–3353 (2023). https://doi.org/10.18653/v1/2023.acl-long.188. https://aclanthology.org/2023.acl-long.188
26. Vaswani, A., et al.: Attention is all you need. In: Proceedings of the 31st Conference on Neural Information Processing Systems (NIPS 2017), Long Beach, CA, USA (2017). https://api.semanticscholar.org/CorpusID:13756489
27. Wettig, A., Gao, T., Zhong, Z., Chen, D.: Should you mask 15% in masked language modeling? In: Proceedings of the 17th Conference of the EACL, Dubrovnik, Croatia, pp. 2985–3000 (2023). https://doi.org/10.18653/v1/2023.eacl-main.217. https://aclanthology.org/2023.eacl-main.217
28. Wu, Y., Fang, K., Zhang, D., Wang, H., Zhang, H., Chen, G.: TLM: token-level masking for transformers. In: Proceedings of the 2023 Conference on Empirical Methods in Natural Language Processing (EMNLP), Singapore, pp. 14099–14111 (2023). https://doi.org/10.18653/v1/2023.emnlp-main.871. https://aclanthology.org/2023.emnlp-main.871
29. Zeng, G., Zhang, P., Lu, W.: One network, many masks: towards more parameter-efficient transfer learning. In: Proceedings of the 61st Annual Meeting of the ACL, Toronto, Canada, pp. 7564–7580 (2023). https://aclanthology.org/2023.acl-long.418/
30. Zhang, X., Li, C., Yu, D., Davidson, S., Yu, Z.: Filling conversation ellipsis for better social dialog understanding. In: Proceedings of the AAAI Conference on AI, New York, NY, USA, pp. 9587–9595 (2020). https://doi.org/10.1609/aaai.v34i05.6505. https://ojs.aaai.org/index.php/AAAI/article/view/6505

Better Low-Resource Machine Translation with Smaller Vocabularies

Edoardo Signoroni[✉] and Pavel Rychlý

NLP Centre, Faculty of Informatics, Masaryk University, Botanická 68a,
602 00 Brno, Czech Republic
e.signoroni@mail.muni.cz

Abstract. Data scarcity is still a major challenge in machine translation. The performance of state-of-the-art deep learning architectures, such as the Transformers, for under-resourced languages is well below the one for high-resourced languages. This precludes access to information for millions of speakers across the globe. Previous research has shown that the Transformer is highly sensitive to hyperparameters in low-resource conditions. One such parameter is the size of the subword vocabulary of the model. In this paper, we show that using smaller vocabularies, as low as 1k tokens, instead of the default value of 32k, is preferable in a diverse array of low-resource conditions. We experiment with different sizes on English-Akkadian, Lower Sorbian-German, English-Manipuri, to obtain models that are faster to train, smaller, and better performing than the default setting. These models achieve improvements of up to 322% ChrF score, while being up to 66% smaller and up to 17% faster to train.

Keywords: Low-resource · Neural Machine Translation · Tokenization

1 Introduction

End-to-end models based on neural architectures have supplanted most of the traditional natural language processing (NLP) pipeline components. Tokenization, however, is still required for almost all widely used models, which require a preprocessing stage to convert raw text into a sequence of discrete model inputs. This is valid also for machine translation, with neural machine translation (NMT) becoming the dominant approach, due to the performance of model based on encoder-decoder architectures, such as the Transformer [40]. However, these systems still struggle with the *open vocabulary* problem, where rare words [15] not represented in the training data may result in out-of-vocabulary (OOV) unknown tokens at inference time. Several solutions have been proposed to address this issue, with the most commonly used nowadays being data-driven algorithms such as Byte-Pair Encoding (BPE) [35] and SentencePiece [16] which automatically split strings based on frequencies in a training corpus, thus producing a more robust and generalizable vocabulary made of subwords. These subwords units can be compounded during generation to form previously unseen words. While a reasonable and effective improvement over white-space tokenization, subword methods are still incapable to handle complex linguistic structures, such as agglutinative morphology or challenging domains, such as informal text, containing typos, non-standard forms, variation, and the like [4].

To overcome these limitations, other work proposes character-level [18, 25, 39, 41], hybrid [20], and token-free approaches [4, 44], which work directly on the raw text. However, character-level models appear to show mixed results, performing better than subword ones when data and resources are sparse [3, 7], while only comparably or worse in other cases [19]. Moreover, these methods are computationally heavier, since the time complexity of the Transformer's attention layer grows quadratically with input length, which for character models is roughly four times the one of subword models. Thus, at the current moment, subword approaches remain the most employed in NLP and NMT [19], even in a low-resource setting.

Moreover, the choice of tokenization approach and its parameters, such as vocabulary size, has been show to impact the quality of downstream NMT [3, 6, 10, 36]. This is particularly relevant for low-resource conditions, where parallel data is sparse and often noisy [14, 32]. Therefore, to make the most of the available data, it is necessary to understand how tokenization behave in such conditions and how to best apply the current methods through careful selection of the tokenization algorithm and its parameters, such as vocabulary size. While some work [5, 10, 36] already investigated these problems, no definitive picture has been drawn for the optimal setting for these parameters [24]. This further reinforces the idea that, as Mielke et al. (2021) poses it, "there is [not] and likely will never be a silver bullet singular solution for all applications and that thinking seriously about tokenization remains important [...]".

Our paper investigates the impact of vocabulary size on three main aspects of MT models: i. the downstream quality of translations; ii. the size of the model; iii. the training speed. While the first is arguably the most important one, in low-resource scenarios, one can also be limited by the available hardware, thus the importance of fitting smaller and more efficient models. We experiment with three low-resource datasets (Lower Sorbian-German, English-Manipuri, English-Akkadian) to show that these three factors can be combined by reducing the size of the vocabulary.

After this Introduction, Sect. 2 summarizes some related work on tokenization (2.1) and its impact on downstream NMT (2.2). Section 3 describes the methodology of our experiments, while Sect. 4 reports on our results. Section 5 relates some conclusions and the future work that remains to be done.

2 Related Work

In this section, we first introduce the tokenization algorithm we employ in our experiments, then we briefly summarize relevant previous work on the impact of tokenization on NMT performance.

2.1 Tokenization

Sennrich et al. (2016) [35] adapts the BPE compression algorithm [9] to the task of word segmentation, thus the name of the tokenizer. The method initializes the vocabulary with characters, plus an end-of-word symbol. Then it replaces adjacent symbols with a new one representing that pair, iteratively merging all occurrences of the most frequent pair. These character n-grams are then merged together similarly until the desired number of merge operations is reached. The algorithm does not consider pairs that cross word boundaries.

SentencePiece [16] is a software package that implements BPE segmentation algorithms and removes the need for preprocessing, such as pretokenization. SentencePiece trains subword models directly from the raw sentences, by removing the spaces as word boundaries and allowing learned subwords to cross them. This enables segmentation for languages without whitespace-tokenized words, such as Chinese or Japanese. They also change the word continuation symbol @@ of the original BPE to the SentencePiece beginning-of-word _.

2.2 Impact on NMT

Other works investigate the impact of tokenization approaches on NMT. Regarding the vocabulary size, most studies use default values between 32k and 40k tokens, regardless of the characteristics of the dataset. However, Gowda and May (2020) [10] finds that each dataset has its own optimal vocabulary size. Moreover, Sennrich and Zhang (2019) [36] show that a smaller vocabulary size improves NMT quality in simulated low-resource scenarios. This was confirmed by work on African low-resource languages, such as the studies by Martinus and Abbott (2019) [22] and Rajab (2022) [31], and in further experiments by Carrión and Casacuberta (2022) [2] in simulated low-resource conditions and on actual low-resource language pairs. Attempts at finding the best vocabulary automatically, such as the one from Xu et al. (2021) [43], while effective, are still too demanding in terms of computational resources and time.

Recall from the Introduction that several approaches have been proposed to address tokenization, from character-level, to subword-level, and token-free methods. The proponents of these algorithms carry out their own evaluation to validate their approach on downstream NMT quality. Apart from the evaluations in their papers, Libovicky et al. [19] evaluates character-level MT, finding no improvements on subword MT apart from enhanced robustness to source-side noise and larger beam size at generation. Carrión and Casacuberta (2022) [2] shows the effectiveness of "quasi character-level" vocabularies, that is very small vocabularies obtained with subword algorithms run with a desired size of around 350 items. They show these tiny vocabularies to be beneficial when training data is scarce. Edman et al. (2023) [7] compares the performance of character-level the multilingual pretrained ByT5 [44] and subword-level mT5 [45] for MT. After fine-tuning them to a number of languages, they find that character-level models are competitive or superior to subword-level ones in many circumstances, particularly on unseen languages and when data are limited.

3 Methodology

In the following section, we explain our methodology, both in terms of the datasets employed and the experiments we carried out.

3.1 Datasets

Our experiments are carried out on three publicly available low-resource datasets, summarized in Table 1. The datasets are of varying sizes and contain text from different domains. They involve both high-resource languages, such as English and German, and a diverse selection of under-resourced languages.

Lower Sorbian (*Dolnoserbšćina*) is a West Slavic language spoken in Eastern Germany by around 7000 native speakers, most of these belong to the older generations however, and thus the language is severely endangered. It is written in Latin script, with added diacritic. Linguistically, Lower Sorbian has six grammatical cases and has the dual for nouns, pronouns, adjectives, and verbs. It uses no articles. The dataset used in our experiments was curated by the Witaj Sprachzentrum[1] (Witaj Language Centre) [21].

Manipuri (*Meiteilon*) is a Tibeto-Burman language official in the Manipuri state of India and also at federal level. It is spoken natively by 1.8 million people, the Meitei, both in Manipur and in the neighbouring states. It is considered "vulnerable" by the UNESCO. Manipuri has an extensive suffix with limited prefixation and SVO word order. Other features of this language include agglutinative verb morphology, tone, the absence of grammatical person, number, gender, and a prevalence of aspect over tense [27]. Manipuri employs a wide array of writing systems, the official ones being the Meitei script and the Bengali script, which is the writing system used for all the Manipuri data in our experiments. The Latin script is also used. The corpus is a modified version [27] of the work by Haddow and Kirefu (2020) [12], Laitonjam and Ranbir Singh (2021) [17], and Huidrom et al. (2021) [13]. Each split of the dataset consists mostly of news or otherwise informative text.

Akkadian is an extinct East Semitic language that was used in ancient Mesopotamia from the third millennium BCE to the 1st century CE. It is written in the cuneiform script, which has a complex logophonetic nature: its symbols may be used as logograms, determinatives, or phonograms/syllabograms, with each of this function resulting in a different reading of the symbol. The data comes from parts of the ORACC corpus[2] and mostly consists of Neo-Assyrian royal inscriptions and administrative letters. Due to the nature of the source (i.e. clay tablets), data is incomplete and sentences may be truncated. Another issue is the wide stylistic difference between genres [11]. Akkadian is a fusional language with grammatical case and uses a system of consonantal roots.

---

[1] https://www.witaj-sprachzentrum.de/.
[2] https://oracc.museum.upenn.edu/index.html.

Table 1. Summary of the datasets in our experiments. The columns report the languages in the dataset, its original source, and the number of training examples.

Languages	Abbreviation	Dataset	N. of Pairs
English-Akkadian	eng-akk	EvaCun 2023	45269
Lower Sorbian-German	dsb-deu	WMT22 Low-res shared Task	40194
English-Manipuri	eng-mni	WMT23 Indic Shared Task	21287

3.2 Experiments

We conduct our experiment using BPE [35], as implemented in SentencePiece [16]. We learn separated vocabularies for source and target, with a range of six sizes of the vocabulary: 1k, 2k, 4k, 8k, 16k, 32k. It is common to use a vocabulary size of 32k-40k [5,43]

We evaluate the impact of the settings mentioned above by training several models with Fairseq [26]. For both datasets, we train only on the given parallel data, without any additional step apart from tokenization. This was possible because the test sets of the dataset are in-domain. The systems are based on the Transformer architecture [40], but smaller than its *base* version. All models were trained until BLEU score [28] on the validation set did not increase for 10 consecutive epochs or until 12k updates were reached. The training hyperparameters are summarized in Table 2. We did not perform extensive hyperparameter search, since the focus of our study is the size of the subword vocabulary. Instead, we started from the parameters tested by previous research on low-resource machine translation [1,37,38] and then experimented with our own tweaks. All models were trained locally on a single Nvidia A40 GPU.

Table 2. Hyperparameters for the models in our experiments.

enc/dec layers	4
enc/dec embedding dimension	256
enc/dec feed-forward dimension	1024
enc/dec attention heads	4
optimizer	adam
adam-betas	0.9, 0.98
learning rate	1e–4
learning rate scheduler	inverse square root
warmup updates	4000
dropout	0.1
label smoothing	0.1
max tokens	16000

After training, we generate test set translations for each model and obtain sentence-level BLEU [28], ChrF [29], ChrF++ [30], and COMET [33] scores as implemented in Hugging Face `evaluate` library. We employ bootstrap evaluation on 200 batches of 400 test sentences to obtain the final scores.

Mathur et al. (2020) [23] argue for the retirement of BLEU in favour of ChrF++. We keep BLEU scores to allow comparisons with previous research. Sai B. et al. (2023) [34] finds that ChrF++ performs the best among overlap metrics for a selection of Indic languages. The results of recent WMT Metrics shared tasks [8] demonstrate that learned neural metrics are the most optimal. Among these, COMET is the current state-of-the-art, and is widely employed in machine translation studies. However, pretrained neural metrics are unreliable for unseen languages, especially under-resourced ones. Works such as the ones by Sai B. et al. (2023) [34] and Wang et al. (2024) [42] show that fine-tuned COMET models perform better for specific sets of low-resource languages, than baseline models. For these reasons, we chose ChrF as the metric of reference for our experiments.

From the training logs, we retrieved the model size in millions of parameters, and the elapsed time for each epoch in seconds. We then compute a per-epoch average for each model.

4 Results

Table 3. Full results of the experiments. For each language pair and vocabulary size, the first four columns report the BLEU, ChrF, ChrF++, and COMET scores. The last two columns give the size of the models in million of parameters and the average training speed in seconds per epoch.

Source	Target	Voc. Size	BLEU	ChrF	Chrf++	COMET	Params	sec/epoch
eng	**akk**	1k	**28.41**	**34.425**	**29.782**	**0.932**	7.885	27.987
		2k	25.389	26.764	23.656	0.919	8.393	23.200
		4k	23.878	20.438	18.562	0.918	9.398	**21.514**
		8k	24.294	19.530	17.878	0.917	11.373	21.622
		16k	22.155	12.472	12.294	0.848	15.104	21.857
		32k	23.641	13.603	13.206	0.881	21.115	25.844
dsb	**deu**	1k	19.867	35.755	35.462	0.485	7.889	15.123
		2k	20.798	38.011	37.493	0.505	8.397	13.627
		4k	24.337	42.621	42.110	0.54	9.417	**13.219**
		8k	**26.932**	45.434	44.851	**0.571**	11.434	13.313
		16k	26.360	**45.468**	**45.019**	0.563	15.387	13.402
		32k	26.670	45.248	44.979	0.564	22.784	15.807
eng	**mni**	1k	10.824	36.171	32.412	0.643	7.885	16.452
		2k	12.538	38.952	35.180	0.664	8.395	14.905
		4k	**14.12**	**40.878**	**37.213**	**0.678**	9.400	**14.203**
		8k	2.663	10.604	9.400	0.412	11.366	14.63
		16k	2.845	11.090	9.930	0.408	15.03	14.839
		32k	3.019	9.685	8.733	0.391	20.363	17.276

The results of our experiments are summarized in Table 3. Through this section, we will refer to the models we trained using their vocabulary size, the only parameter that differs among them. We will compare the default setting of 32k to the best performing model for each language pair.

For English-Akkadian the best model overall is the 1k with 34.425 ChrF (+20.822), 7.885M parameters (−13.23M parameters), and an average per-epoch train time of 27.987 s (+2.143). It is thus 153.07% better than the default, while being 62.66% smaller. It is, however, 8.29% slower to train, due to the high percentage of single character tokens in the vocabulary.

For Lower Sorbian-German, the 8k, 16k, and 32k models achieve around 45 ChrF. The 8k, however, is significantly smaller with 11.434M parameters (−11.35M), and faster to train with 13.313 s per-epoch (−2.49). Overall, the 8k is 49.82% smaller and 15.78% faster than the 32k, while performing similarly to bigger models. Moreover, while training of the 16k and the 32k ended due to early stopping, the 8k reached the maximum numbers of updates while train loss was still decreasing. Further training may have led to a better model also in terms of translation quality.

For English-Manipuri we found the best model to be the 4k. It reaches a ChrF score of 40.878 (+31.193), with 9.4M parameters and an average train time of 14.203 s per epoch. These numbers make the 4k 322.08% better, 53.86% smaller, and 17.79% faster than the default 32k. It is important to note that in the case of English-Manipuri the models from 8k to 32k failed to reach meaningful representations during training early on, despite training being attempted multiple times with different random seeds. This further reinforces the idea that vocabulary size is a crucial parameter that has to be tuned in order to obtain functional machine translation models in low-resource scenarios.

5 Conclusions

Our paper explored how the choice of tokenization vocabulary size influences machine translation quality in low-resource scenarios. We experimented with three datasets of different sizes and domains, containing both high (English, German) and under-resourced languages (Akkadian, Manipuri, Lower Sorbian). The results of our work show that a careful selection of the vocabulary size is a crucial step when developing machine translation systems for under-resourced languages. The choice of an optimal vocabulary size can indeed lead to much better and more efficient machine translation models. Conversely, including too many items in the vocabulary, as is the case with the commonplace value size of 32k, can be detrimental or even prevent meaningful training. While there appears to be no "one-fits-all" setting for the vocabulary size for low-resource datasets, since it may depend on its features, our results show that the best setting most probably rests on the smaller side of the spectrum.

6 Limitations and Future Work

Admittedly, our work has some limitations.

First, the sample of languages and datasets is tiny with respect to the 7000+ living languages of the world. Faced with the clear impossibility of covering all of them, we tried to focus on datasets relevant to real world use-cases, such as information gathering from English to an under-resourced language (English-Manipuri), language and knowledge preservation (Lower Sorbian-German); or that are challenging for MT due to linguistic and data characteristics (English-Akkadian).

Second, the scope of our experiments is also limited with respect to the architectures and training parameters of the model, which could influence the downstream quality of the translation. The focus of the present work is the impact of tokenization parameters on the model's output, we thus decided to use only the architecture which in our previous experience generally worked best in a low-resource setting.

Lastly, we were not able to perform a deep qualitative analysis of the output, and we had to rely on automated metrics to estimate the quality of the translation. We tried to offset this setback by employing several metrics, and by selecting the most reliable according to previous research.

Future work will involve a deeper look at the structure of the datasets, its internal variability and sequence length, and how it influences the optimal size of the vocabulary. We also plan to extend our analysis to the impact of vocabulary composition, with the aim to further reduce the size and train times of the models.

Acknowledgments. The work described herein has been supported by the Ministry of Education, Youth and Sports of the Czech Republic, Project No. LM2023062 LIN-DAT/ CLARIAH-CZ and by Lexical Computing CZ s.r.o.

References

1. Araabi, A., Monz, C.: Optimizing transformer for low-resource neural machine translation. In: Proceedings of the 28th International Conference on Computational Linguistics, pp. 3429–3435. International Committee on Computational Linguistics, Barcelona (Online) (2020). https://doi.org/10.18653/v1/2020.coling-main.304. https://aclanthology.org/2020.coling-main.304
2. Carrión-Ponz, S., Casacuberta, F.: On the effectiveness of quasi character-level models for machine translation. In: Proceedings of the 15th biennial conference of the Association for Machine Translation in the Americas, vol. 1: Research Track, pp. 131–143. Association for Machine Translation in the Americas, Orlando (2022). https://aclanthology.org/2022.amta-research.10
3. Cherry, C., Foster, G., Bapna, A., Firat, O., Macherey, W.: Revisiting character-based neural machine translation with capacity and compression. In: Proceedings of the 2018 Conference on Empirical Methods in Natural Language Processing, pp. 4295–4305. Association for Computational Linguistics, Brussels (2018). https://doi.org/10.18653/v1/D18-1461, https://aclanthology.org/D18-1461

4. Clark, J.H., Garrette, D., Turc, I., Wieting, J.: Canine: pre-training an efficient tokenization-free encoder for language representation. Trans. Assoc. Comput. Linguist. **10**, 73–91 (2022)
5. Ding, S., Renduchintala, A., Duh, K.: A call for prudent choice of subword merge operations in neural machine translation. In: Proceedings of Machine Translation Summit XVII: Research Track, pp. 204–213. European Association for Machine Translation, Dublin (2019). https://aclanthology.org/W19-6620
6. Domingo, M., García-Martínez, M., Helle, A., Casacuberta, F., Herranz, M.: How much does tokenization affect neural machine translation? In: Computational Linguistics and Intelligent Text Processing: 20th International Conference, CICLing 2019, La Rochelle, France, 7–13 April 2019, Revised Selected Papers, Part I, pp. 545–554. Springer, Heidelberg (2019). https://doi.org/10.1007/978-3-031-24337-0_38
7. Edman, L., Toral, A., van Noord, G.: Are character-level translations worth the wait? an extensive comparison of character- and subword-level models for machine translation (2023)
8. Freitag, M., et al.: Results of WMT22 metrics shared task: stop using BLEU – neural metrics are better and more robust. In: Proceedings of the Seventh Conference on Machine Translation (WMT), pp. 46–68. Association for Computational Linguistics, Abu Dhabi (Hybrid) (2022). https://aclanthology.org/2022.wmt-1.2
9. Gage, P.: A new algorithm for data compression. C Users J. **12**(2), 23–38 (1994)
10. Gowda, T., May, J.: Finding the optimal vocabulary size for neural machine translation. In: Findings of the Association for Computational Linguistics: EMNLP 2020, pp. 3955–3964. Association for Computational Linguistics (2020). https://doi.org/10.18653/v1/2020.findings-emnlp.352. https://aclanthology.org/2020.findings-emnlp.352
11. Gutherz, G., Gordin, S., Sáenz, L., Levy, O., Berant, J.: Translating Akkadian to English with neural machine translation. PNAS Nexus **2**(5), pgad096 (2023). https://doi.org/10.1093/pnasnexus/pgad096
12. Haddow, B., Kirefu, F.: Pmindia – a collection of parallel corpora of languages of India (2020)
13. Huidrom, R., Lepage, Y., Khomdram, K.: EM corpus: a comparable corpus for a less-resourced language pair Manipuri-English. In: Proceedings of the 14th Workshop on Building and Using Comparable Corpora (BUCC 2021), pp. 60–67. INCOMA Ltd., Online (2021). https://aclanthology.org/2021.bucc-1.8
14. Joshi, P., Santy, S., Budhiraja, A., Bali, K., Choudhury, M.: The state and fate of linguistic diversity and inclusion in the NLP world. In: Proceedings of the 58th Annual Meeting of the Association for Computational Linguistics, pp. 6282–6293. Association for Computational Linguistics, Online (2020). https://doi.org/10.18653/v1/2020.acl-main.560. https://aclanthology.org/2020.acl-main.560
15. Koehn, P., Knowles, R.: Six challenges for neural machine translation. In: Proceedings of the First Workshop on Neural Machine Translation, pp. 28–39. Association for Computational Linguistics, Vancouver (2017). https://doi.org/10.18653/v1/W17-3204. https://aclanthology.org/W17-3204
16. Kudo, T., Richardson, J.: SentencePiece: a simple and language independent subword tokenizer and detokenizer for neural text processing. In: Proceedings of the 2018 Conference on Empirical Methods in Natural Language Processing: System Demonstrations, pp. 66–71. Association for Computational Linguistics, Brussels (2018). https://doi.org/10.18653/v1/D18-2012. https://aclanthology.org/D18-2012

17. Laitonjam, L., Ranbir Singh, S.: Manipuri-English machine translation using comparable corpus. In: Proceedings of the 4th Workshop on Technologies for MT of Low Resource Languages (LoResMT2021), pp. 78–88. Association for Machine Translation in the Americas, Virtual (2021). https://aclanthology.org/2021.mtsummit-loresmt.8
18. Lee, J., Cho, K., Hofmann, T.: Fully character-level neural machine translation without explicit segmentation. Trans. Assoc. Comput. Linguist. **5**, 365–378 (2017)
19. Libovický, J., Schmid, H., Fraser, A.: Why don't people use character-level machine translation? In: Findings of the Association for Computational Linguistics: ACL 2022, pp. 2470–2485. Association for Computational Linguistics, Dublin (2022). https://doi.org/10.18653/v1/2022.findings-acl.194. https://aclanthology.org/2022.findings-acl.194
20. Luong, M.T., Manning, C.D.: Achieving open vocabulary neural machine translation with hybrid word-character models. In: Proceedings of the 54th Annual Meeting of the Association for Computational Linguistics, vol. 1: Long Papers, pp. 1054–1063. Association for Computational Linguistics, Berlin, Germany (2016). https://doi.org/10.18653/v1/P16-1100. https://aclanthology.org/P16-1100
21. Weller-di Marco, M., Fraser, A.: Findings of the WMT 2022 shared tasks in unsupervised MT and very low resource supervised MT. In: Proceedings of the Seventh Conference on Machine Translation (WMT), pp. 801–805. Association for Computational Linguistics, Abu Dhabi (Hybrid) (2022). https://aclanthology.org/2022.wmt-1.73
22. Martinus, L., Abbott, J.Z.: A focus on neural machine translation for African languages (2019)
23. Mathur, N., Baldwin, T., Cohn, T.: Tangled up in BLEU: Reevaluating the evaluation of automatic machine translation evaluation metrics. In: Proceedings of the 58th Annual Meeting of the Association for Computational Linguistics, pp. 4984–4997. Association for Computational Linguistics, Online (2020). https://doi.org/10.18653/v1/2020.acl-main.448. https://aclanthology.org/2020.acl-main.448
24. Mielke, S.J., et al.: Between words and characters: a brief history of open-vocabulary modeling and tokenization in NLP. arXiv preprint arXiv:2112.10508 (2021)
25. Neubig, G., Watanabe, T., Mori, S., Kawahara, T.: Substring-based machine translation. Mach. Transl. **27**(2), 139–166 (2013). http://www.jstor.org/stable/42628800
26. Ott, M., et al.: fairseq: a fast, extensible toolkit for sequence modeling. In: Proceedings of the 2019 Conference of the North American Chapter of the Association for Computational Linguistics (Demonstrations), pp. 48–53. Association for Computational Linguistics, Minneapolis (2019). https://doi.org/10.18653/v1/N19-4009. https://aclanthology.org/N19-4009
27. Pal, S., et al.: Findings of the WMT 2023 shared task on low-resource Indic language translation. In: Koehn, P., Haddow, B., Kocmi, T., Monz, C. (eds.) Proceedings of the Eighth Conference on Machine Translation, pp. 682–694. Association for Computational Linguistics, Singapore (2023). https://doi.org/10.18653/v1/2023.wmt-1.56. https://aclanthology.org/2023.wmt-1.56
28. Papineni, K., Roukos, S., Ward, T., Zhu, W.J.: Bleu: a method for automatic evaluation of machine translation. In: Proceedings of the 40th Annual Meeting of the Association for Computational Linguistics, pp. 311–318. Association for Computational Linguistics, Philadelphia (Jul 2002). https://doi.org/10.3115/1073083.1073135. https://aclanthology.org/P02-1040

29. Popović, M.: chrF: character n-gram F-score for automatic MT evaluation. In: Proceedings of the Tenth Workshop on Statistical Machine Translation, pp. 392–395. Association for Computational Linguistics, Lisbon (2015). https://doi.org/10.18653/v1/W15-3049. https://aclanthology.org/W15-3049
30. Popović, M.: chrF++: words helping character n-grams. In: Proceedings of the Second Conference on Machine Translation, pp. 612–618. Association for Computational Linguistics, Copenhagen (2017). https://doi.org/10.18653/v1/W17-4770. https://aclanthology.org/W17-4770
31. Rajab, J.: Effect of tokenisation strategies for low-resourced southern african languages. In: 3rd Workshop on African Natural Language Processing (2022). https://openreview.net/forum?id=SpMeq5M48W9
32. Ranathunga, S., Lee, E.S.A., Prifti Skenduli, M., Shekhar, R., Alam, M., Kaur, R.: Neural machine translation for low-resource languages: a survey. ACM Comput. Surv. **55**(11) (2 2023). https://doi.org/10.1145/3567592
33. Rei, R., Stewart, C., Farinha, A.C., Lavie, A.: COMET: a neural framework for MT evaluation. In: Proceedings of the 2020 Conference on Empirical Methods in Natural Language Processing (EMNLP), pp. 2685–2702. Association for Computational Linguistics, Online (2020). https://doi.org/10.18653/v1/2020.emnlp-main.213. https://aclanthology.org/2020.emnlp-main.213
34. Sai B, A., et al.: IndicMT eval: a dataset to meta-evaluate machine translation metrics for Indian languages. In: Rogers, A., Boyd-Graber, J., Okazaki, N. (eds.) Proceedings of the 61st Annual Meeting of the Association for Computational Linguistics, vol. 1: Long Papers, pp. 14210–14228. Association for Computational Linguistics, Toronto (2023). https://doi.org/10.18653/v1/2023.acl-long.795. https://aclanthology.org/2023.acl-long.795
35. Sennrich, R., Haddow, B., Birch, A.: Neural machine translation of rare words with subword units. In: Proceedings of the 54th Annual Meeting of the Association for Computational Linguistics, vol. 1: Long Papers, pp. 1715–1725. Association for Computational Linguistics, Berlin (2016). https://doi.org/10.18653/v1/P16-1162. https://aclanthology.org/P16-1162
36. Sennrich, R., Zhang, B.: Revisiting low-resource neural machine translation: a case study. In: Proceedings of the 57th Annual Meeting of the Association for Computational Linguistics, pp. 211–221. Association for Computational Linguistics, Florence (2019). https://doi.org/10.18653/v1/P19-1021. https://aclanthology.org/P19-1021
37. Signoroni, E., Rychlý, P.: MUNI-NLP systems for Lower Sorbian-German and Lower Sorbian-Upper Sorbian machine translation WMT22. In: Proceedings of the Seventh Conference on Machine Translation (WMT), pp. 1111–1116. Association for Computational Linguistics, Abu Dhabi (Hybrid) (2022). https://aclanthology.org/2022.wmt-1.109
38. Signoroni, E., Rychly, P.: MUNI-NLP systems for low-resource Indic machine translation. In: Koehn, P., Haddow, B., Kocmi, T., Monz, C. (eds.) Proceedings of the Eighth Conference on Machine Translation, pp. 959–966. Association for Computational Linguistics, Singapore (2023). https://doi.org/10.18653/v1/2023.wmt-1.91. https://aclanthology.org/2023.wmt-1.91
39. Tay, Y., et al.: Charformer: fast character transformers via gradient-based subword tokenization. In: International Conference on Learning Representations (2022). https://openreview.net/forum?id=JtBRnrlOEFN
40. Vaswani, A., et al.: Attention is all you need. In: Proceedings of the 31st International Conference on Neural Information Processing Systems, NIPS 2017, pp. 6000–6010. Curran Associates Inc., Red Hook (2017)

41. Vilar, D., Peter, J.T., Ney, H.: Can we translate letters? In: Proceedings of the Second Workshop on Statistical Machine Translation, pp. 33–39. Association for Computational Linguistics, Prague (2007). https://aclanthology.org/W07-0705
42. Wang, J., et al.: Afrimte and africomet: enhancing comet to embrace under-resourced African languages (2024)
43. Xu, J., Zhou, H., Gan, C., Zheng, Z., Li, L.: Vocabulary learning via optimal transport for neural machine translation. In: Proceedings of the 59th Annual Meeting of the Association for Computational Linguistics and the 11th International Joint Conference on Natural Language Processing, vol. 1: Long Papers, pp. 7361–7373. Association for Computational Linguistics, Online (2021). https://doi.org/10.18653/v1/2021.acl-long.571. https://aclanthology.org/2021.acl-long.571
44. Xue, L., et al.: ByT5: Towards a token-free future with pre-trained byte-to-byte models. Trans. Assoc. Comput. Linguist. **10**, 291–306 (2022)
45. Xue, L., et al.: mT5: a massively multilingual pre-trained text-to-text transformer. In: Proceedings of the 2021 Conference of the North American Chapter of the Association for Computational Linguistics: Human Language Technologies, pp. 483–498. Association for Computational Linguistics, Online (2021). https://doi.org/10.18653/v1/2021.naacl-main.41. https://aclanthology.org/2021.naacl-main.41

Bella Turca: A Large-Scale Dataset of Diverse Text Sources for Turkish Language Modeling

Duygu Altinok[✉]

Deepgram Research, San Francisco, USA
duygu.altinok@deepgram.com

Abstract. In recent studies, it has been demonstrated that incorporating diverse training datasets enhances the overall knowledge and generalization capabilities of large-scale language models, especially in cross-domain scenarios. In line with this, we introduce Bella Turca: a comprehensive Turkish text corpus, totaling 265GB, specifically curated for training language models. Bella Turca encompasses 25 distinct subsets of 4 genre, carefully chosen to ensure diversity and high quality. While Turkish is spoken widely across three continents, it suffers from a dearth of robust data resources for language modelling. Existing transformers and language models have primarily relied on repetitive corpora such as OSCAR and/or Wiki, which lack the desired diversity. Our work aims to break free from this monotony by introducing a fresh perspective to Turkish corpora resources. To the best of our knowledge, this release marks the first instance of such a vast and diverse dataset tailored for the Turkish language. Additionally, we contribute to the community by providing the code used in the dataset's construction and cleaning, fostering collaboration and knowledge sharing.

Keywords: Turkish corpus · Turkish dataset · Turkish language modeling · Turkish NLP

1 Introduction

Recent advancements in general-purpose language modeling have showcased the effectiveness of training large-scale models on extensive text corpora for various downstream applications [5,25,26,29]. As the field continues to push the boundaries of language model training, there is a growing demand for high-quality, massive text datasets [18].

To meet the data requirements for language modeling, earlier large language models (LLMs) heavily relied on the Common Crawl dataset as their primary or sole source of training data [5,26]. While training on the Common Crawl has proven effective, recent studies have demonstrated that dataset diversity contributes to improved generalization capabilities in downstream tasks [27].

Furthermore, large language models have shown the ability to acquire knowledge effectively in new domains with relatively small amounts of domain-specific training data [5,27]. These findings suggest that by combining numerous smaller high-quality datasets from diverse sources, we can enhance the cross-domain knowledge and downstream generalization of models compared to those trained on only a limited set of data sources.

Addressing the need for dataset diversity, the Pile dataset was introduced and made available to the community [11]. The Pile consists of 22 diverse and high-quality datasets, including genres such as books, web crawls, Arxiv papers, and more, totaling 825GB. Several successful LLMs [4,22,29,39,40] have been trained on the Pile, many of which have outperformed competing models with similar architectures.

Following the success of the Pile, the trend of utilizing diverse datasets continued. Although not openly released, MassiveText [9] was compiled by Google as a collection of large English-language text datasets from multiple sources, including web pages, books, news articles, and code, to support projects like Gopher [9] and Chinchilla [15]. While the training corpus is not published, the PaLM 2 [33] paper describes their dataset as comprising a diverse set of sources, including web documents, books, code, mathematics, and conversational data. The open-source LLM, LLaMa [17], is available, although its training corpus is not openly accessible. However, it is described in [36] as a mixture of several sources, including books, web crawls, Arxiv papers, GitHub code, and Wikipedia. Recently, an open-source counterpart to LLaMa, called the RedPajama dataset [8], has been created. BLOOM [3] is powered by the extensive 1.6TB ROOTS dataset [16], which is not only openly released but also accompanied by data cleaning, deduplication, and document filtering code. Lastly, Dolma, another massive dataset comprising 3 trillion tokens from multiple resources, has been released [30]. All of these developments indicate that the future direction for LLMs involves not only utilizing more data but also emphasizing clean, curated, and diverse datasets.

When it comes to non-English models and datasets, the availability of diverse datasets for many languages is still limited, and as a result, there are not many large language models trained on such datasets. However, there are several mid-size language models pretrained on diverse datasets for certain non-English languages. For instance, the Finnish BERT model introduced in [37] is trained on news, online discussions, and internet crawl data. The Brazilian Portuguese transformer model BERTimbau [31] is trained on brWaC [38], which is a curated dataset from the web, ensuring diversity and quality through filtering and cleaning. The Spanish BERT model introduced in [7] utilizes training corpora from various sources, including United Nations and Government journals, TED Talks, subtitles, and news stories. Finally, FlauBERT [20], a French transformer model, is trained on a corpus consisting of 24 sub-corpora gathered from different sources, covering diverse topics and writing styles, ranging from formal texts such as Wikipedia and books to random text crawled from the internet (e.g.,

Common Crawl). These examples demonstrate that even mid-size models can benefit from dataset diversity.

However, when it comes to Turkish models, the available resources are scarce and not very diverse. The first and most widely used Turkish transformer model, BERTurk [28], was trained on the OSCAR corpus [1] and several OPUS corpora [35]. Later, the authors trained an ELECTRA model on mC4. Another recent work by [19] trained BERT models of varying sizes (small, medium, and large) and described their corpus as a diverse dataset totaling over 75GB in size. However, this dataset is not openly accessible, and details regarding the compilation, cleaning, and other aspects are unknown. Another recent work by [21] focused on training a medical domain BERT but only utilized BERTurk's dataset, which is 35GB in size. Similarly, a recent work by [32] trained a RoBERTa model [23] for Turkish using the OSCAR and mC4 datasets.

As evident from the situation described, the absence of easily accessible, large, and diverse corpora presents a challenge for researchers. This pushes them to either rely on existing limited corpora such as OSCAR and/or mC4, or engage in repetitive efforts to compile corpora without the benefit of knowledge sharing. In summary, the availability of only two significant corpora, OSCAR and mC4, is quite unfortunate for several reasons. Firstly, the lack of diversity is a major drawback, as both corpora are derived from web crawls. Secondly, the size of these corpora is also disappointing, as indicated by the previously mentioned papers, where transformers trained on these resources were limited to just 35GB of data.

To address these limitations, we present a large-scale, diverse, and curated dataset to the Turkish NLP research community. Taking inspiration from the Pile dataset, which is large and diverse, we have compiled several sub-corpora from diverse genres, including existing datasets and newly created ones. We introduce new datasets from genres such as books, academic papers, a curated web crawl, forum conversations, and customer reviews about commercial products. We have also processed the OSCAR datasets and mC4 to ensure quality and included them in our compilation.

We also provide publicly accessible preprocessing code for the constituent datasets of Bella Turca[1], ensuring reproducibility and enabling others to create similar datasets.

Our main contributions can be summarized as follows:

---

[1] The dataset's name, "Bella Turca," represents a blend of Latin and Italian, signifying "Beautiful Turkish Language." Originally, our intention was to have a name in Latin. However, we opted to incorporate "Bella" from Italian, a Romance language, in order to emphasize the elegance, richness, and aesthetic appeal of the Turkish language. For the second word, we sought a term that would harmonize with "Bella." We decided against "lingua Turcica" to avoid excessive length. Instead, we substituted "person" with "language," as it posed no harm, resulting in either "Turca" or "Turcus," denoting "Turkish person" in the feminine or masculine form, respectively. We chose "Turca" for two reasons: it rhymes with "Bella," and the author, being female, favored the feminine word form. In essence, we believe that only a language as rich as Latin can truly extol the richness and beauty of the Turkish language.

- We present a Turkish-language corpus for language modeling, totaling 265.4GB and consisting of 31.6 billion tokens. This corpus combines 25 distinct subsets sourced from four different genres.
- We introduce 21 new datasets, which we believe will be of independent interest to researchers working in the field.
- Our corpus is the first of its kind for the Turkish language. It is a large-scale dataset made openly available, accompanied by comprehensive documentation and provided with code for reproduction purposes.
- Above all, our intention is to inspire other researchers in the Turkish NLP community to undertake similar work. The primary motivation behind compiling this paper is to encourage and motivate all Turkish NLP researchers to contribute to the field.

Our dataset is conveniently accessible through its dedicated Huggingface repository[2]. Additionally, the dataset curation codes, which were used in the process of creating and organizing the dataset, are available on our Github repository[3]. This ensures easy and convenient access to both the dataset itself and the resources used for its curation.

Table 1. Genre and size details of Bella TurcaâĂŹs collections.

Collection	Genre	UTF-8 bytes (GB)	Documents (millions)	Unicode words (billions)	Mean Document Size (KB)
AcademiaCrawl	academic	4.44	0.19	0.50	17.19
Books	books	14.6	93.4	1.36	0.21
CleanOSCAR	web pages	50	20.3	6.1	2.58
CleanMC4	web pages	172.7	76.4	20.5	2.49
CraftedCrawl	web pages	5.06	1.39	0.55	3.37
ForumChats	forum	18.6	3	2.6	12.8

2 Bella Turca Datasets

Bella Turca consists of 25 sub-corpora categorized into 4 genres/collections, as outlined in Table 1. The specific details of each sub-corpora can be found in their respective collection sections. It is important to note that while we use the English translations of the dataset names to cater to a wider audience, the original dataset names in the Huggingface and Github repositories are in Turkish. For a more comprehensive understanding of the types of data included in the datasets, please refer to Appendix A for sample excerpts.

---

[2] https://huggingface.co/datasets/turkish-nlp-suite/BellaTurca.
[3] https://github.com/turkish-nlp-suite/bella-turca-cleaners.

2.1 AcademiaCrawl

The AcademiaCrawl collection comprises five datasets: Articles, Academic-Abstracts, Medical-Articles, Medical-Abstracts, and Bilkent-Writings. Among these, Bilkent-Writings is obtained from the creative writings of Turkish 101 and Turkish 102 courses[4] at Bilkent University between 2014 and 2018.

The remaining four datasets are crawled from various sources. The Academic-Abstracts dataset is compiled from two different resources, namely YÖK Açık Erişim[5] and Dergipark[6]. Both organizations, YÖK and TÜBİTAK-Dergipark, are government-affiliated and aim to provide high-quality research papers and journals on their respective websites. Of the total abstracts, 309,169 are sourced from YÖK Açık Erişim, while 188,106 are sourced from Dergipark.

The Articles, Medical-Articles, and Medical-Abstracts datasets are solely obtained from Dergipark. The sizes of each dataset can be found in Table 2. It was particularly challenging to find medical articles and abstracts, resulting in a comparatively smaller amount. Specifically, we were able to scrape data from 85 medicine journals in Dergipark, while there were 1,871 journals available for non-medical sciences.

Table 2. Subsets of the academic crawl collection.

Dataset	Size	Documents	Words (millions)
Articles	3.3GB	147,961	392.34
Academic-Abstracts	924MB	497,275	86.92
Medical-Articles	142MB	16,592	16.58
Medical-Abstracts	36.6MB	21,318	3.39
Bilkent-Writings	30MB	6,451	3.67

The academic text within the AcademiaCrawl collection covers a wide range of topics, spanning from scientific disciplines to sociological subjects. Consequently, this diversity of topics leads to a rich and varied vocabulary within the dataset. Moreover, academic texts are subjected to rigorous scrutiny by journals, peers, and thesis advisors, ensuring a high level of quality and credibility.

2.2 Books

Books play a crucial role in language modeling, as demonstrated by the initial English BERT model and its effectiveness [12]. Therefore, including a books

---

[4] https://stars.bilkent.edu.tr/turkce/.
[5] The Council of Higher Education, https://acikerisim.yok.gov.tr/acik-erisim.
[6] Offered by TÜBİTAK, The Scientific and Technological Research Council Of Türkiye, https://dergipark.org.tr.

dataset is essential in creating a diverse language dataset. Recognizing this significance, we have collected approximately 80,000 books from various sources for the Bella Turca dataset.

These books are sourced from free e-book platforms such as the Culture Ministry's e-book platform[7] and library websites like Altkitap[8]. We have organized the books into four distinct collections based on their subjects. The collections include purely literary books, books with a focus on philosophy and sociology, scientific books, and textbooks.

The textbooks in particular are primarily obtained from Open Education[9], a platform that offers free educational resources online for different levels of education, including middle school, high school, and college. The sizes of each dataset in the Books collection can be found in Table 3.

Table 3. Breakdown of the Books collection and the size of each subset.

Dataset	Size	Documents	Words (billions)
Books1	10.5 GB	73,356,502	1.08
Books2	2.2 GB	6,882,043	0.1
Books3	1.9 GB	13,074,892	0.18
TextBooks	14.5 MB	87,513	0.0015

Each document within the Bella Turca dataset is structured with a book ID and a single sentence. The decision to organize the dataset at the sentence level was made to address potential parsing errors that may occur during the extraction of text from PDFs. Extensive filtering was applied to the sentences to ensure the reliability and quality of the dataset. Section 3 provides additional insights into how the dataset was carefully curated and processed.

2.3 CleanOSCAR

This collection of the Bella Turca dataset is derived from existing resources, namely the Turkish partitions of OSCAR-2019[10], OSCAR-2021[11], and OSCAR-2023[12]. This portion of the dataset has undergone thorough cleaning and extensive filtering to ensure its quality and reliability. A significant portion of the web crawl data included in our dataset is sourced from this particular collection.

Table 4 provides information on the sizes of the sub-datasets within this OSCAR-derived collection.

---

[7] https://ekitap.ktb.gov.tr/.
[8] https://tr.wikipedia.org/wiki/Altkitap.
[9] https://www.anadolu.edu.tr/en/open-education.
[10] https://huggingface.co/datasets/oscar-corpus/OSCAR-2109.
[11] https://huggingface.co/datasets/oscar-corpus/OSCAR-2201.
[12] https://huggingface.co/datasets/oscar-corpus/OSCAR-2301.

Table 4. Breakdown of the CleanOSCAR collection and the size of each subset.

Dataset	Size	Documents	Words (billions)
clean-OSCAR-2109	18GB	8,472,809	2.22
clean-OSCAR-2201	17.6GB	6,642,315	2.20
clean-OSCAR-2301	14.4GB	5,193,341	1.75

2.4 CleanMC4

The subset of Bella Turca is obtained from an existing corpus called mC4. The majority of the web crawl data in our dataset originates from this subset. After undergoing filtering and cleaning processes, the total size amounts to 172.7 GB, comprising 76.4 million documents and encompassing 20.5 billion words.

2.5 CraftedCrawl

CraftedCrawl is a meticulously curated collection of high-quality web crawl data from carefully selected websites, encompassing articles, journals, and magazines. The goal of CraftedCrawl is to ensure the inclusion of valuable and extensive textual content, with a preference for lengthy articles. To achieve this, specific topics were targeted, including travel, news, culture, fairy tales and folk stories, movie critiques, popular science, service/product complaints, fashion and self-care, popular technology, popular culture, and literature.

Table 5. Specific topics and corresponding datasets in the CraftedCrawl collection.

Dataset	Size	Documents	Words (millions)
Literature	72.9MB	74,529	8.08
Popular Culture	535MB	146,315	64.34
Popular Technology	584MB	160,680	46.92
Fashion	33.4MB	8,993	3.87
Complaints	23.6MB	23,923	8.09
Popular Science	101MB	23,424	11.52
Cinema	135MB	36,888	16.23
Folk stories	9.08MB	2,621	0.10
Culture	508MB	113,118	59.29
Travel	181MB	33,174	21.88
News	2.88GB	747,574	315

Each sub-corpus, except for the News sub-corpus, is sourced from a minimum of 50 high-quality domains. The diversity of content within these sub-corpora allows the models to capture the nuances of modern Turkish culture in detail, making this sub-corpus highly valuable. The distribution of topics within the CraftedCrawl collection can be found in Table 5.

Notably, the News sub-corpus, originally named "Havadis," is a special dataset as it represents the first large-scale Turkish news crawl ever conducted. It includes data crawled from 11 distinct newspaper websites, namely CNN Türk, Habertürk, Hürriyet, Milliyet, NTV haber, Posta, Sabah, Sözcü, Star, T24, and Takvim.

2.6 ForumChats

The ForumChats dataset is collected from 12 distinct forum websites that are known for their high quality and moderation. These forums cover a wide range of topics including technology, culture, life, science, and more. This dataset comprises around 3 million documents, with each document representing a forum thread. For further details about this dataset, please refer to its corresponding Huggingface repository[13].

3 Creating and Curating the Data

In the process of curating the Bella Turca dataset, we utilized a modular design approach. Since each document within the different collections had its own unique structure and presented specific challenges, we recognized the need for a flexible and adaptable curation process. Hence, we employed a plug-and-play approach, selecting and implementing the most suitable modules for each collection.

By adopting this modular design, we were able to address the specific curation requirements of each collection more effectively. This approach allowed us to tailor the data cleaning, filtering, and processing tasks to the characteristics and challenges presented by each dataset (Fig. 1).

Our toolkit has the following main modules:

- **Language filtering.** In order to create a monolingual corpus, a language filter is essential. We utilized PyCLD2[14] due to its efficiency. Depending on the document genre, we applied filtering at the document, paragraph, or sentence level. However, we did not perform any filtering on datasets that were already pre-filtered for Turkish-only documents, such as OSCAR.
- **Structural parsing.** Certain types of documents, such as books and scientific articles, may contain sections that do not provide significant semantic information, such as forewords and references. For these documents, we aimed to extract the relevant textual content while excluding such sections. On the other hand, for documents like blogs and web articles, we removed sections like page headers and site visitor comments/messages.

---
[13] https://huggingface.co/datasets/turkish-nlp-suite/ForumSohbetleri.
[14] https://pypi.org/project/pycld2/.

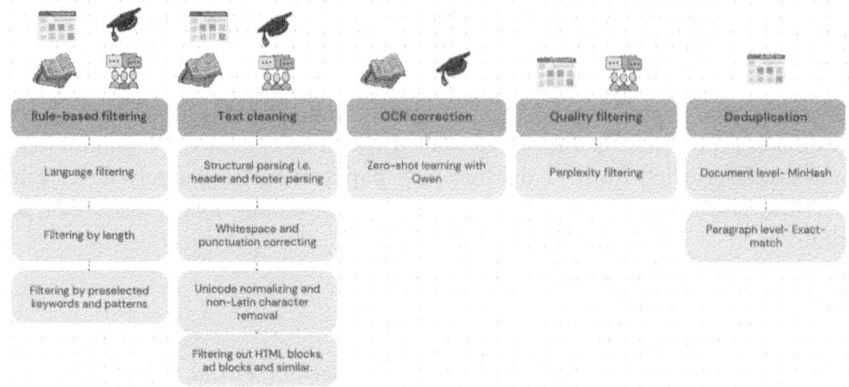

Fig. 1. Overview of the text preprocessing modules.

- **Quality filtering.** Removing text classified as "low quality" is a common practice, although there is no widely agreed-upon definition or standardized approach for automated tools.[15] In the case of web-based datasets, we performed extensive filtering at both the document and sentence levels. This module comprises two submodules: document filtering and sentence filtering. The following items provide further details on these submodules.
- **Document filtering.** In the case of web-based genres, our toolkit applies several filters to remove undesirable documents. These filters include:
  - Eliminating documents with an insufficient number of sentences (less than 5 sentences).
  - Removing documents containing pejorative or sexually explicit content.
  - Filtering out documents that resemble advertisements.
  - Discarding documents that appear to be written for search engine optimization (SEO) purposes, characterized by excessive repetition of keywords.
  - Removing documents with an excessive number of non-word characters.
  - Excluding documents with a high proportion of non-word characters.
  - Filtering out documents containing excessive profanity.

  Each removal rule is implemented as a class method within the toolkit. For the book genre, we specifically addressed extraction errors from PDF files, which occasionally introduced anomalous non-graphical characters. Additionally, we addressed issues with incorrectly placed inter-word spaces by utilizing a 3-gram model for their elimination.

---

[15] The term "quality filter," although commonly used in literature, may not accurately depict the result of filtering a dataset. The term "quality" can be interpreted as a judgment on the informativeness, comprehensiveness, or other subjective characteristics valued by humans. However, it is important to note that the filters employed in Bella Turca and other language model endeavors are based on criteria that inherently carry ideological implications [13].

Another consideration for web crawls and books is the presence of religious texts that may include archaic and Arabic words derived from religious literature. To address this, we implemented a "too much Arabic" filtering method based on the count of Arabic characters and a word list of religious terms.
- **Sentence filtering.** Depending on the genre, our toolkit applies filters to remove low-quality sentences, including:
    • Eliminating sentences that are too short, containing fewer than 6 words.
    • Removing sentences that lack punctuation and capitalization.
    • Filtering out sentences with excessive punctuation errors, such as an excessive number of punctuation marks or incorrect placement of punctuation.
    • Excluding sentences with unusually long words, which often indicate spacing errors in PDF extraction.
  The specific threshold for determining what is considered "too many" depends on the genre. For web genres, we utilized the trafilatura package [2] for crawling, while for text extraction from PDFs, we employed the PyMuPDF package[16].
- **Content filtering.** Research has shown that large pretraining corpora can contain personally identifiable information [10], which language models are capable of reproducing during inference [6]. In Bella Turca, we have implemented content filtering to identify and remove sensitive information using a list of regular expressions for detecting email addresses, addresses, and phone numbers. We replace each identified entity with a corresponding entity token, such as <EMAIL>, <ADR>, and <PHONE_NUM>.
  It is important to note that while we filtered out documents containing excessive profanity, we did not employ checks specifically for removing toxic content. This dataset serves text generation purposes as well as general natural language understanding (NLU) tasks. We intentionally allowed toxic content to persist in certain collections to enable researchers to study and address the issue of toxic content. Two collections, ForumChat and CustomerTrends, may contain diverse and potentially toxic language due to the nature of forum discussions. However, other collections are highly clean in terms of toxic content, as they were collected in a targeted manner.
- **Text cleaning.** Text cleaning is an essential component of any language pipeline. We have incorporated a general cleaning module and added or removed specific submodules based on the genre. The general cleaning steps include:
    • Normalizing space and punctuation characters by mapping Unicode character variations to standard whitespace and punctuation characters.
    • Filtering out Unicode blocks that appear after the extended Latin-Extended-B block, while considering the ranges of emoji and emoticon characters.
    • Normalizing characters with circumflex accents (which are no longer officially used in the writing language) to their regular versions (e.g., â -> a).

---

[16] https://pypi.org/project/PyMuPDF/.

- Filtering out HTML blocks.
- Filtering out Chinese, Korean, and Arabic characters.
- Applying fixes to punctuation marks before and after them.

For certain genres like News, we filter out emoji and emoticon characters as they are not crucial for the content. However, for genres such as customer reviews, these characters may carry important information, so we retain them.

- **Deduplication.** The code utilized in our work was adapted from the extensive deduplication pipeline described in the methodology outlined by [24]. This pipeline combines document-level MinHash deduplication and sub-document exact-match deduplication techniques, enabling the efficient identification and elimination of duplicate content both within individual documents and across multiple documents. To handle web crawl genre collections, we employed collectionwise MinHash deduplication to identify duplicates at the document level, while for each document, we performed paragraphwise exact-match deduplication to address duplicates at a more granular level.
- **Quality filtering.** To further enhance the quality, we employed perplexity filtering on the web crawl genre collections. This process involved utilizing the KenLM library [14] to identify and filter out documents with perplexity scores that significantly exceeded the average.
- **OCR correction.** In the pipeline for the AcademiaCrawl and Books collections, OCR correction was deemed necessary. The Books collection exhibited a noticeable number of character-level errors, making OCR correction crucial to maintain a satisfactory level of quality. On the other hand, the AcademiaCrawl collection had only occasional mistakes in certain documents, making it usable without OCR correction. However, incorporating OCR correction significantly enhanced the overall quality of the AcademiaCrawl collection. To carry out this task, we utilized the open-source LLM Qwen-7B [34] and employed few-shot learning techniques for both collections. The processing time for these two collections was 24 h for AcademiaCrawl and 8 h for Books, utilizing a single NVIDIA A100 80G GPU.

Next, we will proceed to provide a comprehensive analysis of each collection, elucidating the specific details pertaining to each one.

AcademiaCrawl. In the AcademiaCrawl collection, the key focus was on structural parsing and language filtering. Since the journal name could be in Turkish while the article itself might be in English, it was necessary to handle such cases. Additionally, the article titles often contained scientific terms that could adversely affect the Turkish language score. To address these issues, the first ten lines of the abstract were examined, and a language filter was applied to identify and filter out English articles.

Language filtering was performed in conjunction with structural parsing, as some abstracts included an English version alongside the Turkish content. In certain cases, the English part appeared after the Turkish content, so a check was made for contiguous chunks of English sentences. If found, the abstract and/or the article were cropped accordingly.

As this collection was obtained from PDF files, Furthermore, we implemented a filtering process to identify and exclude documents that still contained errors resulting from the PDF-to-text extraction process. While not the primary objective, improving OCR accuracy remained a significant focus throughout our work.

Since this collection was crawled from PDF files, one of our significant objectives was to address and rectify any OCR errors. To achieve this, we employed our pipeline's OCR correction module. Furthermore, we implemented a filtering process to identify and exclude documents that still contained errors resulting from the PDF-to-text extraction process. These errors were identified using a comprehensive set of manual rules, such as excessive occurrences of single digits or short syllable-like sequences.

Regarding text cleaning, the exclusion of Chinese and Korean characters filtering method was not applied in this corpus. These characters were considered integral to the articles, as they often appeared in named entities or terms written in their original language.

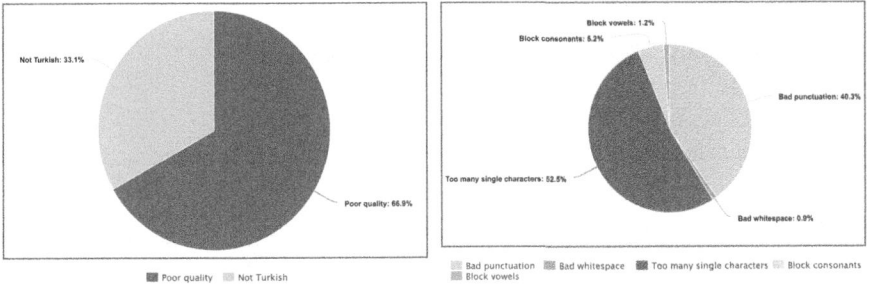

Fig. 2. Statistics on the reasons for document filtering in the Books collection.

Books. In the Books collection, our primary focus was on addressing PDF-to-text extraction errors. We developed an extensive set of handmade regex rules to identify and classify these errors, referred to as "quality mistakes." Documents with a high number of quality mistakes were eliminated, as it typically indicates poor PDF-to-text conversion quality.

The surviving documents were then segmented into sentences, and another round of filtering was applied using quality rules. These rules encompassed the following criteria:

– Excessive punctuation mistakes
– Excessive whitespace mistakes
– Excessive consecutive single characters (e.g., "a b o o k")
– Long blocks of consonants or vowels (which are highly unlikely in Turkish)

Additionally, sentences that were not in Turkish were also removed. Figure 2 illustrates the percentage distribution of different mistake types that led to the removal of sentences. The majority of sentences were eliminated due to poor quality issues.

Correcting the OCR mistakes constitutes a significant aspect of cleaning this collection. In our endeavor, we explored multiple multilingual language models (LLMs), and it was only the Qwen model family that demonstrated satisfactory performance in this regard. However, the process of locating pretrained models, especially for Turkish language resources, posed a notable challenge. The scarcity of available resources further complicated the task of creating Turkish language resources.

CleanOSCAR and CleamMC4. Extensive filtering was applied to the CleanOSCAR and CleanMC4 collections due to the diverse range of content found in web documents. Both entire documents and individual sentences within the documents underwent thorough filtering. Sentences with fewer than 5 words were excluded, as well as non-Turkish sentences.

Various criteria were employed to discard entire documents, including:

- Inadequate length
- Punctuation mistakes
- Non-Turkish language
- Presence of sexually explicit content
- Identification as an advertisement or SEO-optimized page
- Excessive length of words (character count >50), indicating potential spacing mistakes
- Absence of comma, period, exclamation marks, or question marks
- Lack of capital letters
- Excessive usage of archaic religious terms
- Inclusion of Arabic characters

Figure 3 presents the percentage distribution of the different filtering reasons for documents within both collections.

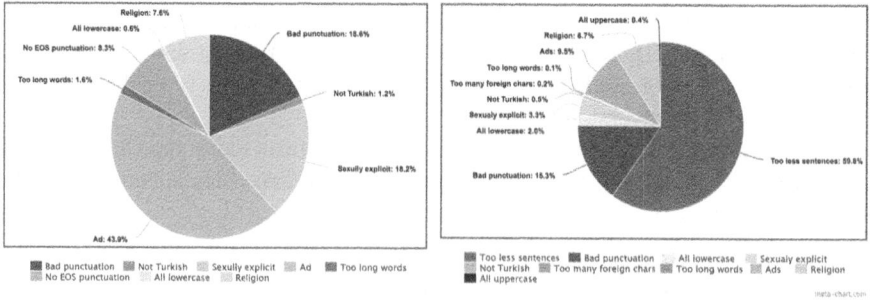

Fig. 3. Percentage distribution of the various reasons for filtering documents in CleanOSCAR (left) and CleanMC4 (right) collections.

CraftedCrawl. The URLs in the CraftedCrawl collection were deliberately selected from known sources of high-quality content. Consequently, the cleaning process for this collection was minimal. We performed text cleaning and conducted some structural parsing to remove article/blog headers, user comments/reviews, and similar elements. Remarkably, we encountered negligible document losses during the filtering stage.

ForumChats. The ForumChats dataset is quite diverse and exhibits a wide range of text types. Our initial step involved parsing out non-content text, such as phrases like "leave an answer," "quote this message," or "userX quoted." During the structural parsing phase, we also filtered out excessively short threads and messages. Following that, we applied our standard text cleaning pipeline, which preserved emoji and emoticon characters. It is worth noting that this dataset encompasses various text categories, including technical forum discussions, technological product reviews, and personal names. We deliberately refrained from applying any further filtering to maintain the dataset's wild and heterogeneous nature, preserving its semantic richness.

4 Ethical and Legal Concerns

All the web crawls conducted for data collection were performed on publicly available content, and we consistently respected the guidelines provided in the website's robots.txt file and any legal warnings issued by the website owners. However, we acknowledge that large web crawls may inadvertently include copyrighted material. Currently, there are no reliable or scalable tools to detect copyrighted material within a corpus of this magnitude. We refer interested readers to our public position on AI and fair use[17].

We are aware that datasets derived from extensive web crawls may contain inaccuracies, toxic language, hate speech, personally identifiable information (PII), and other forms of harmful content. While we have made diligent efforts to curate this dataset with these concerns in mind, we believe that risk mitigation should be approached through multiple avenues, including careful consideration of licenses and access controls.

5 Conclusion

In this paper, we have introduced our new, diverse, and large-scale dataset specifically designed for training Turkish language models. We have provided an overview of the structure and content of the individual collections comprising our dataset, as well as insights into the curation process. We hope that our work will inspire more researchers to contribute to the development of Turkish language resources by providing their own corpus offerings.

---

[17] https://www.regulations.gov/comment/COLC-2023-0006-8762.

Disclosure of Interests. The authors have no competing interests to declare that are relevant to the content of this article. The authors did not receive support from any organization for the submitted work.

A Data Samples

Below are two random, unbiased samples from each of Bella Turca's collections, which were selected from the train split. To accommodate the page length, we made some adjustments, particularly in cropping certain samples, such as articles.

A.1 AcademiaCrawl

Metformin güvenilir bir oral antidiyabetik olmakla birlikte yüksek dozda alınmasıyla ortaya çıkabilen laktik asidoz ve böbrek yetmezliği tablosu ölümcül olabilir. Bu çalışmada intihar amacıyla metformin alan olguların klinik seyri tartışılmıştır. Eylül 2009-Mayıs 2017 tarihleri arasında kliniğimizde takip edilen, intihar amaçlı yüksek doz metformin alan olgular değerlendirildi. Demografik verileri, biyokimya sonuçları, klinik seyirleri ve tedavileri dosyalarından kaydedildi. Toplam 15 olgunun 10 tanesi kadın, 5 tanesi erkekti. Medyan yaş kadınlarda 34 yıl (18-68), erkeklerde 38 yıl (23-58) saptandı. Alınan medyan metformin dozu, kadınlarda 19,5 g (8,5-30,0), erkeklerde 24,0 g (14,0-50,0) bulundu.

Bu çalışmada, örgütsel güvenin, örgütsel bağlılık ve işten ayrılma niyeti ile yakın bir ilişkiye sahip olduğu varsayılarak, aralarındaki ilişkinin gücünün ve yönünün karşılaştırmalı olarak analiz edilmesi ve sonuçların ortaya konulması amaçlanmıştır. Varsayımın test edilmesi amacıyla nicel bir yöntem olan meta analiz uygulanmış, etki büyüklüğü değeri elde edebilmek amacıyla, Pearson (r) ve örneklem (n) değerleri kullanılmıştır. Yapılan taramada 150 çalışmaya ulaşılmış, seçim kıstaslarına uygun olan 91 çalışma analize dahil edilmiştir. Örgütsel güven ile örgütsel bağlılık arasında ES=0.55 değerinde güçlü ve pozitif, örgütsel güven ile işten ayrılma niyeti arasında ES= -0.42 değerinde, orta ve negatif bir ilişki elde edilmiştir. Alt boyutlar açısından bakıldığında, örgüte ve yöneticiye duyulan güvenin, duygusal bağlılık ile pozitif yönde ve işten ayrılma niyeti ile negatif yönde daha güçlü bir etkiye sahip olduğu anlaşılmıştır. Güven gerek kişilerarası ilişkilerde gerekse örgütsel seviyede, etkileşimin ve rekabetin gittikçe daha karmaşık, çok boyutlu ve yoğun bir hale geldiği günümüz şartlarında, daha çok önem verilen, kazanılması ve sürdürülmesi için çaba sarf edilen soyut bir varlık haline gelmiştir. Kişisel olarak ilişkilerin daha sağlıklı ve sürdürülebilir bir hale gelmesini kolaylaştıran güven, örgütsel başarının devamında da etkili bir araç durumundadır.

A.2 Books

Eski Avusturyanın milliyetler arasındaki mücadelesine katılma fırsatım oldu.

Korenin geleneksel giysisi olan Hanbok, yüzyıllardan bu yana kadın ve erkekler için aynı biçimiyle gelmiş, iklim ve kültüre çok iyi uyduğu için, jeogori ve çimanın boyu dışında değişmeden kalmıştır.

A.3 CleanOSCAR and CleanMC4

Bu makalede Microsoft Generic Report üzerinden Unix sunuculara ait CPU performansı raporunun nasıl alınacağından bahsedilecektir. Unix sunucuların izlenmesi için gerekli Management Pack yüklendikten sonra default olarak Universal Linux Monitoring raporları Report sekmesine gelmektedir. İsmi her ne kadar Microsoft ile başlasa da Group ve Unix kuralı eşleşiyorsa istediğiniz çeşitlilikte rapor almanız mevcut. Daha sonra raporu oluşturacak group, Rule ve tarih aralıkları belirlenir. Group kısmına linux sunucuları içeren grubu tanımlıyoruz Rules kısmına da Total CPU değerini veren Universal Unix Rule ekliyoruz. Kısımda kural tanımlarken aşağıdaki kuralı seçiyorum. Rapor detaylandırılmak istenilirse 2. sekmeden yeni bir seri oluşturup kuralı ile beraber rapora eklenebilir. Tarih aralığı belirtildikten sonra raporu çalıştırıyoruz. Bu raporu zenginleştirmek için ilgili objeleri eklenebilir. Ancak MP ile birlikte zaten zengin içerikli raporlar geliyor. Bu yönetimi sadece belirli bir kuralın detaylı raporunu almak için kullanabilirsiniz.

Güney Kıbrıs Rum Yönetimi (GKRY) sözde mültecilerin Kuzey Kıbrıs Türk Cumhuriyeti (KKTC) üzerinden Güney Kıbrısa girişini engellemek amacıyla Yeşil Hat olarak anılan KKTC-GKRY sınırına yaklaşık... Belarus Devlet Başkanı Lukaşenko, Brükselin, Belarus-Avrupa Birliği sınırındaki göçmen akışını kendisinin organize ettiği yönündeki suçlamalarını da reddetti. Ancak BBCye mülakat veren Lukaşenko, Biz Slavız, kalp... Fransanın kuzeydeki Dunquerke bölgesinde bulunan Grande-Synthe kentinde güvenlik güçleri, İngiltereye ulaşmak isteyen binden fazla göçmenin kaldığı kampı dağıttı. Operasyon, Dunquerke mahkemesinin Grande-Synthe topraklarında bulunan arazilerden... Belarus Devlet Başkanı Aleksander Lukaşenko ile Almanya Başbakanı Angela Merkel bir telefon görüşmesi yaptı. Yaklaşık 50 dakika süren görüşmede Belarus-Polonya sınırındaki göçmen krizinin yanı sıra... İspanyaya bağlı Kanarya Adaları hükümeti, İspanyol sahil güvenlik görevlilerinin Pazar günü Gran Canaria adası açıklarında denizde sürüklenen bir teknede sekiz Afrikalı göçmenin cesetlerini bulduklarını açıkladı... Ukrayna başta Polonya olmak üzere Belarus ile sınırı olan Litvanya ve Letonya gibi AB ülkelerinde yaşanan göçmen krizinin Ukraynayı da etkilemesi ihtimaline karşı önlem aldı...

A.4 CraftedCrawl

Kaynakların sürdürülebilir kullanımı, iklim kriziyle mücadele ve ormanların korunmasının giderek önem kazandığı günümüzde Atlas dergisi de yeni bir adım atıyor ve Eylül 2023 sayısından itibaren okurlarının karşısına yepyeni bir kağıtla çıkıyor. Derginin iç sayfalarında kullanılan yeni kağıdın en önemli özelliği tamamen geri dönüştürülmüş malzemeden elde edilmesi, ayrıca geliştirilmiş standartları ve gramajıyla derginin görsel kalitesini daha da yükseltmesi. Alman Leipa firması ürettiği kağıtlarda hammadde olarak yüzde 100 atık kağıt kullanıyor, firma bu yolla yılda bir buçuk milyon tondan fazla geri kazanılmış kağıdı tekrar değerlendiriyor. Atlas'ın Leipa tarafından üretilen yeni kağıdı, sürdürülebilir ve yenilikçi üretim tekniğiyle şimdiye kadar birçok doğa koruma belgesi almaya hak kazandı. Derginin kağıdı Blauer Engel ve Uluslararası Orman Yönetim Konseyi, Orman Sertifikasyonu Onaylama Programı, EU Ecolabel gibi sertifikaların sahibi oldu.

Binbir Tv yazarları Işınla Bizi Scotty ve Uzun çorap, 2017'nin dizilerini değerlendirdi. Uç: Hımmm. Güzel bir bakış açısı. 2017 yılını şöyle bir düşündüğümde "Aman süper" diyemediğimi fark ettim. Fakat tek tek dizilerin üstünden geçerken belki unuttuklarımı hatırlarım, fikrim değişir. Bana diziler bakımından çok güçlü bir yıl gibi gelmedi. Bu yıla damga vuran diziye gelince... Dizilere şöyle bir bakıyorum, bir sürü kısa ömürlü dizi olmuş. Hem de çok. Senenin ilk çeyreğinde başlayıp halen süren 5 dizi var. İstanbullu Gelin, Fazilet Hanım ve Kızları, Fi, Payitaht ve Yeni Gelin...

A.5 CustomerTrends

Kıkırdak ve burun delemedim ama kulak çok kolay delindi süperrr. Kendi işini kendi halletmek isteyenler için süper bir çözüm :) ağrısız kendim kulağımı deldim o kadar hissetmedim ki deldiğimi anlamadım ancak herkesin acı eşiği farklıdır.

İade iade. Kumaşı inanılmaz kalitesiz bu parayı hak etmiyor diye düşünüyorum. Üzerine biraz daha ekleyerek mangodan daha kaliteli bir çanta alınabilir. çok büyük degil daha büyük beklemiştim kumaşı çok güzel rengi çok güzel kaliteli ama en olarak daha büyük duruyor fotoğrafta ama büyük çanta tabii yine de asla fiyatını hak etmeyen bir urun rengi göründüğü gibi değil daha soluk. fotoğraftaki gibi duruşu formu değil. bir bez çantadan biraz daha hallice. hayal kırıklığına uğradım. fotoğrafları ile yanıltıcı.

A.6 ForumChats

Sa arkadaslar... 1 GB RAM 512 MB ekran kartim var su anda windows 7 ultimate kullanıyorum ve format atcam sizce windows 8 mi 7 mi yoksa XP mi... Memnunsan 7 kullanmaya devam et, tekrar temiz olarak Windows 7 kurulumu yap yani, eğer ram belleğini artırma şansın varsa, Windows 10 a da geçebilirsin. Bence 7 den memnunsan ve kasmadan donmalar yaşamadan kullanıyorsan devam et ama sıkıntılar yaşıyorsan xp ye dön. çünkg ram miktarın az ve diğer sistemlerde sıkıntı çıkarabilir Geçtim teşekkür ederim çözüldü sorun.

Sistemin gücüne dayalı iş yapıyorsan tv ye bağla kasaya yüklen derim. AIO ların geliştirilmesi parça eklenmesi zahmetli fiyatlarıda donanımına oranla yüksek goluyor. Tv de de bakış mesafesi açısı kişiden kişiye değişiyor ama boyun ve göz ağrılarına neden olabiliyor Denemiş ve memnunsan sistem de güçlü olsun dersen tv ile gidebilirsin.

References

1. Abadji, J., Ortiz Suarez, P., Romary, L., Sagot, B.: Towards a Cleaner Document-Oriented Multilingual Crawled Corpus, January 2022. arXiv e-prints arXiv:2201.06642
2. Barbaresi, A.: Trafilatura: a web scraping library and command-line tool for text discovery and extraction. In: Proceedings of the Joint Conference of the 59th Annual Meeting of the Association for Computational Linguistics and the 11th International Joint Conference on Natural Language Processing: System Demonstrations, pp. 122–131. Association for Computational Linguistics (2021), https://aclanthology.org/2021.acl-demo.15
3. BigScience Workshop et al.: Bloom: A 176b-parameter open-access multilingual language model (2023)
4. Black, S., et al.: Gpt-neox-20b: an open-source autoregressive language model (2022)
5. Brown, T.B., et al.: Language models are few-shot learners (2020)
6. Carlini, N., Ippolito, D., Jagielski, M., Lee, K., Tramer, F., Zhang, C.: Quantifying memorization across neural language models (2023)
7. Cañete, J., Chaperon, G., Fuentes, R., Ho, J.H., Kang, H., Pérez, J.: Spanish pre-trained bert model and evaluation data (2023)
8. Computer, T.: Redpajama: An open source recipe to reproduce llama training dataset, April 2023. https://github.com/togethercomputer/RedPajama-Data
9. DeepMind: Scaling language models: Methods, analysis & insights from training gopher (2022)
10. Elazar, Y., et al.: What's in my big data? (2024)
11. Gao, L., et al.: The pile: an 800gb dataset of diverse text for language modeling (2020)
12. Gunasekar, S., et al.: Textbooks are all you need (2023)
13. Gururangan, S., et al.: Whose language counts as high quality? measuring language ideologies in text data selection. In: Goldberg, Y., Kozareva, Z., Zhang, Y. (eds.) Proceedings of the 2022 Conference on Empirical Methods in Natural Language Processing, pp. 2562–2580. Association for Computational Linguistics, Abu Dhabi, United Arab Emirates, December 2022. https://doi.org/10.18653/v1/2022.emnlp-main.165, https://aclanthology.org/2022.emnlp-main.165
14. Heafield, K.: KenLM: Faster and smaller language model queries. In: Callison-Burch, C., Koehn, P., Monz, C., Zaidan, O.F. (eds.) Proceedings of the Sixth Workshop on Statistical Machine Translation, pp. 187–197. Association for Computational Linguistics, Edinburgh, Scotland, July 2011. https://aclanthology.org/W11-2123
15. Hoffmann, J., Borgeaud, S., et al.: Training compute-optimal large language models (2022)
16. Hugo Laurençon et al.: The bigscience roots corpus: A 1.6tb composite multilingual dataset (2023)
17. Touvron, H., et al.: Llama 2: Open foundation and fine-tuned chat models (2023)
18. Kaplan, J., et al.: Scaling laws for neural language models (2020)
19. Kesgin, H.T., Yuce, M.K., Amasyali, M.F.: Developing and evaluating tiny to medium-sized turkish bert models (2023)
20. Le, H., et al.: FlauBERT: Unsupervised language model pre-training for French. In: Calzolari, N., et al. (eds.) Proceedings of the Twelfth Language Resources and Evaluation Conference, pp. 2479–2490. European Language Resources Association, Marseille, France, May 2020. https://aclanthology.org/2020.lrec-1.302

21. Lepikhin, D., et al.: Gshard: Scaling giant models with conditional computation and automatic sharding (2020)
22. Lieber, O., Sharir, O., Lenz, B., Shoham, Y.: Jurassic-1: Technical details and evaluation. Tech. rep., AI21 Labs, August 2021
23. Liu, Y., et al.: Roberta: A robustly optimized bert pretraining approach (2019)
24. Penedo, G., et al.: The refinedweb dataset for falcon llm: Outperforming curated corpora with web data, and web data only (2023)
25. Radford, A., Wu, J., Child, R., Luan, D., Amodei, D., Sutskever, I.: Language models are unsupervised multitask learners (2019). https://api.semanticscholar.org/CorpusID:160025533
26. Raffel, C., et al.: Exploring the limits of transfer learning with a unified text-to-text transformer (2023)
27. Rosset, C.: Turing-nlg: A 17-billion-parameter language model by microsoft (2019)
28. Schweter, S.: Berturk - bert models for turkish, April 2020. https://doi.org/10.5281/zenodo.3770924. https://doi.org/10.5281/zenodo.3770924
29. Shoeybi, M., Patwary, M., Puri, R., LeGresley, P., Casper, J., Catanzaro, B.: Megatron-lm: Training multi-billion parameter language models using model parallelism (2020)
30. Soldaini, L., et al.: Dolma: an open corpus of three trillion tokens for language model pretraining research (2024)
31. Souza, F., Nogueira, R., Lotufo, R.: Bert models for Brazilian portuguese: pretraining, evaluation and tokenization analysis. Appl. Soft Comput. **149**, 110901 (2023)
32. Tas, N.: Roberturk: Adjusting roberta for Turkish (2024)
33. Team, P.: Palm 2 technical report (2023)
34. Team, Q.: Qwen technical report. arXiv preprint arXiv:2309.16609 (2023)
35. Tiedemann, J.: Parallel data, tools and interfaces in OPUS. In: Proceedings of the Eight International Conference on Language Resources and Evaluation (LREC'12). European Language Resources Association (ELRA), Istanbul, Turkey, May 2012
36. Touvron, H., et al.: Llama: open and efficient foundation language models (2023)
37. Virtanen, A., et al.: Multilingual is not enough: Bert for finnish (2019)
38. Wagner Filho, J.A., Wilkens, R., Idiart, M., Villavicencio, A.: The brWaC corpus: a new open resource for Brazilian Portuguese. In: Calzolari, N., et al.: (eds.) Proceedings of the Eleventh International Conference on Language Resources and Evaluation (LREC 2018). European Language Resources Association (ELRA), Miyazaki, Japan, May 2018. https://aclanthology.org/L18-1686
39. Wang, B., Komatsuzaki, A.: GPT-J-6B: A 6 Billion Parameter Autoregressive Language Model, May 2021. https://github.com/kingoflolz/mesh-transformer-jax
40. Zhang, S., et al.: Opt: Open pre-trained transformer language models (2022)

Evaluation Metrics in LLM Code Generation

Kai Hartung[1]([✉]) [ID], Sambit Mallick[1], Sören Gröttrup[1], and Munir Georges[1,2]

[1] Technische Hochschule Ingolstadt, Ingolstadt, Germany
{kai.hartung,sambit.mallick,soren.grottrup,munir.georges}@thi.de
[2] AIMotion Bavaria, Ingolstadt, Germany

Abstract. The advanced capabilities of large language models can also be seen in their increasing use in the automatic generation of programming code. Although models are generally getting better and better, there are very few metrics that can be used to evaluate the quality of the generated code. In particular, this evaluation becomes challenging without dependence on good reference data in the form of tests or alternative solutions. In this paper, we explore both existing and new approaches to evaluate generated python code. These approaches can be classified into two categories: similarity-based and reference independent. The similarity-based approaches involve examining the code's syntax-tree structure and embeddings and comparing them to reference code from the dataset. On the other hand, the reference independent approaches utilize static code analysis metrics used to assess human-written code. These metrics include maintainability and adherence to style guidelines. We examine these metrics on the example of several state-of-the-art code generation models to test their validity. Based on our results, the independent metrics seem to be the most promising approaches for future research.

Keywords: Code Generation · Large Language Models · Evaluation

1 Introduction

With the rise of large language models, code generation becomes an increasingly feasible task, and models like the GPT family [22,24] show impressive feats on these tasks. However, code differs in structure from natural language (NL), and the question, what makes good code cannot be answered the same way as for NL. This raises the question, how to effectively evaluate generations in this context.

Commonly used metrics, such as unit testing or similarity-based measures are what one might call reference-dependent. They require additional information from the dataset, beyond simple labels. For testing, this means they depend on well-defined tests, while similarity-based metrics require code samples as reference to compare generated code with. This means the quality of the measurements relies on the quality of the reference. Liu et al. [18] have shown that it is not trivial to provide sufficient test coverage even for relatively simple tasks. On the other hand, similarity-based metrics only become relevant, if they manage to capture what is actually intended to be measured in the comparison, and for previous attempts it has been argued, that they don't [9,26]. Both of these issues will only become more prevalent if the code generated by models becomes more comprehensive, and therefore more problematic to test and to compare.

In this study, we attempt to compare both, existing and new metrics, in how well they capture quality of code. As an approximation to the usability of similarity-based metrics, we test how well they predict if generated code will pass provided test cases or not and compare this to reference-independent metrics. As examples of those, we explore the use of static code analysis metrics like maintainability, cyclomatic complexity and adherence to coding styles on the example of python. These should give an assessment of the quality of the generated code, independent of additional references provided by the dataset.

2 Related Work

The predominant metric for code generation is whether or not test cases provided with a task can be passed. As pass@n it reports whether any of generations for a task pass the provided test cases. The metric itself is binary, and the report for a whole dataset is the percentage of tasks passed [6,12,17,21,27,29]. This metric provides the most straight forward quality assessment of code. But a downside is, that test cases need to be available in a dataset in the first place and the code needs to be executed to be tested. This opens up the danger of executing harmful code, if generated samples are not manually controlled first. In addition the provided test cases need to fit the generated code. This means, the method signature to be tested needs to be known to the model when generating, or the tests need to be adapted to the generated code.

Several static evaluation alternatives have been used to examine the generated code, which don't require execution. Among the more commonly used is the BLEU score [23,26–28]. It is a classic machine translation metric, which measures n-gram-wise token overlap between prediction and reference. However, this does not apply as well to code as it does to NL [26]. The BLEU score only captures surface level similarities and cannot account for the underlying tree structure of code. It also overvalues token similarities for example for variable names, which can differ between implementations while functionally being exactly the same. When considering token matches as evaluation for CodeT5 [28], the authors go a step further and also consider an Exact Match score, the binary measure whether generated code perfectly matches the desired reference.

The CodeBLEU score is an attempt to alleviate the shortcomings of BLEU by accounting for structural aspects in addition to the token level view of BLEU [26]. But, it has been shown to not perform better than the standard BLEU score [9].

For the evaluation of PHI-1 [12], the authors used two more evaluation approaches. On one hand, they embedded code snippets to be compared using a

Codegen model [21], and calculated the L2-distance between the embeddings as similarity metric. On the other hand, the authors computed a tree edit distance for the abstract syntax trees of the code.

Except for pass@n, the above mentioned metrics all attempt to find similarity between generation and references. A complete different approach to assessing code quality comes up, when considering the quality of human written code. For example, in [5] the authors explore what makes code readable for humans and observe that readability is correlated with software quality, while the authors of [4] find that adherence to python coding style guidelines is positively correlated with higher-valued contributions.

3 Datasets

We perform code generation experiments on two datasets, each including a task description in NL, a ground truth python code sample to fit the task, and a set of test cases to validate the functionality of the code. An overview of the dataset sizes can be seen in Table 1.

Table 1. Dataset Overview

Dataset	Data Set Size	Test Set Size
mbpp (sanitized)	427	257
HumanEval	164	164

The first dataset is the Mostly Basic Python Problems Dataset (mbpp) released by Google [3]. As the name implies, the coding tasks are intended to be solvable by entry-level programmers and with the use of native python libraries only. The tasks have been written by crowd sourced workers, who also provided the ground-truth solutions tests with the tasks. Mbpp contains a sanitized subset where instances with ambiguous task descriptions or misaligned tests were filtered out. For our experiments, we use this sanitized version of the dataset.

The HumanEval dataset [6] has been created specifically for the evaluation of the Codex model and therefor consists only of its test split. The coding problems, solutions and tests are hand-written. In this dataset, the task description is given the form of a docstring for a function signature, making it viable for both code completion as well as NL-based code generation testing.

4 Methods

We compare 4 large language models (Phi, Code Llama, Magicoder and GPT) that are trained to produce python code from NL descriptions. After processing the results obtained from these models, we evaluate each model's capability to generate Python code with different metrics.

4.1 Models

To get an impression of the expressivity of the metrics studied in this work, we apply a range of Large Language Models already trained to generate code from NL description. The models generate code for the two datasets described in Sect. 3 and the metrics are applied on this generated code. In addition we evaluate the ground truth samples provided by the data sets and report their results as 'oracle'. The models we use are the derivatives of the Llama2 Model: CodeLlama-7B and CodeLlama-7B-py which is focused on python specifically [27], the Phi-models Phi-1 [12] and Phi-2 [16], Magicoder [29] and the GPT models [24] in the versions 3.5, 4 and 4-turbo [22].

To run the models, we consider five prompting variations, with a different focus of the task. The default setup *p-0* uses the task from the dataset and an example test case as prompt. For the MBPP dataset, each of the following four prompts begins with "Generate python code according to the description below." and for the Human Eval dataset with "Complete the following python code". The further additions are:

p-1: It should pass the provided test cases.
p-2: It should pass the provided test cases. Make sure to adhere to the pep8 style guidelines.
p-3: It should pass the provided test cases. It should pass the provided test cases. Make sure to write code with low complexity.
p-4: It should pass the provided test cases. It should pass the provided test cases. Make sure it adheres to the pep8 style guidelines and to write code with low complexity.

For MBPP tasks, the first of the respective sample's provided test cases is appended to the prompt, when provided test cases are mentioned. For human eval tasks, the task already includes test cases, so no further tests are included.

For Magicoder and the GPT models, we use additional system prompts:

Magicoder: You are an exceptionally intelligent coding assistant that consistently delivers accurate and reliable responses to user instructions.
GPT: You are GPT, an exceptionally intelligent coding assistant that consistently delivers accurate and reliable responses to user instructions.

For the remaining models, preliminary tests showed no positive effect of adding a system prompt, and for the Phi models it deteriorated performance, so we don't use system prompts for those. These prompts are joined with the tasks using each model's prompt template and given as input to the models.

4.2 Metrics

To evaluate the quality of generated code, we propose different metrics, which can be split into the two groups: similarity-based and reference independent metrics.

pass@1. The most important metric is whether test cases for a given task can be passed. We apply this as pass@1, where a single task is passed if all provided test cases for the task are passed.

BLEU Score. We also tried running BLEU and CodeBLEU scores, but found no new results of note for these scores, so we are omitting the results on those.

Tree-Edit Distance (TED). Another level to compare the similarity of code samples is the graph structure of the syntax tree of the code. The minimum-edit distance between the syntax trees has been used in the evaluation of the phi-1 model [12]. We apply this metric using the `networkx` library [2,13].

Isomorphic Distance. A similar approach to look at the overall tree structure is to search for tree isomorphisms. Two graphs G and H are isomorphic if there is a bijective mapping of nodes, such that for each connecting pair of nodes in G, there is also an edge connecting the mapped pair of nodes in H. As such, the isomorphism allows a binary statement, about whether two trees are isomorphic or not. To be able to use it as a more gradual metric (*iso-score*), we traverse the subtrees of each tree to test whether those are isomorph and then report the percentage of the nodes of both trees, which are part of an isomorphic subtree. Leaf nodes are excluded from the test, as each leaf node is automatically isomorphic to each other leaf node. This score ranges from 0 to 1, where 1 is a complete isomorphism of the generated code and ground truth, whereas 0 would be no overlap at all. To the best of our knowledge an approach like this has not yet been used to explore code similarity.

In addition, we can report an isomorphic depth, i.e., the depth of the isomorphic node closest to the root of the parse tree, normalized by the maximum depth of the tree. This gives an impression if the isomorphic similarity stems more from many isomorph subtrees close to the leaves, or from fewer larger subtrees close to the root. To use the depth as a metric (*iso-level*) with 1 denoting the highest value, i.e., the trees are completely isomorphic, we compute the mean of both generation and ground truth, and define:

$$iso\text{-}level = 1 - \frac{1}{2}\left(\frac{d_{gen}}{max\_d_{gen}} + \frac{d_{truth}}{max\_d_{truth}}\right),$$

where d_{gen} and max_d_{gen} resp. d_{truth} and max_d_{truth} denote the isomorphic depth and maximum depth of the generation resp. ground truth.

Cosine Similarity (CosSim). Another approach to asses similarities between two samples is to compute their distance in an embedding space. For NL this can be done for example using sentence BERT models to embed text and compute a distance like the cosine distance [25]. To be able to capture the nature of the text being code, we use the CodeBERT model [10], which has been trained on both code and NL data. A similar approach has been used in the evaluation of the Phi-1 model [12]. Here, the authors use the Codegen-350M model to compute embeddings and calculate the L2-distance as the similarity measure.

Task-Similarity (TaskSim). As a first independent metric, we compute a similarity between the description of the task and the generated code. This is derived from the CosSim approach and uses the exact same method, except instead of comparing to the oracle code, the generation is compared to the task description. For this, we use the CodeBERT model, as it has been trained both on code and NL to perform code search, and as such should be able to handle and compare NL and code. While similarity of embeddings has been used before to assess code [12], similarity between task description and code appears not to have been considered yet.

Maintainability (MI) and McCabe Number. A different approach to code evaluation is to look at the coding style. A classical evaluation focus for human written code is to inspect how accessible it is for human readers and how well it can be maintained. The Maintainability score MI [8,15] is intended to measure the ease of maintaining a piece of code and is used in software development environments such as, for example, Visual Studio [1]. It ranges from 0 to 100 with a higher value marking better code. MI comprises four components: the Cyclomatic Complexity G [19], the Halstead Volume V [14], the number of lines in the code L, and the percentage of comment lines C. It is constructed as a polynomial regression model using the components as variables to predict software engineers' subjective assessment of maintainability in software systems [8]. The formula we use is based on the description from the Radon library [20]:

$$MI = max(0, 100\frac{171 - 5.2\ln V - 0,23G + 16.2\ln L + 50\sin{(\sqrt{2.4C})}}{171})$$

The Cyclomatic Complexity or *McCabe* number [19] counts the number of decision blocks or control flow statements in the code, representing how many paths there are through the code. The number is intended as an indicator for how many test cases are necessary to be defined to capture all possible paths. It has a minimum value of one, in case there are no decision blocks.

The Halstead Volume is one of several Halstead Metrics [14] and it relates the program vocabulary size η to its length N as:

$$V = N \; log_2 \; \eta$$

There is a discussion on the reliability of these metrics and how well they really predict the maintenance effort of code [7]. One particular argument is, that they are susceptible to code size and behave differently for different orders of size. This can be somewhat alleviated by normalizing over code size (*MI-norm, McCabe-norm*). Since the tasks and code snippets of the used datasets are internally relatively consistent and overall rather small, the metrics should be able to give a reliable estimate with each dataset, however, this does make the metrics partially dependent on the datasets.

PEP 8. As we are evaluating the generation of python code specifically, we also test for the adherence of the generations to the PEP 8 standards [11]. The PEP 8 style guide is intended to improve readability of code and to achieve this, focuses on consistency. This can give insights into the accessibility of the code for human readers. We use the Flake8 linting tool[1] and count the occurrences of each type of error message as they are coded by Flake8, and the included add-ons pycodestyle[2].

PEP 8 Types. The resulting PEP8 score is open-ended in range and therefore more difficult to normalize independently. Therefore, we also consider a second less detailed variation, where we count the percentage of flake8 error message types, which are observed at all.

5 Results

The results for all experiments can be seen in Tables 2 and 3, split into compilability, passing, syntax tree- and embedding-based (Table 2) metrics, and maintenance and PEP8 metrics (Table 3). The results shown include only the best performing prompt variation for each model, as judged by the pass@1 score.

Table 2 shows that for pass@1, the GPT models notably outperform most other models, though Magicoder comes close and even outperforms GPT-3.5 on

Table 2. Code execution as well as tree-based and embedding-based metric Overview. The table shows only the best prompt variants per model with respect to pass@1. p-n stands for the prompt variant which achieved the reported results (see Sect. 4.1).

dataset	model	compiled	pass@1	iso-score	iso-level	TED	CosSim	TaskSim
mbpp	oracle	1	0.98	–	–	–	–	0.92
mbpp	codellama7b p-4	0.85	0.4	0.03	0.15	115.77	0.85	0.91
mbpp	codellama7b-py p-4	0.84	0.4	0.02	0.14	113.64	0.85	0.91
mbpp	gpt-35 p-0	1	0.79	0.05	0.25	109.99	0.84	0.92
mbpp	gpt-4 p-0	1	0.84	0.05	0.25	110.69	0.86	0.92
mbpp	gpt-4-turbo p-0	1	0.8	0.03	0.23	156.6	0.84	0.93
mbpp	magicoder p-1	0.99	0.72	0.07	0.31	95.4	0.86	0.92
mbpp	phi1 p-3	1	0.32	0.04	0.63	198.54	0.82	0.88
mbpp	phi2 p-0	0.95	0.54	0.06	0.35	130.42	0.88	0.93
he	oracle	1	1	–	–	–	–	0.99
he	gpt-35 p-0	0.96	0.54	0.05	0.33	155.47	0.82	0.97
he	gpt-4 p-0	0.99	0.72	0.05	0.32	144.82	0.82	0.97
he	gpt-4-turbo p-3	1	0.74	0.04	0.37	172.79	0.86	0.98
he	magicoder p-0	1	0.57	0.06	0.55	182.89	0.82	0.97
he	phi1 p-0	1	0.35	0.01	0.23	200.0	0.84	0.96
he	phi2 p-0	1	0.37	0.06	0.56	186.85	0.82	0.97

---
[1] https://flake8.pycqa.org/en/latest/.
[2] https://pycodestyle.pycqa.org/en/latest/index.html.

Table 3. Maintenance metric and style adherence Overview. The table shows only the best prompt variants per model with respect to pass@1. p-n stands for the prompt variant which achieved the reported results (see Sect. 4.1).

dataset	model	MI	MI-norm	McCabe	McCabe-norm	PEP8	PEP8 types
mbpp	oracle	0.72	17.92	2.53	0.49	1.37	1.97e−2
mbpp	codellama7b p-4	0.71	18.66	6.35	1.88	0.95	1.69e−2
mbpp	codellama7b-py p-4	0.72	20.66	6.47	2.06	1.0	1.67e−2
mbpp	gpt-35 p-0	0.77	18.84	9.43	2.05	0.59	0.82e−2
mbpp	gpt-4 p-0	0.8	24.09	11.17	3.05	0.84	1.08e−2
mbpp	gpt-4-turbo p-0	0.88	17.24	13.24	2.2	1.15	1.79e−2
mbpp	magicoder p-1	0.88	29.9	7.94	2.24	0.44	0.59e−2
mbpp	phi1 p-3	0.64	2.55	15.61	0.61	1.04	1.79e−2
mbpp	phi2 p-0	0.76	19.56	8.52	1.98	0.98	1.74e−2
he	oracle	0.74	6.69	7.63	0.53	1.07	1.54e−2
he	gpt-35 p-0	0.72	12.98	10.42	1.46	0.57	1.01e−2
he	gpt-4 p-0	0.76	16.64	12.95	2.49	0.66	1.08e−2
he	gpt-4-turbo p-3	0.81	12.01	13.14	1.61	1.71	1.52e−2
he	magicoder p-0	0.69	4.24	9.95	0.52	1.46	1.62e−2
he	phi1 p-0	0.59	1.57	18.48	0.48	1.04	1.8e−2
he	phi2 p-0	0.67	3.8	14.23	0.63	1.57	1.78e−2

Table 4. Results from the regression models. Multivariate linear regression, using all other metrics to predict pass@1. All metrics have been z-normalized to have a mean of 0 and a standard deviation of 1. Oracle instances are excluded as outliers in the similarity-based univariate models and in the multivariate model. The table is sorted by the multivariate effects value.

metric	Univariate		Multivariate	
	r^2	effect	r^2	effect
MI	**0.61**	0.16	**0.84**	0.22
McCabe-norm	0.42	0.18		0.11
iso-level	0.23	−0.13		0.07
compiled	0.15	0.11		0.05
PEP8	0.04	−0.05		0.02
CosSim	0.19	0.12		0.02
TaskSim	0.01	−0.02		0.00
McCabe	0.03	−0.05		−0.07
PEP8 types	0.3	−015		−0.08
iso-score	0.00	0.01		−0.15
TED	0.20	−0.12		−0.21
MI-norm	0.34	0.16		−0.25

the Human Eval dataset. Overall the table shows that both the datasets and the metrics disagree on which model or group of models performs best. On mbpp, magicoder has the highest iso-score, the lowest tree-edit-distance and with GPT-4 the highest Cosine Similarity. But the phi models perform best, when judged by iso-level and phi-2 has the highest Task Similarity together with GPT-4. On the HumanEval dataset, the GPT models show the best tree-edit-distance and the best embedding distances. However it is also noticable that the iso-score and the embedding similarities show very little variance across models, so that it is hard to say there is an actual best model.

Among the reference-independent metrics in Table 3, only one metric has such a low variance. That is the PEP8 types metric, for which all models achieve very low values and make no notable difference to the oracle. Generally, it can be seen in this table, that the oracle code does not perform much better than the models, if at all. But it can be seen again, that the metrics differ in which models perform best. While the GPT models achieve high maintainability, their complexity as defined by the McCabe norm is generally the highest.

We further explored the relation of each dependent and independent metric to the pass@1 score, respectively, via a linear regression analysis. For better comparability each predictor variable is z-normalized to have a mean of 0 and a standard deviation of 1. For each of these scaled metrics, we perform a univariate linear regression, to test how well they predict pass@1. The resulting r^2 score for each variable is shown in Table 4. For all similarity-based metrics, the oracle instances are excluded as outliers as they achieve maximal similarity by definition. Most of the metrics have only very little relation to pass@1. The highest score is achieved by the Maintainability MI with 0.61. The next strongest predictors are the normalized Maintainability MI-norm and normalized McCabe number.

Finally, multivariate regression models are fitted for each subset of variable combinations to predict pass@1. To compare the model variants, we evaluate by ten-fold cross validation and the results for the model with the best r^2-score can be seen in Table. 4. Here too, the oracle instances are included as outliers. This model utilizes all of the variables considered in this study and with 0.84 it achieves a much higher r^2 score than any of the univariate models. Maintainability (MI) shows the largest effects both normalized and original, but in inverse directions. The weakest effects come from the embedding-based scores CosSim and TaskSim and PEP8.

6 Discussion

When viewed as single predictors, none of the considered metrics come close to predicting pass@1 reliably enough to be applicable as a workaround for the use of unit tests. The similarity-based metrics show worse results than the reference-independent metrics, specifically the static code analysis-based ones. This means that, as they are now, the similarity-based metrics, are not useful. None of them capture similarity in a manner that makes it inherently useful to know that samples are similar. For the considered metrics, especially the embedding representations but also the tree-level views are to vague or to abstract in their representation for the comparison to mean much by itself. But when considering these comparisons it should also be considered, that similarities might become more predictive if more references per sample were given. Though this would again make the construction of datasets more expensive.

The reference-independent metrics suffer less in that regard. But even though their correlation with pass@1 is generally higher, it is not enough to promise reliable predictions. However, the static code analysis based metrics Maintainability, McCabe norm and PEP8 adherence still have value in that their values make useful statements on their own. The PEP8-style metrics address the style of the code and with that the readability for human readers. Here PEP8 shows more variance than PEP8 types (see Table 3) and has a more interpretable value, which makes it a more useful metric on its own. Maintainability gives a good summary on maintainability, however for more detailed interpretation it is best to also look at its components like the McCabe norm. The McCabe norm is maybe the most clearly interpretable value, as it attempts to directly measure the amount of tests necessary to adequately cover a piece of code. As such it could even serve as a simple measure of reliability of pass@n scores when compared to the actual number of tests given for a task. Compilability does not predict pass@1 either and once a model becomes good enough to generate mostly parsable output, it ceases to be a useful metric on its own. Task similarity suffers from the same issue as the similarity metrics, as it both badly predicts pass@1, and conveys no useful information on its own. In summary, the static code analysis based metrics show the most promise as they convey useful information about the generated code. But they can not serve as the only metrics, as they don't provide critical information like the chance of generated code to fulfil the requested task.

The predictive power of the metrics for pass@1 increases notably, when joint in a multivariate regression. Even though some variables appear useless as single predictors in Table 4, the multivariate model with the best r^2 score is one that includes all of the variables. It is also notable that the strength of effect of each variable in the multivariate model does not mirror the correlations as estimated by the univariate models. So, even if a metric by itself is not useful to evaluate the code, it might still be worth to consider including it in the construction of a joint metric or evaluation model.

7 Conclusion

Our paper presents the analysis of the evaluation study conducted on 4 large language models and their variations and 13 evaluation metrics for code quality to determine how reliably the metrics evaluate the quality of code generated for a NL task description. The evaluation metrics included the commonly used pass@1, four similarity-based metrics, which compare generated code to references from the dataset, and six reference-independent metrics. Of the similarity metrics Cosine Similarity and tree-edit-distance have been used before, but the isomorphism-based metrics are by our design. However, none of these showcase much use in our current study. The similarity metrics might gain from the inclusion of more references and the embedding-based approaches specifically might also gain from training embedding models more specifically on such comparison tasks. As reference-dependent metrics, we proposed the embedding-based Task Similarity with no success and the use of the static code analysis approaches Maintainability, McCabe number and adherence to PEP8 standards. The Maintainability and McCabe norm appear most promising for future use, as they make a clear statement on their own and show the strongest correlation with the pass@1 score. In addition, we found that joining all metrics in a multivariate regression model can predict pass@1 better than any single one could, which suggests the use of a composite approach to derive evaluation from a combination of metrics. While the PEP8 metrics didn't provide much information in the context of this study, it would be useful to further look into the specific types of violations made by specific models to explore different model behaviours. Overall, we only tested on a small number of datasets and future work might consider extending on this aspect as well as exploring different languages.

Acknowledgments. This study received funding from the Bayrisches Staatsministerium für Wissenschaft und Kunst under the grant number: H.2-F1116.IN/47/2.

Disclosure of Interests. The authors have no competing interests to declare that are relevant to the content of this article.

References

1. Code metrics values. https://learn.microsoft.com/en-us/visualstudio/code-quality/code-metrics-values?view=vs-2022. Accessed 13 Feb 2023
2. Abu-Aisheh, Z., Raveaux, R., Ramel, J.Y., Martineau, P.: An exact graph edit distance algorithm for solving pattern recognition problems. In: 4th International Conference on Pattern Recognition Applications and Methods 2015, Lisbon, Portugal (2015). https://doi.org/10.5220/0005209202710278. https://hal.science/hal-01168816
3. Austin, J., et al.: Program synthesis with large language models. arXiv preprint arXiv:2108.07732 (2021)
4. Bafatakis, N., et al.: Python coding style compliance on stack overflow. In: 2019 IEEE/ACM 16th International Conference on Mining Software Repositories (MSR), pp. 210–214. IEEE (2019)
5. Buse, R.P., Weimer, W.R.: Learning a metric for code readability. IEEE Trans. Softw. Eng. **36**(4), 546–558 (2009)
6. Chen, M., et al.: Evaluating large language models trained on code. arXiv preprint arXiv:2107.03374 (2021)
7. Chowdhury, S., Holmes, R., Zaidman, A., Kazman, R.: Revisiting the debate: are code metrics useful for measuring maintenance effort? Empir. Softw. Eng. **27**(6), 158 (2022)
8. Coleman, D., Ash, D., Lowther, B., Oman, P.: Using metrics to evaluate software system maintainability. Computer **27**(8), 44–49 (1994)

9. Evtikhiev, M., Bogomolov, E., Sokolov, Y., Bryksin, T.: Out of the BLEU: how should we assess quality of the Code Generation models? J. Syst. Softw. **203**, 111741 (2023)
10. Feng, Z., et al.: Codebert: a pre-trained model for programming and natural languages. arXiv preprint arXiv:2002.08155 (2020)
11. Guido van Rossum, Barry Warsaw, Alyssa Coghlan: PEP 8 - Style Guide for Python Code (2001). https://peps.python.org/pep-0008/
12. Gunasekar, S., et al.: Textbooks Are All You Need (2023). https://arxiv.org/pdf/2306.11644v2.pdf. _eprint: 2306.11644
13. Hagberg, A.A., Schult, D.A., Swart, P.J.: Exploring network structure, dynamics, and function using networkX. In: Varoquaux, G., Vaught, T., Millman, J. (eds.) Proceedings of the 7th Python in Science Conference, Pasadena, CA, USA, pp. 11–15 (2008)
14. Halstead, M.H.: Elements of Software Science (Operating and programming systems series). Elsevier Science Inc. (1977)
15. Heitlager, I., Kuipers, T., Visser, J.: A practical model for measuring maintainability. In: 6th International Conference on the Quality of Information and Communications Technology (QUATIC 2007), pp. 30–39 (2007). https://doi.org/10.1109/QUATIC.2007.8. https://ieeexplore.ieee.org/stamp/stamp.jsp?tp=&arnumber=4335232
16. Li, Y., Bubeck, S., Eldan, R., Giorno, A.D., Gunasekar, S., Lee, Y.T.: Textbooks are all you need ii: phi-1.5 technical report (2023)
17. Li, Y., et al.: Competition-level code generation with alphacode. Science **378**(6624), 1092–1097 (2022)
18. Liu, J., Xia, C.S., Wang, Y., ZHANG, L.: Is your code generated by ChatGPT really correct? rigorous evaluation of large language models for code generation. In: Oh, A., Neumann, T., Globerson, A., Saenko, K., Hardt, M., Levine, S. (eds.) Advances in Neural Information Processing Systems, vol. 36, pp. 21558–21572. Curran Associates, Inc. (2023)
19. McCabe, T.: A complexity measure. IEEE Trans. Softw. Eng. **SE-2**(4), 308–320 (1976). https://doi.org/10.1109/TSE.1976.233837. https://ieeexplore.ieee.org/stamp/stamp.jsp?tp=&arnumber=1702388
20. Lacchia, M.: Introduction to Code Metrics (2020). https://radon.readthedocs.io/en/latest/intro.html
21. Nijkamp, E., et al.: Codegen: an open large language model for code with multi-turn program synthesis. arXiv preprint arXiv:2203.13474 (2022)
22. Achiam, J., et al.: OpenAI: Gpt-4 technical report (2023)
23. Papineni, K., Roukos, S., Ward, T., Zhu, W.J.: Bleu: a method for automatic evaluation of machine translation. In: Isabelle, P., Charniak, E., Lin, D. (eds.) Proceedings of the 40th Annual Meeting of the Association for Computational Linguistics, pp. 311–318. Association for Computational Linguistics, Philadelphia (2002). https://doi.org/10.3115/1073083.1073135. https://aclanthology.org/P02-1040
24. Radford, A., Narasimhan, K., Salimans, T., Sutskever, I., et al.: Improving language understanding by generative pre-training (2018)
25. Reimers, N., Gurevych, I.: Sentence-bert: sentence embeddings using siamese bert-networks (2019). https://arxiv.org/abs/1908.10084
26. Ren, S., et al.: CodeBLEU: a method for automatic evaluation of code synthesis (2020). https://arxiv.org/pdf/2009.10297.pdf. _eprint: 2009.10297
27. Rozière, B., et al.: Code Llama: open foundation models for code (2023). https://arxiv.org/pdf/2308.12950.pdf. _eprint: 2308.12950

28. Wang, Y., Wang, W., Joty, S., Hoi, S.C.: Codet5: identifier-aware unified pretrained encoder-decoder models for code understanding and generation. arXiv preprint arXiv:2109.00859 (2021)
29. Wei, Y., Wang, Z., Liu, J., Ding, Y., Zhang, L.: Magicoder: source code is all you need (2023)

Kernel Least Squares Transformations for Cross-Lingual Semantic Spaces

Adam Mištera[1] and Tomáš Brychcín[2]

[1] Department of Computer Science and Engineering, Faculty of Applied Sciences, University of West Bohemia, Pilsen, Czech Republic
amistera@kiv.zcu.cz
[2] NTIS – New Technologies for the Information Society, Faculty of Applied Sciences, University of West Bohemia, Pilsen, Czech Republic
brychcin@kiv.zcu.cz

Abstract. The rapid development in the field of natural language processing (NLP) and the increasing complexity of linguistic tasks demand the use of efficient and effective methods. Cross-lingual linear transformations between semantic spaces play a crucial role in this domain. However, compared to more advanced models such as transformers, linear transformations often fall short, especially in terms of accuracy. It is thus necessary to employ innovative approaches that not only enhance performance but also maintain low computational complexity.

In this study, we propose Kernel Least Squares (KLS) for linear transformation between semantic spaces. In our comprehensive analysis involving three intrinsic and two extrinsic experiments across six languages from three different language families and a comparative evaluation with nine different linear transformation methods, we demonstrate the superior performance of KLS. Our results show that the proposed method significantly improves word translation accuracy, thereby standing out as the most efficient method for transforming only the source semantic space.

Keywords: cross-lingual transformations · kernels · linear transformations · semantic spaces

1 Introduction

Semantic spaces are based on the *Distributional Hypothesis* [18], which states that the meaning of a word is determined by its surroundings, so words that occur in similar contexts will also have similar meanings. Based on this hypothesis, several different semantic models were consequently developed [5, 25, 28].

The natural next step was the development of methods that would allow semantic spaces to be transformed between each other, or to create a unified space across several different languages. We can divide these methods for cross-lingual transformations into two basic groups, the supervised and unsupervised methods. Supervised methods are most often based on linear transformations [2,6,13,19,22,26] and only need to build a dictionary containing a few thousand words [31]. The second group consists of unsupervised methods [3,11,21] that produce their own dictionary based on internal similarities in the given spaces. In this case, the dictionary is created only during the transformation and is of very high quality compared to the supervised approach.

Cross-lingual semantic spaces find application in many different NLP tasks. For instance, they can be used for sentiment analysis [1,27], document classification [20], or syntactic dependency parsing across languages [16]. Recently, more complex models based on the transformer architecture [30], such as BERT models [12,32], have been increasingly used for these tasks. However, experiments show that semantic spaces in certain cases can still achieve comparable quality at significantly lower computational cost [29].

In this paper, we present the *Kernel Least Squares* (KLS) method, a new approach proposed to improve the accuracy and efficiency of cross-lingual semantic transformations. Unlike traditional methods, the KLS method uses kernel-based techniques to capture nonlinear relationships in the data, which promises to significantly improve word translation accuracy and overall performance on a variety of language tasks.

The paper is organized as follows: Sect. 2 provides a detailed analysis of previous work in the field of cross-lingual transformations. In Sect. 3 we describe the kernels and our proposed transformation KLS. Section 4 presents the experimental setup, while Sect. 5 presents the measured results. Finally, we conclude in Sect. 6.

2 Cross-Lingual Transformations Between Semantic Spaces

A cross-lingual linear transformation between semantic spaces can be defined as:

$$\mathbf{Y} = \mathbf{T}^{x \to y} \mathbf{X}, \tag{1}$$

where matrices $\mathbf{X} \in \mathbb{R}^{m \times d}$ and $\mathbf{Y} \in \mathbb{R}^{m \times d}$ are constructed using the dictionary $D^{x \to y}$ of word pairs (w^x, w^y), where x is the source language and y is the target language. Symbol m represents the size of the used dictionary, and d represents the dimension of the semantic space. In the following text, we present nine different cross-lingual transformations that were used for comparison in the experiments with our proposed transformation. They all differ in how they estimate the matrix $\mathbf{T}$.

The first of these transformations, the *Least Squares Transformation* [26], minimizes the total squared differences between paired word vectors in two languages aligned according to a bilingual dictionary. This approach offers an analytical solution through singular value decomposition (SVD). It uses a transformation matrix derived via the Moore-Penrose pseudoinverse, which provides a direct estimation for mapping between semantic spaces.

Orthogonal Transformation [2] extends the least squares approach by enforcing the orthogonality of the transformation matrix to preserve the angles between the word vectors of the transformed semantic space. This method also uses SVD to compute the optimal transformation matrix.

The next method, *Canonical Correlation Analysis* [13] tries to maximize the correlation between projected vectors of two sets from different languages, finding optimal basis vectors for each set. This method uses canonical directions to project words into a shared semantic space, enhancing multilingual semantic performance.

Ranking Transformation [22] employs a max-margin hinge loss to reduce hubness in high-dimensional spaces, optimizing the alignment of words across languages by adjusting their ranks. The goal is to ensure the correct alignment of words across languages by optimizing their relative ranks within the semantic space.

Orthogonal Ranking Transformation [6] adds orthogonal constraints to the Ranking Transformation to address asymmetry problem in cross-lingual mappings. It uses two max-margin loss functions to optimize transformation matrix towards being *nearly orthogonal*, thus preserving the monolingual performance.

Geometry-aware Multilingual Mapping [19] aligns semantic spaces using orthogonal transformations and *Mahalanobis* metrics allowing for more efficient learning of similarity measures. This method optimizes both the source and target semantic spaces to achieve more accurate semantic alignment.

Vector Mapping [3] is an unsupervised method that iteratively refines the transformation matrix without the need for labeled parallel data, which is ideal for resource-limited languages. It starts by aligning semantic spaces and refines the mapping using a self-learning algorithm focusing on the internal structures of the semantic spaces.

Multilingual Unsupervised and Supervised Embeddings [11] is an unsupervised method that uses adversarial training to align semantic spaces without the need for bilingual dictionaries. It improves the alignment quality by using a discriminator that tries to distinguish between transformed and real word vectors of the target semantic space.

Kernel Canonical Correlation Analysis [4] is method based on the previously mentioned canonical correlation analysis, but enhances it by using kernels. By using kernels, the nonlinear relationship between two languages can be captured using a linear transformation. The obtained transformation matrix can thus be used in the same way as in the previous transformations.

3 Proposed Transformation

In the proposed *Kernel Least Squares Transformation* method, we build on the first proposed transformation, the *Least Squares Transformation* method, but greatly improve it by using *kernels*. The use of kernels allows us to significantly enhance the accuracy of the transformations by facilitating the capture of complex, high-dimensional relationships in the data. Since kernels are an essential part of our transformation, we will briefly introduce them in more detail in the following Sect. 3.1.

3.1 Kernel Function

The use of a *kernel* or *kernel function* $\kappa : \mathcal{S} \times \mathcal{S} \to \mathbb{R}$ in the proposed transformation allows us to avoid the explicit mapping that is necessary for linear learning algorithms to model a nonlinear function. We define the kernel between any two data inputs a_i and a_j from the space $\mathcal{S} = \{a_1, \ldots, a_n\}$ as:

$$\kappa(a_i, a_j) = \langle \varphi(a_i), \varphi(a_j) \rangle, \qquad (2)$$

where $\varphi : \mathcal{S} \to \mathcal{F}$ represents a mapping to a feature space $\mathcal{F}$, and the function $\langle \cdot, \cdot \rangle$ denotes a generalized inner product in the feature space, that can be used for both vectors and matrices. This transformation of inputs a_i and a_j into a higher-dimensional space is essential for identifying latent structures in semantic spaces and enhancing transformations.

To further explain the use of kernels in the transformation, it is useful to introduce the concept of *kernel matrix* or *Gram matrix*. This $n \times n$ matrix, denoted as $\mathbf{K}$, is created by evaluating the chosen kernel function for each pair of data points in a given data set. The entries of the matrix are given as:

$$\mathbf{K}_{ij} = \langle \varphi(a_i), \varphi(a_j) \rangle = \kappa(a_i, a_j). \qquad (3)$$

The kernel matrix is symmetric since $\mathbf{K}_{ij} = \mathbf{K}_{ji}$, and contains the pairwise comparisons within the space.

Table 1. Kernel types.

Kernel Function	Kernel Formula
LK	$\kappa(a_i, a_j) = a_i^T a_j$
PK	$\kappa(a_i, a_j) = (a_i^T a_j + c)^d$
RBF	$\kappa(a_i, a_j) = \exp\left(-\gamma \|a_i - a_j\|_2^2\right)$
CSK	$\kappa(a_i, a_j) = a_i^T a_j / (\|a_i\|_2 \|a_j\|_2)$

Several commonly used types of kernels are defined in the Euclidean space $\mathbb{R}^d$. Choosing the right kernel type is crucial for the final performance of the transformation, as different kernels can significantly affect the resulting quality of transformed semantic space and thus the results of individual experiments. In our experiments, we employ four different types of kernels, namely, *Linear Kernel* (LK), *Polynomial Kernel* (PK), *Radial Basis Function Kernel* (RBF), and *Cosine Similarity Kernel* (CSK). The formulas for their calculation are given in Table 1. Each of these kernels offers distinct advantages and is key to the success of the semantic space transformation. Most of these kernels, especially RBF, are commonly encountered in the context of SVM algorithms, which make extensive use of them to find nonlinear decision boundaries [10]. However, in the field of NLP we often see polynomial kernels of lower degrees [10,14].

3.2 Kernel Least Squares Transformation

The optimization criterion is defined by Eq. 4, where $\mathbf{K}$ denotes the kernel matrix computed from the input data $\mathbf{X}$, encapsulating the similarities between data points. The symbol $\hat{\mathbf{T}}^{x \to y}$ indicates the optimal transformation matrix.

$$\hat{\mathbf{T}}^{x \to y} = \underset{\mathbf{T}^{x \to y}}{\arg\min} \left\| \mathbf{Y} - \mathbf{K}\mathbf{T}^{x \to y} \right\|_2^2. \qquad (4)$$

If we take the derivative of the equation with respect to $\hat{\mathbf{T}}^{x \to y}$, then the analytic solution is given as $\hat{\mathbf{T}}^{x \to y} = \mathbf{K}^{-1}\mathbf{Y}$. However, in practice, to increase stability and ensure that the $\mathbf{K}$ matrix is invertible, we add the regularization term λ. The final equation with this term is then given by:

$$\hat{\mathbf{T}}^{x \to y} = (\mathbf{K} + \lambda \mathbf{I})^{-1} \mathbf{Y}. \qquad (5)$$

Here $\lambda > 0$ is a scalar value that adjusts the regularization strength, and $\mathbf{I}$ represents the identity matrix.

Subsequently, we can transform arbitrary word vector $\mathbf{x}$ from source semantic space to target semantic space with the following equation

$$\hat{y} = \kappa(\mathbf{x}, \mathbf{X})\hat{\mathbf{T}}^{x \to y}. \qquad (6)$$

Essentially, we create a new kernel matrix for the transformed vector $\mathbf{x}$ that contains the calculated similarities of this vector to all vectors from the input data $\mathbf{X}$ in the given feature space. Therefore, this equation can be used to easily transform each word vector from the source to the target semantic space.

The whole process of the proposed transformation method then involves several key steps. First, we construct a kernel matrix $\mathbf{K}$ with the chosen kernel function $\kappa(\mathbf{x}_i, \mathbf{x}_j)$, which we evaluate for each pair of vectors $\mathbf{x}_i, \mathbf{x}_j \in \mathbf{X}$. We then calculate the optimal transformation matrix $\hat{\mathbf{T}}^{x \to y}$ that achieves the smallest values of the loss function as given in Eq. 4. In the last step, it is necessary to compute a kernel $\kappa(\mathbf{x}, \mathbf{X})$ between all $\mathbf{x}$ from the source semantic space $\mathcal{S}^x$ and matrix $\mathbf{X}$, enabling the transformation of the word vectors according to Eq. 6. After these steps, the transformation is complete and the semantic space can then be used in any subsequent downstream tasks.

3.3 Preprocessing and Postprocessing

A necessary step to ensure optimal transformation quality is the appropriate configuration and application of preprocessing and postprocessing techniques to the semantic spaces. Both semantic spaces are therefore first column-wise centered, followed by unit vector normalization, as shown in [2]. Normalizing the semantic spaces ensures that the vectors contribute equally to the transformation and their length does not affect the mapping.

Our research showed that applying the above steps to the source semantic space after the transformation has been performed can further improve the measured results. For this reason, in Sect. 5, we present the results of methods with this setup, i.e. postprocessing. In the following Sect. 4, we present the experiments used to evaluate the quality of the proposed transformation and the measured results.

4 Experiments

In total, we conducted five different experiments, three intrinsic ones, namely *Word Translation* (WT), *Cross-lingual Word Analogies* (WA), and *Cross-lingual Word Similarities* (WS), and two extrinsic ones, *Topic Classification* (TC), and *Sentiment Analysis* (SA). The individual experiments are described in more detail in Sect. 4.2. All experiments were performed in both directions, i.e. from the source semantic space to the target space and vice versa. This allowed us to more accurately assess the resulting quality of individual transformations, since some transformations may perform very well in the first direction and fail in the opposite direction due to the asymmetry problem [6].

Nine additional linear transformations have been tested, including seven supervised methods, namely *Least Squares Transform* (LS), *Orthogonal Transform* (OT), *Canonical Correlation Analysis* (CCA), *Ranking Transform* (RT), *Orthogonal Ranking Transformation* (ORT), *Geometry-aware Multilingual Mapping* (GEOMM), and *Kernel Canonical Correlation Analysis* (KCAA) and two unsupervised methods, namely *Vector Mapping* (VM) and *Multilingual Unsupervised and Supervised Embeddings* (MUSE). The evaluation is conducted on six languages from different language families, namely Czech (Cs), Croatian (Hr), English (En), German (De), Italian (It), and Spanish (Es).

4.1 Experimental Setup

To create the transformation matrices $\mathbf{X} \in \mathbb{R}^{m \times d}$ and $\mathbf{Y} \in \mathbb{R}^{m \times d}$, where $m = 20,000$ as recommended in [6], we first translated the top $50,000$ words from the source language into the target language using *Google Translate*. We then divided the created bilingual dictionary into train and test subsets, reserving the latter for later use in the word translation task. For all tested languages we employed semantic spaces with dimension size $d = 300$ pre-trained on a corpus combined from *Common Crawl* and *Wikipedia* [15].

All transformations were performed with the recommended settings and with the previously mentioned dictionary size. However, only the source semantic space was transformed for the experiments, while the target space remained unchanged to ensure a fair comparison, except for the MUSE and KCAA methods, which do not support this constraint. The RT and ORT methods were configured to use five negative samples, with the parameter $\gamma = 0.275$. The VM and MUSE transformations were run in completely unsupervised mode without a provided train dictionary. Preprocessing was applied to all methods before transformation, and the process was repeated after the transformation was completed.

4.2 Evaluation Metrics

Word Translation. In this experiment, words from the source semantic space are transformed into the target semantic space using a transformation matrix, and the success of finding correct translations is tested by checking whether the k-nearest neighbors of the transformed word vector contain an accurate translation. Two different settings of the number of neighbors were used to evaluate the accuracy of translations, specifically $k = 1$ and $k = 5$. The test part of the dictionary was used to evaluate results. This experiment was performed for all language pairs.

Cross-lingual Word Analogies. This experiment serves for evaluating cross-lingual word analogies using a dataset divided into semantic and syntactic categories [7]. The experiment involves predicting a word that completes an analogy based on relationships between word pairs across languages. To find the target word that best matches the analogy, vector operations are used. As in the previous case, this experiment was conducted in two settings ($k = 1$ and $k = 5$). This experiment was performed for all language pairs.

Cross-lingual Word Similarities. In this intrinsic experiment, we evaluate cross-lingual word similarities using three datasets with predefined word pair similarities in multiple languages, namely *RG-65* [9], *SemEval2017* [8], and *WS353* [9]. Each dataset uses a scale to measure the similarity between word pairs, with scores provided by human annotators. The experiment evaluates the quality of transformations by comparing the cosine similarity of the word vectors with provided scores using Pearson's correlation coefficient, where a higher correlation means a more accurate transformation.

Topic Classification. First extrinsic evaluation involves topic classification using the RCV2 Reuters dataset [23], where documents are categorized into four predefined categories in *English*, *German*, and *Spanish*. For classification evaluation, we train two different models based on *Convolutional Neural Networks* (CNN) and *Long Short-Term Memory* (LSTM) architectures, which contain significantly fewer parameters than transformers. The experiment tests the classification accuracy using *F-measure*. The experiment was divided into four subtasks. Firstly, the CNN was trained on the transformed source semantic space, and the classification was evaluated on the unmodified target space. Then, the experiment was repeated with training on the target space and evaluation on the source space. The same experiments were also performed for the LSTM.

Sentiment Analysis. Finally, a sentiment analysis experiment for movie reviews was conducted using two datasets for *Czech* and *English*, divided into positive and negative sentiments. The datasets, CSFD [17] and IMDB [24], involve $90,000$ and $50,000$ reviews respectively. Similar to the previous classification task, this experiment employs CNN and LSTM models, with performance evaluated using the *F-measure*. As in the previous case, this experiment was divided into four subtasks.

5 Results

Table 2 shows the complete results measured in our experiments for all tested methods for transforming semantic spaces. The numbers in the columns correspond to the measured results for the individual experiments discussed in Sect. 4.2. As shown in the table, the proposed method with the *polynomial kernel* achieved the best result among all the transformations tested with an average score across all task equal to 0.655. The proposed method particularly excels at the word translation task, where it achieved the highest value. With a value of 0.452, it outperforms the second-best method by more than three percent, representing a significant improvement for this category. The second place was also taken by the proposed method, but this time with the *radial basis function kernel*, achieving an average score 0.645, which is comparable to the result of the ORT method. The KCCA transformation, which also uses kernels, performed especially well in the topic classification and sentiment analysis tasks, where it achieved the best scores.

Table 2. Overall results of all experiments. For each experiment, we show the average across all languages tested in the experiment and across both directions of transformation. The last column, denoted as AVG, contains the average score across all experiments performed in both directions.

	WT	WA	WS	TC	SA	AVG
LS	0.393	0.559	0.685	0.753	0.786	0.635
OT	0.382	0.569	0.694	0.747	0.784	0.635
CCA	0.385	**0.570**	0.691	0.752	0.786	0.636
RT	0.363	0.547	0.691	0.716	0.766	0.617
ORT	0.419	0.561	**0.727**	0.729	0.786	0.644
GEOMM	0.413	0.530	0.732	0.725	0.778	0.636
VM	0.336	0.537	0.698	0.755	0.784	0.622
MUSE	0.281	0.489	0.675	0.748	0.747	0.588
KCCA	0.329	0.517	0.669	**0.798**	**0.811**	0.625
KLS+LK	0.393	0.559	0.685	0.756	0.785	0.636
KLS+CSK	0.393	0.559	0.685	0.752	0.783	0.634
KLS+RBF	0.418	0.567	0.706	0.746	0.786	0.645
KLS+PK	**0.452**	0.563	0.719	0.748	0.793	**0.655**

The first two kernels, specifically the *linear kernel* and the *cosine similarity kernel*, achieved comparable results to the ordinary least squares method, especially in the first three categories. The results suggests that these kernels therefore do not introduce any additional information to the transformation that would improve it in the result. This not only shows that the linear kernel and the cosine similarity kernel are completely equivalent when using unit normalization of semantic space vectors, but also highlights the importance of an appropriate choice of preprocessing and postprocessing techniques for semantic spaces.

The *radial basis function kernel* performed overall very well, improving on the previous two kernels in all measured categories. It also performed the best of all the kernels tested in the cross-lingual word analogy category. Despite its popularity and frequent use, however, it has not achieved the best results in these experiments overall. This may be caused by the overly complex nature of the kernel, as the feature space of this kernel has infinitely many dimensions, which may reduce the quality of the resulting transformation.

The best performing kernel was the *polynomial kernel*. We conducted several experiments with different settings of the polynomial degree for this kernel, among which the kernel with polynomial of degree $d = 6$ performed the best. The final scores of the measured experiments gradually decreased with increasing and decreasing polynomial degree, respectively.

Table 3. Average results of the KLS method over all experiments for all language pairs.

	Cs	De	En	Es	Hr	It
Cs	1.000	0.478	0.664	0.449	0.424	0.462
De	0.504	1.000	0.699	0.612	0.440	0.535
En	0.681	0.683	1.000	0.696	0.554	0.645
Es	0.459	0.574	0.689	1.000	0.415	0.614
Hr	0.451	0.411	0.510	0.388	1.000	0.402
It	0.484	0.515	0.645	0.623	0.437	1.000

The Table 3 shows the results for each language pair for the KLS method. In the table we can see that transformations to or from the English semantic space yield the best results. This is an expected behavior, since English is the most resource-rich language and thus its semantic spaces are generally of high quality. To improve the results of individual tasks, it is therefore worth transforming the semantic space into such a space.

6 Conclusion

In this paper, we introduced a new transformation method, *Kernel Least Squares*, which uses kernels to improve the quality of the transformation and the resulting semantic space. Our findings show that this newly proposed method outperforms existing transformations that focus only on modifying the source semantic space, on average across all languages and experiments. It also completely dominates in the word translation task compared to all tested transformations. At the same time, *Kernel Canonical Correlation Analysis* achieved the best scores for the topic classification and sentiment analysis tasks. Thus, it can be concluded that kernel transforms are definitely a promising direction.

The main advantage of the proposed method is significantly lower computational complexity compared to more advanced models based on transformer architecture. In addition, the newly proposed method provides opportunities for further improvement by designing custom kernels that can further optimize the resulting transformation.

Acknowledgments. This work has been partly supported by grant No. SGS-2022-016 Advanced methods of data processing and analysis. Computational resources were provided by the e-INFRA CZ project (ID:90254), supported by the Ministry of Education, Youth and Sports of the Czech Republic.

References

1. Abdalla, M., Hirst, G.: Cross-lingual sentiment analysis without (good) translation. In: Proceedings of the Eighth International Joint Conference on Natural Language Processing, vol. 1: Long Papers), pp. 506–515. Asian Federation of Natural Language Processing, Taipei (2017). https://aclanthology.org/I17-1051
2. Artetxe, M., Labaka, G., Agirre, E.: Learning principled bilingual mappings of word embeddings while preserving monolingual invariance. In: Proceedings of the 2016 Conference on Empirical Methods in Natural Language Processing, pp. 2289–2294. Association for Computational Linguistics, Austin (2016). https://aclweb.org/anthology/D16-1250
3. Artetxe, M., Labaka, G., Agirre, E.: A robust self-learning method for fully unsupervised cross-lingual mappings of word embeddings. In: Proceedings of the 56th Annual Meeting of the Association for Computational Linguistics, vol. 1: Long Papers, pp. 789–798. Association for Computational Linguistics, Melbourne (2018). https://doi.org/10.18653/v1/P18-1073. https://aclanthology.org/P18-1073
4. Bai, X., Cao, H., Zhao, T.: Improving vector space word representations via kernel canonical correlation analysis. ACM Trans. Asian Low-Resour. Lang. Inf. Process. **17**(4) (2018). https://doi.org/10.1145/3197566
5. Bojanowski, P., Grave, E., Joulin, A., Mikolov, T.: Enriching word vectors with subword information. Trans. Assoc. Comput. Linguist. **5**, 135–146 (2017). https://transacl.org/ojs/index.php/tacl/article/view/999
6. Brychcín, T.: Linear transformations for cross-lingual semantic textual similarity. Knowl.-Based Syst. **187**, 104819 (2020)
7. Brychcín, T., Taylor, S., Svoboda, L.: Cross-lingual word analogies using linear transformations between semantic spaces. Expert Syst. Appl. **135**, 287–295 (2019)
8. Camacho-Collados, J., Pilehvar, M.T., Collier, N., Navigli, R.: Semeval-2017 task 2: Multilingual and cross-lingual semantic word similarity. In: Proceedings of the 11th International Workshop on Semantic Evaluation (SemEval-2017), pp. 15–26. Association for Computational Linguistics, Vancouver (2017). http://www.aclweb.org/anthology/S17-2002
9. Camacho-Collados, J., Pilehvar, M.T., Navigli, R.: A framework for the construction of monolingual and cross-lingual word similarity datasets. In: Proceedings of ACL, no. 2, pp. 1–7 (2015)

10. Chang, Y.W., Hsieh, C.J., Chang, K.W., Ringgaard, M., Lin, C.J.: Training and testing low-degree polynomial data mappings via linear svm. J. Mach. Learn. Res. **11**(48), 1471–1490 (2010). http://jmlr.org/papers/v11/chang10a.html
11. Conneau, A., Lample, G., Ranzato, M., Denoyer, L., Jégou, H.: Word translation without parallel data. arXiv preprint arXiv:1710.04087 (2017)
12. Devlin, J., Chang, M.W., Lee, K., Toutanova, K.: BERT: pre-training of deep bidirectional transformers for language understanding. In: Proceedings of the 2019 Conference of the North American Chapter of the Association for Computational Linguistics: Human Language Technologies, vol. 1 (Long and Short Papers), pp. 4171–4186. Association for Computational Linguistics, Minneapolis (2019). https://doi.org/10.18653/v1/N19-1423. https://aclanthology.org/N19-1423
13. Faruqui, M., Dyer, C.: Improving vector space word representations using multilingual correlation. In: Proceedings of the 14th Conference of the European Chapter of the Association for Computational Linguistics, pp. 462–471. Association for Computational Linguistics, Gothenburg (2014). http://www.aclweb.org/anthology/E14-1049
14. Goldberg, Y., Elhadad, M.: splitSVM: fast, space-efficient, non-heuristic, polynomial kernel computation for NLP applications. In: Moore, J.D., Teufel, S., Allan, J., Furui, S. (eds.) Proceedings of ACL-08: HLT, Short Papers, pp. 237–240. Association for Computational Linguistics, Columbus (2008). https://aclanthology.org/P08-2060
15. Grave, E., Bojanowski, P., Gupta, P., Joulin, A., Mikolov, T.: Learning word vectors for 157 languages. In: Proceedings of the International Conference on Language Resources and Evaluation (LREC 2018) (2018)
16. Guo, J., Che, W., Yarowsky, D., Wang, H., Liu, T.: Cross-lingual dependency parsing based on distributed representations. In: Proceedings of the 53rd Annual Meeting of the Association for Computational Linguistics and the 7th International Joint Conference on Natural Language Processing, vol. 1: Long Papers, pp. 1234–1244. Association for Computational Linguistics, Beijing (2015). http://www.aclweb.org/anthology/P15-1119
17. Habernal, I., Ptáček, T., Steinberger, J.: Sentiment analysis in Czech social media using supervised machine learning. In: Proceedings of the 4th Workshop on Computational Approaches to Subjectivity, Sentiment and Social Media Analysis, pp. 65–74. Association for Computational Linguistics, Atlanta (2013). https://aclanthology.org/W13-1609
18. Harris, Z.: Distributional structure. Word **10**(23), 146–162 (1954)
19. Jawanpuria, P., Balgovind, A., Kunchukuttan, A., Mishra, B.: Learning multilingual word embeddings in latent metric space: a geometric approach. Trans. Assoc. Comput. Linguist. **7**, 107–120 (2019). https://doi.org/10.1162/tacl_a_00257
20. Klementiev, A., Titov, I., Bhattarai, B.: Inducing crosslingual distributed representations of words. In: Proceedings of COLING 2012, pp. 1459–1474. The COLING 2012 Organizing Committee, Mumbai (2012). http://www.aclweb.org/anthology/C12-1089
21. Lample, G., Conneau, A., Ranzato, M., Denoyer, L., Jégou, H.: Word translation without parallel data. In: International Conference on Learning Representations (2018). https://openreview.net/forum?id=H196sainb

22. Lazaridou, A., Dinu, G., Baroni, M.: Hubness and pollution: Delving into cross-space mapping for zero-shot learning. In: Proceedings of the 53rd Annual Meeting of the Association for Computational Linguistics and the 7th International Joint Conference on Natural Language Processing, vol. 1: Long Papers, pp. 270–280. Association for Computational Linguistics, Beijing (2015). http://www.aclweb.org/anthology/P15-1027
23. Lewis, D.D., Yang, Y., Russell-Rose, T., Li, F.: Rcv1: a new benchmark collection for text categorization research. J. Mach. Learn. Res. **5**, 361–397 (2004)
24. Maas, A., Daly, R.E., Pham, P.T., Huang, D., Ng, A.Y., Potts, C.: Learning word vectors for sentiment analysis. In: Proceedings of the 49th Annual Meeting of the Association for Computational Linguistics: Human Language Technologies, pp. 142–150 (2011)
25. Mikolov, T., Chen, K., Corrado, G., Dean, J.: Efficient estimation of word representations in vector space. CoRR arxiv: 1301.3781 (2013)
26. Mikolov, T., Le, Q.V., Sutskever, I.: Exploiting similarities among languages for machine translation. CoRR arxiv:1309.4168 (2013)
27. Mogadala, A., Rettinger, A.: Bilingual word embeddings from parallel and non-parallel corpora for cross-language text classification. In: Proceedings of the 2016 Conference of the North American Chapter of the Association for Computational Linguistics: Human Language Technologies, pp. 692–702. Association for Computational Linguistics, San Diego (2016). http://www.aclweb.org/anthology/N16-1083
28. Pennington, J., Socher, R., Manning, C.: Glove: global vectors for word representation. In: Proceedings of the 2014 Conference on Empirical Methods in Natural Language Processing (EMNLP), pp. 1532–1543. Association for Computational Linguistics, Doha (2014). http://www.aclweb.org/anthology/D14-1162
29. Přibáň, P., Šmíd, J., Mištera, A., Král, P.: Linear transformations for cross-lingual sentiment analysis. In: Text, Speech, and Dialogue: 25th International Conference, TSD 2022, Brno, Czech Republic, 6–9 September 2022, Proceedings, pp. 125–137. Springer, Heidelberg (2022). https://doi.org/10.1007/978-3-031-16270-1_11
30. Vaswani, A., et al.: Attention is all you need. Adv. Neural Inf. Process. Syst. **30** (2017)
31. Vulić, I., Korhonen, A.: On the role of seed lexicons in learning bilingual word embeddings. In: Proceedings of the 54th Annual Meeting of the Association for Computational Linguistics, vol. 1: Long Papers, pp. 247–257. Association for Computational Linguistics, Berlin (2016). http://www.aclweb.org/anthology/P16-1024
32. , Zhuang, L., Wayne, L., Ya, S., Jun, Z.: A robustly optimized BERT pre-training approach with post-training. In: Proceedings of the 20th Chinese National Conference on Computational Linguistics, pp. 1218–1227. Chinese Information Processing Society of China, Huhhot (2021). https://aclanthology.org/2021.ccl-1.108

Unsupervised Extraction of Morphological Categories for Morphemes

Abishek Stephen[✉], Vojtěch John, and Zdeněk Žabokrtský

Faculty of Mathematics and Physics, Institute of Formal and Applied Linguistics,
Charles University, Malostranské Náměstí 25, 11800 Praha, Czechia
{stephen,john,zabokrtsky}@ufal.mff.cuni.cz

Abstract. Words in natural language can be assigned to specific morphological categories. For example, the English word 'apples' can be described using morphological labels like N;PL. The conditional probabilities on such word forms given the labels would reveal for English that the morpheme 's' is present almost always when the label N;PL appears. This indicates that the morphological properties of a word can be traced to its morphemes. We do not have any data resource that associates morphemes with morphological categories. We use UniMorph schema and datasets for universal morphological annotation as a source of morphological categories and morpheme segmentation. We align morphemes (or exponents) with the corresponding morphological categories based on the UniMorph schema for 12 languages. Given the multilingual nature of the task, we utilize unsupervised methods based on the ΔP measure and IBM Models as we test out the effectiveness of alignment methods used in statistical machine translation. Our results indicate that IBM Models accurately capture the alignment asymmetries between morphemes and morphological categories under non-trivial alignment settings.

Keywords: Morphemes · UniMorph · IBM Models

1 Introduction

An inflectional paradigm links a set of word forms to structural descriptions expressed in terms of a stem carrying a lexical meaning and some formal expression of one or more morphological categories [7]. The different inflectional paradigms can be viewed as a byproduct of the morphological categorization of morphemes. For example, the English word *watches* can be analyzed as *watches* N;PL or *watches* V;PRS;3;SG. In the word-based theories of morphology [1,2], the morphological categories are assigned to the word as a whole and not to individual morpheme segments which is not the case for morpheme-based theories [4,11]. However, the association of morphological categories to morphemes is a non-trivial task in NLP because of challenges posed by affix ordering and composition of morphemes in a word [10].

Induction of morphological rules from plain text corpora as performed in Unsupervised Learning of Morphology (ULM) [7] is made possible potentially due to the difference in substring frequencies reflected in recurrent morphological formations [10]. In a text corpora for Italian, one can find the substring -*amo* to be more frequent than any other substring of the same length. This substring is a morpheme encoding for the morphological categories V;PRS;1;PL and such morphologically potent substrings or morphemes can be identified through frequency asymmetries under ULM.

The modern approaches to NLP utilizing subwords do not rely on linguistic information much but for fruitful cross-lingual experiments and better interpretability of neural network-based architectures taking a morpheme-based approach should be given a higher preference. But in the absence of hand-annotated morphological categories such supervised or unsupervised methods always involve a trade-off and fall flat on sounding linguistically competent.

On the other hand, let's say we have morphological categories for words segmented into morphemes, how simple is the task of mapping the categories to the segmented morphemes? Consider the Latin word *dominus* 'Lord' which can be realized to have the tags[1] N;NOM;SG;M. We need to know the segmentation of *dominus* to start aligning the morphemes to categories. And even then let's say given the segmentation *domin|us* how well can we assign the morphological categories to the morphemes *domin* and *us* using statistical methods which is also linguistically valid?

In our study, we focus on such mapping or alignment experiments and evaluation metrics and illustrate the challenges in extracting morphological categories to morphemes in an unsupervised fashion for 12 languages from UniMorph that are freely available in the official GitHub repository[2].

2 Motivation and Related Work

Earlier works on morphology induction have used semantic vectors for identifying or grouping affixes based on their morphological properties [14], and unsupervised learning of morphological segmentation using minimum description length [6]. These approaches have hinted at computationally reasonable explanations for the presence of specific morphological characteristics encoded in the morphemes.

In recent times, it has been shown that learning morphological inflection patterns from labeled data is an important challenge in correctly analyzing and generating morphological inflections [9]. Morphological encoder-decoder models for this experiment at SIGMORPHON 2016 Shared Task benefited immensely by using morphological tags. This strongly indicates the need for having a good quality morpheme to tag mapping as training data for similar NLP tasks. We believe that our alignment methods are capable of curating such datasets.

Word-level word embeddings have been shown to capture information about semantic classes of derivational relations between words, despite not having any information about the orthography or morphological makeup of the words in

---

[1] Here, the morphological tags should be interpreted as a bundle or list of morphological categories assigned to a word.
[2] https://github.com/unimorph.

Czech [12]. The availability of morphological information can be supplemented to similar tasks to study derivational networks as the presence of such linguistic information can boost the results. Additionally, cross-lingual comparisons largely improve through morpheme-level alignments based on morphological features because the morphological complexities reduce considerably by moving from the word-level to the morpheme-level parsing [5].

3 Data

For our experiments we exploit UniMorph [3]. UniMorph is a collection of annotated morphological datasets in a universal schema that allows an inflected word from any language to be defined by its lexical meaning, typically carried by the lemma and the inflectional form in terms of a bundle of morphological categories. The UniMorph schema currently comprises 23 dimensions of meaning and over 212 morphological categories for 169 languages. The dimensions of meaning are expressed as morphological categories, like person, number, tense, and aspect.

For example, the Spanish word *habló* is represented by its lemma *hablar* and the morphological tag set FIN;IND;PFV;PST;3;SG. The tags are as follows:

- FIN indicates a finite verb.
- IND indicates the indicative mood.
- PFV indicates the perfective aspect.
- PST indicates the past tense.
- 3 indicates third person.
- SG indicates the singular number.

This tag set yields a mapping, which associates the entire inflected word with its meaning, without indicating the morpheme segments within *habló* or how the UniMorph schema features are distributed among those morphemes. These features or categories remain the same across languages thus enabling direct cross-lingual comparisons. As far as the segmentations go, UniMorph adopts a canonical morphological segmentation which keeps the morphemes in its canonical form [8]. For example, the Czech word *stojím* 'I am standing' is segmented as *stát|m* with the tag set V;PFV|IND;PRS;1;SG.

4 Experiments

One of the main problems in devising methods of aligning morphological categories to morphs is the scarcity of gold data. In the typical scenario, we have words tagged with morphological categories; quite often, we also have morphological segmentation of some words or at least enough segmentation data to induce reasonably reliable morphological segmentation on the tagged corpus. On the other hand, resources combining segmentation and morphological categories of the segments (morphemes) are much scarcer and tend to be generated automatically, which makes using them as gold data problematic. As a result, we have used segmented and tagged datasets for 12 languages from UniMorph as they are as close to gold data as we can get because UniMorph's segmentation files contain parallel lists of pairs of morpheme segments and morphological categories. It should be noted however that the task is made simpler by the UniMorph segmentation format - only inflectional affixes are segmented off. An overview of these can be found in the Table 1. The morphological complexity of the languages differs in several respects. Firstly, the average number of distinct viable analyses

Table 1. UniMorph segmentation data analysis. ind = individual, apw = average number of analyses per word, mpw = average number of morphs per word, fpm = average number of features per morph

Language	Items	w. forms	ind. morphs	ind. features	apw	mpw	fpm
cat	158922	14979	14365	23	10.61	2.19	2.11
ces	816956	33348	33313	32	24.50	1.89	2.44
deu	490331	39275	44702	38	12.48	1.77	2.31
eng	649594	396772	140187	16	1.64	1.34	1.49
fin	1786163	51048	56735	40	34.99	2.30	1.50
fra	453228	52711	50494	24	8.60	1.95	2.35
hun	1049755	19591	19884	52	53.58	2.27	1.63
ita	712021	89763	78820	23	7.93	1.91	2.39
lat	1416437	26927	53282	38	52.60	2.40	2.43
rus	1321025	36387	34602	38	36.30	3.10	2.37
spa	1289324	65565	62009	35	19.66	2.42	2.72
swe	131599	12508	12917	27	10.52	2.15	1.92

Table 2. ΔP based alignment in the Basic setting

Language	WL-acc	ML-acc	WL-acc-multimorph	ML-acc-multimorph
cat	12.07	76.50	1.60	73.36
ces	16.20	69.21	1.48	63.27
deu	24.64	71.93	0.45	63.85
eng	66.05	83.00	0.00	58.83
fin	14.48	52.83	8.06	49.54
fra	13.28	78.44	0.53	76.14
hun	2.96	62.11	0.96	61.47
ita	13.01	76.76	0.38	75.04
lat	8.07	74.68	1.74	73.53
rus	6.15	52.23	1.51	51.19
spa	9.31	73.91	1.08	71.94
swe	21.88	67.74	2.08	59.45

Table 3. ΔP based alignment in the Position setting

Language	WL-acc	ML-acc	WL-acc-multimorph	ML-acc-multimorph
cat	12.08	75.16	1.62	73.33
ces	16.20	69.46	1.47	62.75
deu	24.53	71.90	0.45	63.85
eng	66.05	83.00	0.00	58.83
Fin	14.48	52.63	8.06	49.13
fra	11.87	77.16	3.69	76.99
hun	2.70	62.04	0.70	61.40
ita	13.01	76.76	0.38	75.25
lat	7.43	75.16	1.07	74.30
rus	6.14	50.47	1.50	49.40
spa	8.50	73.59	0.55	72.21
swe	21.88	69.06	2.08	61.73

Table 4. ΔP based alignment in the Combined setting

Language	WL-acc	ML-acc	WL-acc-multimorph	ML-acc-multimorph
cat	10.75	77.85	0.14	76.12
ces	15.44	76.06	0.51	71.85
deu	24.96	76.06	0.82	68.49
eng	66.05	83.00	0.00	58.83
Fin	6.98	50.40	0.00	47.14
Fra	11.42	78.15	0.02	77.11
Hun	3.15	61.60	1.11	60.55
ita	12.40	78.31	0.01	76.61
lat	7.03	78.10	0.97	77.17
rus	4.99	49.09	0.34	47.79
spa	8.50	78.76	0.02	77.07
swe	23.41	72.38	0.09	63.38

of one-morpheme word form is only 1.64 for English but over 50 for Hungarian. In UniMorph segmentation data, each word is segmented into morphemes and each morpheme has assigned some morphological categories. If the same (orthographic) word is either segmented in multiple ways or the categories are assigned to morphemes in multiple ways, we speak of distinct morphological analyses of the word. Hence, we use the average number of morphological analyses per word as a metric of the inherent ambiguity of the morphological segmentation and classification. The average number of categories per morphemes and morphemes per word also vary rather strongly.

4.1 Methods

For the unsupervised alignment, we have used two sets of methods. Firstly, we have used the ΔP measure, As [13] points out, ΔP measures the strength of the link between two events. For events A, and B, the formula is $P(A|B) - P(A|\neg B)$, i.e. the probability of A in the presence of B minus the probability of A in the absence of B. It may make sense to use also the backward variant of this formula, i.e. $P(B|A) - P(B|\neg A)$. In our methods, the events are the presence of morpheme x and the presence of a morphological category (or feature) y.

Table 5. Accuracies for the Random Morpheme Baseline

Language	WL-acc	ML-acc	WL-acc-multimorph	ML-acc-multimorph
cat	13.25	47.89	3.53	44.49
ces	18.62	56.29	4.38	48.94
deu	29.10	61.42	6.51	49.95
eng	73.03	79.43	20.29	49.95
fin	15.67	46.54	9.32	42.65
fra	15.51	52.58	4.60	48.88
hun	8.99	47.01	7.04	45.95
ita	16.56	53.33	4.78	49.40
lat	8.30	42.94	2.81	40.56
rus	6.31	34.18	1.68	32.08
spa	10.23	43.35	1.91	40.28
swe	25.22	55.25	4.14	42.70

Table 6. Accuracies for the Last Morpheme Baseline

Language	WL-acc	ML-acc	WL-acc-multimorph	ML-acc-multimorph
cat	10.05	65.74	0.00	63.47
ces	14.91	75.02	0.00	70.80
deu	24.19	77.59	0.00	70.95
eng	66.05	83.00	0.00	58.83
fin	6.98	40.96	0.00	36.73
fra	11.40	76.48	0.00	74.67
hun	2.07	52.90	0.00	51.95
ita	12.39	77.37	0.00	75.44
lat	5.65	64.04	0.00	62.58
rus	4.71	31.19	0.00	29.03
spa	8.48	58.50	0.00	56.28
swe	21.88	60.44	0.00	49.49

Table 7. Accuracies for IBM Model 1

Language	WL-acc	ML-acc	WL-acc-multimorph	ML-acc-multimorph
cat	43.47	82.88	39.03	82.18
ces	4.44	60.51	5.08	61.59
deu	27.06	74.77	21.38	72.73
eng	31.05	63.21	32.69	67.83
fin	48.28	79.46	51.03	81.31
fra	51.35	85.63	47.29	84.99
hun	36.07	72.22	36.83	72.69
ita	31.12	82.13	23.72	81.09
lat	20.44	63.83	20.44	63.92
rus	8.47	65.46	8.85	65.37
spa	28.53	82.41	31.19	84.01
swe	33.96	72.09	43.28	78.33

Table 8. Accuracies for IBM Model 3

Language	WL-acc	ML-acc	WL-acc-multimorph	ML-acc-multimorph
cat	41.60	85.19	36.84	84.59
ces	10.30	69.28	12.07	69.67
deu	30.06	78.95	38.86	80.96
eng	22.69	61.28	8.08	62.16
fin	46.93	79.23	50.01	79.91
fra	37.63	85.67	29.61	84.57
hun	40.23	81.49	41.08	81.68
ita	48.20	88.49	42.47	87.85
lat	30.64	70.92	29.29	70.35
rus	5.32	67.90	5.35	67.57
spa	33.47	85.50	30.49	85.71
swe	32.35	80.15	36.46	80.38

There may be many morphological categories strongly statistically connected to morphemes that don't express them. For example, if a root is always present only in words with a given category, which is quite expected, as roots tend to be rare in the lexicon, its category-given-morpheme ΔP would be 1. This kind of error is less probable if we use the backward version of ΔP, i.e. the ΔP of a morpheme given a category. Methods relying on this statistic should align categories to morphemes that occur comparatively often with the categories. Supposing there is only a comparatively small number of rare categories, this should avoid the problems with rare predictors.

Table 9. Accuracies for IBM Model 4

Language	WL-acc	ML-acc	WL-acc-multimorph	ML-acc-multimorph
cat	36.06	83.13	30.68	82.40
ces	2.38	65.53	2.76	65.29
deu	37.24	83.76	48.64	87.14
eng	36.10	68.00	47.59	78.42
fin	48.56	77.82	51.76	78.47
fra	39.99	86.80	32.27	85.78
hun	40.25	81.52	41.10	81.71
ita	46.60	88.14	39.05	87.13
lat	29.81	71.90	28.37	71.36
rus	4.15	66.76	4.12	66.38
spa	23.39	83.68	24.95	84.40
swe	32.34	78.66	36.45	79.86

In our methods using the ΔP measure, we go through all categories assigned to a given word and align each category to the morpheme with the highest score. In two of the three settings, we have used only the morpheme-given category ΔP. In the *Basic* setting (Table 2), we did not distinguish between morphemes at all, while in the *Position* setting (Table 3), we have distinguished initial, internal, and final morphemes (*morpheme-, -morpheme-, -morpheme*). Further, we have used the product of morpheme-to-category and category-to-morpheme ΔP in the *Combined* setting (Table 4). The presented results in Sect. 4.1 illustrate the

Table 10. IBM Model 1 precision considering only the features that were aligned

Language	WL-prec	ML-prec	WL-prec-multimorph	ML-prec-multimorph
cat	68.31	90.79	64.77	90.16
ces	46.42	76.57	37.04	74.00
deu	64.47	87.91	51.55	84.01
eng	96.13	97.03	88.60	93.53
fin	86.09	94.52	85.05	94.28
fra	71.89	92.60	68.28	92.02
hun	70.75	87.40	70.13	87.25
ita	76.59	93.25	73.27	92.63
lat	36.26	76.66	32.45	75.97
rus	22.64	75.44	18.94	74.86
spa	71.44	93.68	68.78	93.48
swe	73.50	90.01	66.23	88.51

accuracy scores on word-level (WL-acc), morpheme-level (ML-acc), word-level accuracies for words composed of multiple morphemes (WL-acc-multimorph), and morpheme-level accuracies for words composed of multiple morphemes (ML-acc-multimorph).

Secondly, we have used the IBM alignment models 1 (lexical translation model), 3 (extra fertility model), and 4 (added relative alignment model)[3] using MGIZA++, a popular tool for building word alignment models[4], for aligning categories to morphemes, with 5 training iterations for all the models. We have used the union as the symmetrization heuristics.

We have implemented two baselines: *Random*, which aligns every feature to a random morpheme (Table 5), and *Last*, which aligns every feature to the last morpheme of a given word (Table 6).

5 Evaluation and Results

As the ΔP based alignment methods align every feature to some morpheme, the natural evaluation metrics are word-level accuracy and alignment-level accuracy. The IBM models (and they only) do not align all categories to morphemes. Therefore, for these models, we report both the accuracy of all alignments (considering the missing alignments as incorrect) and the accuracy of the present alignments only i.e. if a feature was not aligned to any morpheme, it was ignored in the evaluation. Suppose we have a word-feature pair, say the feature is V;IPFV

Table 11. IBM Model 3 precision considering only the features that were aligned

Language	WL-prec	ML-prec	WL-prec-multimorph	ML-prec-multimorph
cat	91.81	97.95	90.89	97.80
ces	74.73	90.22	70.30	88.81
deu	75.84	92.94	67.05	90.93
eng	96.32	97.08	89.16	93.30
fin	81.87	92.28	80.51	91.83
fra	92.00	98.01	90.97	97.83
hun	84.85	95.19	84.53	95.11
ita	92.17	98.03	91.07	97.85
lat	40.69	80.88	37.14	80.12
rus	34.77	80.63	31.59	80.08
spa	94.58	99.18	94.11	99.15
swe	81.52	94.05	76.45	92.58

---

[3] We don't need to resolve the deficiency in IBM models 3 and 4, s.t. we don't use the IBM 5 model; we want to model fertility, so apart from the simplest IBM1 model we use only the alignment models including fertility model, i.e. IBM 3 and 4.
[4] https://github.com/sillsdev/giza-py

and the output of an IBM model was 0-0, i.e. V is aligned to the correct morpheme, while IPFV is not aligned. In such a case, the achieved all-alignments alignment-level accuracy would be 50 %, and word-level accuracy would be 0 %, while the present-alignment, alignment-level, and word-level accuracies would both be 100 %. We have computed all of these metrics both on all the data and only on words with more than one morpheme[5]. In evaluation (but not in training), we ignore multi-word expressions (e.g. the separable affixes in German).

The results are rather disappointing for the ΔP-based methods, as we have overcome the *Last* baseline only for some of the languages. Interestingly, the results seem to correspond to the average number of morphemes per word - the more morphemes there are, the more the *Last* baseline tends to get overcome.

Table 12. IBM Model 4 precision considering only the features that were aligned

Language	WL-prec	ML-prec	WL-prec-multimorph	ML-prec-multimorph
cat	89.52	97.33	88.35	97.13
ces	68.38	86.89	62.84	84.96
deu	98.72	99.38	98.26	99.19
eng	100.00	100.00	100.00	100.00
fin	83.10	91.20	81.84	90.70
fra	91.47	97.90	90.37	97.72
hun	85.05	95.22	84.74	95.14
ita	91.35	97.89	90.12	97.69
lat	41.29	81.86	37.77	81.15
rus	36.22	84.10	33.18	83.62
spa	95.30	99.10	94.86	99.06
swe	81.22	93.81	76.07	92.37

For the IBM models, the results are more promising. IBM model 3 performed better than IBM model 1 (Table 7), achieving morpheme-level accuracies between 67 % for Russian and 88 % for Italian in both the all-words and multi-morpheme-words evaluation settings. In general, for the IBM3 model alignment (Table 8), the difference in morpheme-level accuracy caused by the exclusion or inclusion of one-morpheme words tends to be rather small. IBM4 results (Table 9) are comparable to IBM3; they are more often better than worse. Interestingly, all the IBM models often have comparable or better results in the setting where one-morpheme words are excluded. This is probably caused by the fact that the IBM models do not necessarily align every feature to a morpheme. If we do not consider the omitted affixes, all the IBM models by far outperform both the

---

[5] In UniMorph segmentation data, zero morphemes are not segmented off, hence words with the inflectional ending that exhibits a zero morph are considered monomorphemic words.

Table 13. Average precision across languages. -skip=considering only the features that were aligned

Method	WL-prec	ML-prec	WL-prec-multimorph	ML-prec-multimorph
Base-rand	20.10	51.70	5.90	44.70
Base-last	15.70	63.60	0.00	58.40
DP-basic	17.30	69.90	1.70	64.80
DP-posit	17.10	69.70	1.80	64.90
DP-comb	16.30	71.60	0.30	66.80
IBM1	30.40	73.70	30.10	74.70
IBM3	31.60	77.80	30.10	78.00
IBM4	31.40	78.00	32.30	79.00
IBM1-skip	65.40	88.00	60.40	86.70
IBM3-skip	78.40	93.00	75.30	92.10
IBM4-skip	80.10	93.70	78.20	93.20

baselines (Tables 10,11,12). There are however interesting differences between individual languages - the word-level accuracy for Russian and Latin is by far the worst for all the IBM models. Interestingly, the performance gap between Russian and Latin on one side and all the other languages on the other side is much smaller for the alignment-level accuracies. For Latin, this may be caused by a few systematic errors. For example, the part-of-speech feature V, i.e. verb, is usually not assigned, while e.g. V.PTCP i.e. verb participle, is often assigned to the first morpheme (usually root), even though it should have been assigned to the first affix. The wrong assignment of a verb-derived POS tag to the root instead of the first affix appears in 29.7 % of words in total. This might be a case of inconsistency in the UniMorph annotation, e.g. in Czech, the V.PTCP is always aligned to the root in the data but is (arguably more correctly) aligned to the affixes. Another source of errors may be the fact that the categories may be aligned to several morphemes i.e. one morphological function possibly expressed by several different morphemes. It could be the case that morphemes could have doublets, i.e. two (entirely or slightly) different inflectional endings that have the same morphological interpretation, e.g. doublet endings for adjectives in Slavic languages. In this regard, the alignment might be more precise than the UniMorph data. For example, in Czech, the V.CVB is aligned to both the -v and -še affix (as in sežehnout/v/še). Similarly, IMP, i.e. imperative, is often aligned to both -i (which represents imperative singular) and -ěte (which represents imperative plural), even though in the UniMorph data, the feature is aligned only to i.

The results summarized in Table 13 show that the IBM Models on the morpheme level perform better than the ΔP-based methods and the performance of the IBM Models improve when considering only the features that were aligned.

6 Conclusion

Our experiments show that aligning morphological categories or categories to morphemes from the word-level morphological tags is a non-trivial task with linguistic challenges. ΔP-based methods did not perform well for our baselines. Still, all the IBM Models outperformed under different settings especially when only considering the categories being aligned to the morphemes. The results indicate:

- Not all subwords or substrings are capable of inducing linguistic knowledge but rather only morphemes. The accuracy scores surpassing our baselines indicate that.
- For downstream NLP tasks involving subwords would considerably benefit from morphemes being aligned to morphological categories.
- The IBM models work well in non-trivial settings where the alignment is between a fixed order of morphemes in a word with a flexible grouping of morphological category bundles.

Acknowledgments. This work has been using data, tools and services provided by the LINDAT/CLARIAH-CZ Research Infrastructure (https://lindat.cz), supported by the Ministry of Education, Youth and Sports of the Czech Republic (Project No. LM2023062). This work has been supported by Charles University Research Centre program No. 24/SSH/009; project GA UK No. 101924; and partially supported by SVV project number 260 698. We would like to thank three anonymous reviewers for their very insightful feedback.

Disclosure of Interests. The authors have no competing interests to declare that are relevant to the content of this article.

References

1. Aronoff, M.: Word formation in generative grammar. Linguistic Inquiry Monographs Cambridge, Mass (1976)
2. Aronoff, M.: In the beginning was the word. Language **83**(4), 803–830 (2007)
3. In: Batsuren, K., et al. (eds.) Proceedings of the Thirteenth Language Resources and Evaluation Conference, pp. 840–855. European Language Resources Association, Marseille, France (2022)
4. Borer, H.: Structuring sense: Volume 1: In name only, vol. 1. Oxford University Press (2005)
5. Gamba, F., Stephen, A., Žabokrtský, Z.: Universal feature-based morphological trees. In: Bhatia, A., et al. (eds.) Proceedings of the Joint Workshop on Multiword Expressions and Universal Dependencies (MWE-UD) @ LREC-COLING 2024, pp. 125–137. ELRA and ICCL, Torino, Italia (2024). https://aclanthology.org/2024.mwe-1.17
6. Goldsmith, J.: Unsupervised learning of the morphology of a natural language. Comput. Linguist. **27**(2), 153–198 (2001)
7. Hammarström, H., Borin, L.: Unsupervised learning of morphology. Comput. Linguist. **37**(2), 309–350 (2011)

8. Kann, K., Cotterell, R., Schütze, H.: Neural morphological analysis: encoding-decoding canonical segments. In: Proceedings of the 2016 Conference on Empirical Methods in Natural Language Processing, pp. 961–967 (2016)
9. Kann, K., Schütze, H.: MED: the LMU system for the SIGMORPHON 2016 shared task on morphological reinflection. In: Elsner, M., Kuebler, S. (eds.) Proceedings of the 14th SIGMORPHON Workshop on Computational Research in Phonetics, Phonology, and Morphology, pp. 62–70. Association for Computational Linguistics, Berlin, Germany (2016). https://doi.org/10.18653/v1/W16-2010, https://aclanthology.org/W16-2010
10. Manova, S., Hammarström, H., Kastner, I., Nie, Y.: What is in a morpheme?: theoretical, experimental and computational approaches to the relation of meaning and form in morphology. Word Struct. **13**(1), 1–21 (2020)
11. Marantz, A.: Morphology: Word structure in generative grammar (1992)
12. Musil, T., Vidra, J., Mareček, D.: Derivational morphological relations in word embeddings. In: Linzen, T., Chrupała, G., Belinkov, Y., Hupkes, D. (eds.) Proceedings of the 2019 ACL Workshop BlackboxNLP: Analyzing and Interpreting Neural Networks for NLP, pp. 173–180. Association for Computational Linguistics, Florence, Italy (2019)
13. Schneider, U.: ΔP as a measure of collocation strength. Considerations based on analyses of hesitation placement in spontaneous speech. Corpus Linguis. Linguist. Theory **16**(2), 249–274 (2020)
14. Schone, P., Jurafsky, D.: Knowledge-free induction of morphology using latent semantic analysis. In: Fourth Conference on Computational Natural Language Learning and the Second Learning Language in Logic Workshop (2000)

Introducing LCC's NavProc 1.0 Corpus
Annotated Procedural Texts in the Naval Domain

Michael Mohler[✉][®], Sandra Lee, Mary Brunson, and David Bracewell

Language Computer Corporation, Richardson, TX, USA
{michael,sandra,mary,david}@languagecomputer.com

Abstract. In this work, we introduce the NavProc 1.0 Corpus – a medium-scale, annotated corpus of procedural texts within the naval domain – for use as a first step in modeling procedural structures derived from real-world data sources. In particular, we have rigorously produced annotations of frame semantics (i.e., PropBank-inspired trigger/role links) across verbal, nominal, and adjectival frames. Furthermore, we have annotated 21 distinct types of semantic markers and structural links between textual elements (e.g., frame triggers, entities, modifiers) which, taken together, result in a text-focused graph of semantic elements. Such a graph can be used to derive a more complex procedure structure for use in personnel training, simulation, or collaborative procedure execution. Altogether, this annotation effort has encompassed 158 procedural units composed of 2,316 sentences, 44,459 tokens, and 48,137 distinct span annotations. Furthermore, we describe and report LLM-based extraction scores for use as a baseline in future research using this dataset.

Keywords: dataset · knowledge graph · procedures · frame semantics · textual structure

1 Introduction

Procedural knowledge is ubiquitous. It's what enables people (and other agents) to carry out complex tasks, to make fine-grained distinctions about the world, and to satisfy a variety of constraints including laws, rules, norms, policies, and guidelines. Within the context of this effort, we define "procedural texts" broadly to include any unstructured or semi-structured textual instructions which are meant to guide a reader (human or AI) to carry out a task, to make a determination, or to achieve an objective. Our goal in releasing the NavProc Corpus has been to enable text-based procedure extraction and modeling in real-world domains with an eye towards more readily training personnel to carry out procedures, simulating stored procedures for decision support, and ultimately executing procedures through human-AI collaboration. In particular, we are providing procedure understanding resources in the highly-diverse, Navy-relevant domains of ship maintenance (e.g., "what components to look for when assessing a wood-based vessel") and command and control (C2) (e.g., "what policies to follow when pursuing pirates in international waters"). Procedural texts in these real-world domains incorporate not only step-by-step instructions, but also definitions, suggestions, prohibitions, warnings, and contextual explanations for all of the above.

Significant recent research in identifying and modeling procedural knowledge has taken place in the fields of automated planning and execution [10, 20]. This is due to the fact that procedures are by their nature modular, such that an agent can be taught to carry out simpler tasks in a piecemeal fashion, to apply that procedural knowledge in a variety of scenarios, and thereby to build up a repository of more complex procedures. Researchers in common-sense reasoning [30] have also begun to exploit procedures as both a source of common-sense knowledge and as a test-bed for applying such knowledge to a variety of domains and situations. Procedural knowledge is also critical in the fields of collaborative AI [11] and decision support [25] because AI that understands complex procedural knowledge can aid humans in making better decisions to achieve an objective more quickly, more in line with the appropriate guidelines, or with less risk to those involved.

Table 1. Sample procedure from the corpus with frames (triggers) shown in **bold**. Taken from JFMM V4.

Sentence	Frame-Annotated Text
38	22.4 **Procedure**.
39	a. Upon **completion** of **maintenance** and before **underway** for **submerged operations**, the ship must **place** CAUTION tags on the Main Power Switch or **transmit** keys for the **affected** antennas.
40	The **Amplifying Instructions** for the **CAUTION** tags will **state** - "DO NOT **OPERATE** OR **TRANSMIT** ON THIS ANTENNA UNTIL **COMPLETION** OF PASSIVE **CHECKS** FOLLOWING A **DIVE** TO **TEST** DEPTH. PASSIVE **CHECKS** MUST BE **PERFORMED** PRIOR TO EACH **USE** UNTIL THE DEEP **DIVE** IS **COMPLETE**."
41	b. It is **understood** that in some cases, due to water depth **restrictions**, the deep **dive** may not be **performed** for quite some time.
42	In these cases, the ship should **dive** to the maximum depth **possible** and **conduct** passive **checks**.
43	Provided the **checks** are **satisfactory**, the **CAUTION** tag may be **replaced** with one **stating** - "DO NOT **OPERATE** OR **TRANSMIT** ON THIS ANTENNA FOLLOWING **OPERATIONS** GREATER THAN (**enter** max depth **obtained**). PASSIVE **CHECKS** MUST BE **PERFORMED** PRIOR TO EACH **USE** UNTIL THE DEEP **DIVE** IS **COMPLETED**."
44	c. The tag(s) may be **removed** following **completion** of a deep **dive** to **test** depth and **completion** of **satisfactory** passive **testing performed** following the **system specific** technical manuals or **Maintenance** Index Pages and MRCs to **ensure** the **system** is not **grounded**.

Underlying the problem of understanding natural language procedural texts is the problem of understanding text more generally. Consider the (sub-)procedure shown in Table 1. In order to carry out the procedure provided (for safely testing a submarine antenna), it is necessary to model (a) the basic elements (i.e., events, entities, and relations) in each of the constituent sentences and (b) how these pieces fit together. In the annotated dataset which accompanies this work, we aim to provide a resource to advance research in procedural understanding by filling three gaps associated with existing procedure-focused datasets:

- **Domain Variety** - While existing datasets used in procedure modeling are heavily concentrated in constrained domains such as cooking recipes, materials synthesis, and biological processes, we will provide support for research over procedures in the more complex Navy-relevant fields of ship maintenance and C2.
- **Technical Language** - Existing procedural data is often limited to constrained language forms (e.g., ingredient lists, lists of steps in order); we instead provide procedures in real-world contexts with a significant amount of syntactic complexity, lists, conditions, explanatory content, and overlapping constraints.
- **Structure-Focused** - Our hypothesis is that a major component to understanding (and modeling) procedural language is to identify and organize the underlying structural elements in the text; as such, the annotations in this dataset are focused on identifying frame semantics, structural relationships between frames, semantic markers, and other complex links between textual elements which define and constrain the overall procedures.

We believe that the complex structural and discourse elements exhibitied in these domains are also common in other real-world procedural domains, and that the annotations provided in this dataset can be profitably used in these other domains as well.

The remainder of this paper is organized as follows. In Sect. 2, we discuss related work in identifying, modeling, and leveraging procedures with a secondary focus on the importance of frame semantics and knowledge representation in natural language understanding more broadly. In Sect. 3, we describe our process for selecting and preparing the textual data for annotation. In Sect. 4, we describe our methodology for annotating the various structures within the dataset and provide corpus statistics over these structures. In Sect. 5, we describe initial experiments run over this dataset using several state-of-the-art large-language models (LLMs) to serve as a baseline for future research. In Sect. 6, we draw conclusions about the effectiveness of performing LLM-based extraction over this annotated dataset. Finally, in Sect. 7, we highlight our plans to further expand and enhance the NavProc Corpus with the goal of spurring further research in end-to-end procedure modeling.

2 Related Work

2.1 Semantic Representations over Text Elements

Significant work in the NLP research community over the past several decades has been devoted to defining and identifying structural patterns within natural language texts. One prominent example of such work is the influential PropBank project [17] which takes a domain-independent, semantically light view of frame semantic structure extending from more traditional representations of predicate argument structure. The FrameNet project [7], by contrast, defines role types according to an ontology of typed events or other predicates, including roles which explicitly link frames together (e.g., purpose roles).

At the document level, the Penn Discourse Treebank (PDTB) [18] has provided researchers with tools to identify and define the relationships between clauses both within- and across- sentences with sub-categorizations of causal, temporal, contrastive, and elaborative relationships. The CaTeRS corpus [14] was annotated with causal and temporal relations linking events in order to promote further research in narrative understanding and generation. In a similar vein, the ASER dataset [28] encodes fifteen distinct types of relationships between eventualities (i.e., actions, states, and events), extracted from a variety of web sources (reviews, news, forums, social media, movie subtitles, e-books) using classifiers trained over PDTB annotations. These extractions serve as a source of general purpose knowledge that can be used in a variety of reasoning tasks, including reasoning over procedural texts.

2.2 Procedure Modeling

One common source of procedural knowledge derived from text is in recipes. The Recipe Instruction Semantics Corpus (RISeC) [6] consists of 230 recipes with step-by-step instructions for producing certain dishes. These steps are annotated for frame semantics following the PropBank methodology with additional information to model zero-anaphora – i.e., instances of referents which are typically the products of previous steps, which are common in the cooking domain. Likewise, the related CUTL recipes dataset [19] is annotated with semantic structure derived from PropBank and is focused on identifying and modeling coreference throughout the recipe procedures.

The most widely used dataset of procedural texts to date is the ProPara dataset [13], produced by the Allen Institute for AI. This dataset, consisting of 3,300 sentences in 488 paragraphs, was produced via crowdsourcing by providing workers with a prompt (e.g., "What happens during photosynthesis?") and recording their responses. This dataset was then annotated with the existence and locations of all the main entities (the "participants") at every time step (sentence) throughout the procedure to promote automated reasoning about how entities change during the procedure. This dataset was further enriched with Dependency Graph (DG) annotations that record the dependencies between different steps in a process [3], and influence graphs to support answering "What-if" style counterfactual questions [22]. Research on this dataset has included the use of language resources [2,8], consistency bias [4], pre-trained language models [1], and common-sense knowledge [21]. An additional dataset for modeling entity state transitions was produced by Tandon et al. [23] and motivated by the need for a more open attribute vocabulary compared to ProPara.

The annotation effort which is perhaps most similar to our own is the Materials Science Procedural Text Corpus [15]. This corpus was constructed by sampling from academic literature in the materials science domain, automatically identifying procedural text, and annotating with shallow semantic annotations to support further research [9, 15].

In addition to entity state tracking and procedural extraction, downstream systems operating over procedural data have often been focused on expanding world knowledge. Zhang et al. [29] operate over WikiHow data in an effort to detect goals, constituent steps, and temporal ordering requirements from procedural texts. Likewise, Zhou et al. [31] employs neural step-prediction over WikiHow to derive a hierarchy of related procedures to better model the nesting effect of decomposable actions.

2.3 LLM-Based Extraction

Large Language Models (LLMs) have recently revolutionized text analysis in a variety of generation and classification sub-disciplines including question answering, text summarization, and sentiment analysis. However, relatively few researchers have applied such complex generative models to the related tasks of frame identification and information extraction (IE) over unstructured text. Xu et al. [27] have worked to apply a fine-tuned bilingual chat-centric model (ChatGLM) to a variety of IE benchmarks for named entity recognition (NER), relation extraction (RE), and event extraction (EE), while incorporating domain-specific knowledge through the use of reinforcement learning. Wei et al., [26] have attempted to employ ChatGPT to perform NER, RE, and EE within a zero-shot, two-stage, multi-turn framework. They show that while vanilla ChatGPT (untuned and without examples) does not perform strongly on these IE tasks, breaking the problem into simpler pieces may be an effective strategy.

3 Dataset Preparation

In order to achieve our goal of providing a structural framework for future research in Navy-relevant procedures, we have selected procedural data sources that fall into two categories – (1) naval maintenance and (2) command and control (C2). For the maintenance portion of the dataset, we have selected the publicly-available Joint Fleet Maintenance Manual (JFMM)[1] which is organized into seven volumes ("New Constructions", "Integrated Fleet Maintenance", "Deployed Maintenance", "Tests and Inspections", "Quality Maintenance", "Maintenance Programs", and "Contracted Ship Maintenance"). For the C2 portion of the dataset, we have selected three publicly available documents: (a) the NATOPS General Flight and Operating Instructions Manual,[2] (b) the Sanremo

---
[1] https://www.navsea.navy.mil/Portals/103/Documents/SUBMEPP/JFMM/Searchable_JFMM_Rev_D-1.pdf.
[2] https://www.aviation.marines.mil/Portals/11/CNAF%20M-3710.7%20(2016).pdf?ver=2018-10-03-111623-847.

Handbook on Rules of Engagement,[3] and (c) the Commander's Handbook on the Law of Naval Operations.[4]

Due to the size and complexity of these documents, we have selected a sampling of the procedural content to be made available for structural annotation as follows. For each document, we first perform tokenization, sentence segmentation, and page segmentation over the PDF format using off-the-shelf components. Then, we randomly selected N page segments for further analysis. Within each page segment, we manually identified textual content that was highly (or moderately) procedural and selected sufficient context around that content to define a "procedural unit" for annotation.[5] For the JFMM documents, we selected around five procedural units per volume. For the C2 documents, we selected 10 units per document. Altogether, these represent 65 procedural units with 93 contained sub-units for a total of 158.

Due to the nature of pdf-to-text conversion, we carried out a significant amount of document curation to ensure that tokens and sentences were not improperly split or merged and that inter-textual elements found in the PDF (such as headers, footers, page numbers, and footnotes) were removed to ensure that the flow of the procedural text was not interrupted. Furthermore, documents were curated to explicitly label zones in the document (e.g., headings, content-under-headings, lists, etc.) to aid in structural and procedural modeling.[6] All manually curated token, sentence, and zone information is provided as part of our annotation package along with the raw post-curation texts.

4 Annotation Procedure

The goal of this annotation effort has been to produce a gold-standard structural representation of the procedural documents that can be used as building blocks (e.g., steps, conditions, explanations) in modeling the procedures as a whole. At this time, we have focused our efforts on three types of linguistic structures – **frame structures** (i.e., eventive or stative triggers along with their associated roles), **semantic markers**, and **non-frame structural links**. Semantic markers are used in text to identify semantically important features of the text for use in a variety of downstream knowledge representation tasks (e.g., coreference, logical modeling, sequencing), while non-frame structural links are intended to connect disparate frames and other textual elements (e.g., entities[7]) from the

---

[3] https://iihl.org/wp-content/uploads/2022/12/ROE-HANDBOOK-ENGLISH.pdf.
[4] https://stjececmsdusgva001.blob.core.usgovcloudapi.net/public/documents/NWP_1-14M.pdf.
[5] Note that these procedural units are sometimes short (i.e., a few lines) while others span multiple pages and contain other procedural subunits (e.g. 7.1 with 7.1.1 and 7.1.2).
[6] This curation process was carried out by the lead author using LCC's ATESSA web-based interface for document processing and management.
[7] Note that entities are not explicitly annotated as such in this dataset, but are included implicitly as frame roles and other textual spans.

same sentence or adjacent sentences together into a single, connected structural knowledge graph. These markers and structural link types will be described in more detail in Sect. 4.2.

Annotation was carried out by three native English speakers with a professional background in linguistics between June and October of 2023 using LCC's ATESSA web-based interface for document processing and management. Guidelines for each annotation type were provided up front and revised over the course of the annotation process to account for uncertain or ambiguous cases as a result of feedback from and consensus among the annotation team. Annotation was carried out sentence-by-sentence in multiple passes to ensure annotation coverage was thorough. Additional checks have been performed by a secondary annotator to ensure consistency and coverage. The dataset itself, along with its annotation guidelines and version history, can be found on github.[8]

4.1 Frame Annotation

Following existing research in frame semantics, we define a frame as constituting (a) a **trigger** – a textual span which expresses the semantics of the frame, and (b) zero or more **role spans** which serve to identify frame participants, to ground the frame in time, space, etc., or to specify additional properties associated with the frame (e.g., manner). In annotating frame triggers for this effort, we have chosen to take a broad view of what constitutes a "frame". As highlighted in a recent analysis by Orlando et al. [16], semantic frames can be triggered by verbs, nouns, and (in some cases) adjectives. Verbal frames can be expressed using standard subject-verb-object (SVO) structure (active or passive), infinitives, copulas, relative clauses, and adverbial modifiers. Nominal frames can be expressed through deverbal nominalizations (e.g., an **attack**), role nominalizations (e.g., the **aggressor**), gerunds, eventive nouns (e.g., **storm**), and relational nouns (e.g., **capability**). Furthermore, annotators were instructed to label as frame triggers a variety of complex nouns (e.g., **system, policy, law**) which exhibit a latent frame structure between roles as well as acronyms for which the head noun would itself be a nominal trigger (e.g., UNSC – "United Nations Security **Council**"). Finally, adjectival frame triggers are prototypically expressed in this dataset as deverbal participle modifiers (e.g., a **broken** window; a **moving** vessel) or as expressly stative or eventive adjectives (e.g., **accidental, responsible, possible**). An example of each of these types from the dataset is provided in Table 2.

---

[8] https://github.com/lcc-api/navproc

Table 2. Examples of verbal, nominal, and adjectival frames from the NavProc corpus with triggers shown in **bold**

Type	SubType	Example
Verbal	Active SVO	o. The Sustaining Activity must **report** any violation of the DSS operating limits to NAVSEA...
	Passive SVO	The BRB will be **signed** and dated following the representative's review.
	Copula	a. Project Target **is** a NAVSEA nuclear material procurement program.
	Infinitive	Experienced estimators do not normally need to **break** the work down into as many activities as an inexperienced estimator.
	Rel Clause	...the material and parts **required** for work package execution should be provided by the ship.
	Adv Modifier	...the ship must submit an OPNAV 4790/2 K **following** the normal process and provide a copy to the Surveyor.
Nominal	Deverbal Nom	(10) Remarks - enter any pertinent data, event or condition with regards to the **test** discharge.
	Role Nom	The QAI will immediately inform the appropriate Department Head and **Surveyor** of any discrepancies noted.
	Gerund	This means that detail must be inserted to clarify the **meaning** of the rule.
	Eventive Noun	Annex A provides guidance for drafting ROE for specific tasks including... humanitarian assistance/**disaster** relief, and assistance to civil authorities.
	Relation Noun	Similarly, use of the white flag to gain a military advantage over the **enemy** is unlawful.
	Complex Noun	Legal analysis of intended wartime targets requires traditional **law** of war analysis.
	Frame Acronym	g. (Submarine Force only) Support and participate in the **TYCOM**'s CTRA process.
Adjectival	Part Modifier	Excessive heat chars the wood fibers and removal of the **charred** wood results in an uneven surface.
	Eventive Adj	Since they are not pre-negotiated with the contractor, **constructive** changes are against the law.
	Stative Adj	c. All areas stripped as previously described are now **ready** for sanding before priming.

For the set of roles used for annotation, we have employed a simplified version of the roles defined under the PropBank project [17]. These role categories, shown with counts in Table 3, diverge from traditional PropBank roles in three important ways. First, we have split out the TMP category to better account for different dimensions of temporality (e.g., when something happened, how long it lasted, and how frequently it occurs). Second, we have collapsed a great number of the lesser-used PropBank roles in favor of a structure-focused role (PP) indicating that the role is attached to the trigger via preposition. Annotators were instructed to apply this role label cautiously and to be careful to distinguish when frames are being linked to other frames via preposition (e.g., expressing purpose, temporality, causality) and to prefer annotating such cases as a structure link (see Sect. 4.2). Finally, we have provided annotators with an "UNK" role to serve as a catch-all for roles that should be included, but do not fit into any of the other categories. By far the most common usage for this role category is as a noun adjunct occurring before a nominal frame trigger (e.g., "missile **attack**").[9]. Altogether, we have annotated 13,780 frames with 13,673 annotated roles between them.

4.2 Semantic Marker and Non-Frame Structural Link Annotation

In addition to the frame structure annotations described in Sect. 4.1, annotators were instructed to identify a variety of textual patterns linking elements within the sentence (e.g., frames, entities, modifiers) which do not exhibit a clear frame structure. There are two goals associated with these link annotations and, as such, two categories of links. The first category, shown in Table 4, corresponds to markers on existing sentence elements that indicate their use in a logical, coreferential, or sequential context. For instance, the DistinctMark type may provide explicit information on whether or not an entity is the same entity as previously mentioned in a text. The second category, shown in Table 5, corresponds to links between two sentence elements that are expressed using particular sentence patterns apart from the more typical frame semantic structure. In many cases, these are triggered by prepositions (e.g., "of", "in", "because", "to"), conjunctions (e.g., "and", "or", "but") and other specialized lexemes (e.g., "enough", "including"). Note that two of these (FrameLink and MultiSentLink) correspond closely with the notion of cues in the PDTB annotation framework [18]. It is important to point out that our annotations do not attempt to characterize the semantics of such links (e.g., as causal, temporal, etc.), but only indicate the

Table 3. PropBank-derived roles used in annotation along with their corpus counts.

Role	Definition	Count
A0	The agents, causers, or experiencers; subject of transitives and unergatives	2,877
A1	The patient which undergoes a change or is affected by an action	4,768
A2	The beneficiary of some action, often a transfer	48
LOC	The location of the event including physical, virtual, and textual locations	395
TMP	The time (or time range) at which an event takes place or a state is known	190
DUR	A period of time associated with the predicate	23
FREQ	The frequency with which an event occurs	84
MNR	An expression of manner; also used to identify negative modalities	499
JJPRED	An adjectival predicate associated with the trigger	1,305
PP	Other roles expressly linked to a trigger via prepositions (incl. infinitives and "that" clauses)	2,039
UNK	Any role which does not fall into the above categories	1,445
TOTAL		13,673

---

[9] Noun adjuncts have been observed as a semantically ambiguous structure since the time of the Nombank project [12].

Table 4. Types of markers provided in annotations with the markers **bolded** and other element underlined

Marker Type	Definition	Example	Count
CountMark	The relationship between a count (number or percent) and the thing being counted	(9) Ten lowest voltage cells at end of discharge - list **ten** individual cells with the lowest end of discharge voltage.	173
DistinctMark	A marker indicating whether something is necessarily the same as or different from something mentioned previously	A **separate** PDDI must also be submitted.	399
LogicMark	A marker indicating an existential, universal, exclusive, or majoritarian logical constraint should be applied	**Each** of the inspections may require a unique set of activities and measurement criteria .	398
OrderMark	A marker indicating where something comes in an ordered list (e.g., first, fifth, last, next)	Prior to the **first** operation of engines which have been placed in Inactive Equipment Maintenance.	74

presence of such links. The addition of semantic sub-categorization is therefore left for future work (see Section 7). Altogether, we have produced a total of 7,018 semantic link annotations with a total of 15,333 linked elements associated with them.[10]

4.3 Inter-Annotator Agreement

Due to the sequence labelling nature of the task, Cohen's Kappa is an appropriate choice for representing Inter-Annotator Agreement (IAA) over this annotation set. Instead, we have computed an average pairwise F1-Measure among the annotators by modifying the methodology proposed in Hripcsak et al. [5] as follows. Rather than computing correctness in an all-or-nothing manner, we accept as fully correct any spans (triggers or roles) which share the same "head" token, and spans which are linked with an alternate relation type (e.g., different role name, different semantic link type, etc.) are treated as half-correct. Using these parameters, we calculate IAA for this annotation effort as 0.861, sugesting strong agreement.

5 Experiments

In order to provide baseline numbers for further research over this dataset, we have applied two modern large language models (LLMs) – without any additional

---

[10] Note that Membership and ListContain typically have more than two linked elements.

in-house fine-tuning – to the tasks of (a) identifying instances of frames (SemFrame), markers, or structure links defined above for each sentence (or pair of adjacent sentences) in the dataset and (b) identifying the named roles associated with those frames/markers/links for a given gold-standard instance.[11] In particular, we have made use of (a) the open-source Zephyr-7b model [24] and (b) API calls to the GPT-3.5-Turbo model from Open AI. These LLMs were prompted using seven interchangeable parts as shown in Table 6 – System Specification,

Table 5. Types of non-frame structural links provided in annotations. Link triggers are shown as **bold** while other linked elements are underlined.

Structure Type	Definition	Example	Count
AdjNoun	The relationship between an adjective and the noun that it modifies	When the use of Navy pricing information is not appropriate..., the best source of **current** pricing information is...	652
Appositive	The relationship between two noun phrases which are meant to be identical	The ship should provide... a list of the ship's **Quality Assurance Inspectors** (QAI) to be used during the VR period.	230
Comparison	The relationship between two things which are explicitly being compared along some dimension	Emergent work is performed **at greater cost than** work... completed in other availabilities.	27
ConjunctLink	The relationship between two non-frame words or phrases linked by a conjunction	Determine the character of the operation as involving armed conflict (international **or** non-international) or as falling outside...	811
FrameLink	An explicit relationship between two within-sentence frames (incl. temporal, causal, conjunctive, discourse, conditional)	2. The master or person in charge, **upon** request, makes no claim of nationality or registry for that vessel	1667
GroupExclusion	The relationship between a group and an explicitly excluded member	(2) **With the exception of** lagging, Military Specification parts and material are not available to local contractors in other ports.	19
HyphenVerb	The relationship between a verb (or nominalization) and a non-verb within a hyphenated phrase	63E Energizing **fire** - control radars in the direction of (SPECIFY) permitted.	94
ItCopular	A relationship between a constraint and a phrase set off by a pleonastic 'it' (e.g., it is X that/to Y)	Note: It **is** unlawful to... deliberately cause blindness.	24
ItemConstraint	A relationship between an item in a list and a constraint that applied to only that item	(3) Tended ship CO, **MST OIC** (if applicable), Engineer and Availability Coordinator.	44
ListContain	A relationship linking all the items of an explicit list (usually 3 or more) to the list as a whole	d. Ensure accountability... and establish an auditable system of scheduling, performing **and** reporting accomplishment of MRCs.	375
Membership	A relationship between a conceptual category and a member or instance of that category	The Government uses the sealed bidding and negotiation methods of contracting in awarding contracts, **including** job orders.	124
MultiSentLink	The relationship between two identified frames in adjacent sentences, set off by an explicit cue	For example, it is normally more costly to blast and paint... The weather may **also** impact paint curing times and other time factors considered.	91
NounAdjunct	The relationship between two adjacent noun phrases without an explicit semantic relation	(8) Provide for the restoration of all **PCMS** materials removed as interferences for Other tasks.	1147
PartitiveCollect	The relationship between a collective noun or partitive and the noun being grouped with it	For bottom sheathing to be effective..., the sheathing must... be faired to the level of adjoining **pieces** of sheathing.	17
Possessive	A relationship between two nouns indicating a structural possessiveness (incl. ownership, affiliation, part-whole, attribute)	The **scope** of the inspection will be as specified in paragraph 4.3.2.c of this chapter.	484
PrepModifier	The relationship between something and a prepositional phrase modifier outside of an explicit frame structure	a. To prevent overheating of the wood filler blocks and wood cap blocks **in** sea chests, the wood must be insulated from the metal fittings.	176
SuffCondition	A relationship between an item and the capability that is (not) available with the amount of the item available	Use only **enough** heat as required to blister the paint.	4

---

[11] Note that the full role name specification will be included with the annotation guidelines on github.

Task Description,[12] Input Format (a single sentence or adjacent sentences), Output Format (sentence marked up with <>), Example Prep, Examples (20), and finally the Query for a given sentence.[13] For a given target (frame/marker/link), each sentence is categorized as either having one-or-more instances of the target or no instances. Examples for the target are selected randomly, such that, (a) a minimum number of positive/negative examples were reached (8 in these experiments) and (b) the percentage of positive/negative examples were as close to the corpus percentages for the target as possible.

6 Discussion

In general, the results of Table 7 show that both models (but especially the GPT model) exhibit a much higher recall compared to their precision across all types. Only those types which can be considered common (SemFrame, ListCont, FrameLink, ConjLink, LogicMark) were able to be identified with high precision. This suggests that an approach which reduces the likelihood of the LLM applying

Table 6. Sample Prompt Specification for LogicMark identification

System Specification	You are a helpful text analysis machine.
Task Description	You are tasked with identifying markers indicating logical information such as existence, non-existence, proportion, universality, and existentiality among objects or events. These are often indicated with the terms: 'all','none','no','any','some','each','every','most'. This does NOT include periodicity markings (e.g.,'every three weeks').
Input Format	You will be provided with a single English-language sentence, all contained on a single line.
Output Format	For each sentence received, you will indicate the full textual extent of the relation including the logical marker and the entity or event being highlighted on a single line where the full sentence is repeated with triggers indicated using <> as markers – e.g., <trigger>.
Example Prep	20 examples will follow to indicate the expected formats.
Example 1-20	USER: Identify all triggers in the following sentence:"d. Ensure that each ship with an FCA has all the necessary standards, documentation and trained personnel to maintain current certification."
	ASSISTANT: d. Ensure that <each> ship with an FCA has <all> the necessary standards , documentation and trained personnel to maintain current certification.
Query	USER: Identify all triggers in the following sentence: "The Atlantic and Pacific Fleets will use the Consolidated TMDE Readiness Assessment (CTRA) Program."

---

[12] Note that the prompt differs primarily in the Task Description. The full prompts used will be made available on github prior to publication.
[13] In an initial pilot experiment, we explored other options for input (incl. one-token-per-line) and output (incl. one annotation per line, CoNNL format) and performed a grid search to find the optimal number of examples.

Table 7. Precision/Recall/F-Measure of LLM-based extraction experiments for triggers and roles across 22 relation/frame types (and micro/macro averages).

Type	SemFrame	CountMark	DistinctMark	LogicMark	OrderMark	AdjNoun
Zephyr-Trig	65%/15%/24%	5.1%/54%/9.3%	8.0%/31%/13%	14%/55%/23%	2.7%/58%/5.2%	7.7%/27%/12%
GPT-3.5T-Trig	65%/57%/61%	5.7%/58%/10%	11%/71%/19%	27%/88%/41%	3.2%/85%/6.2%	7.2%/47%/13%
Zephyr-Role	3.0%/1.0%/1.4%	55%/51%/53%	86%/83%/84%	70%/68%/69%	68%/70%/69%	74%/69%/71%
GPT-3.5T-Roles	3.7%/2.8%/3.2%	62%/56%/59%	86%/83%/85%	72%/70%/71%	76%/75%/75%	79%/75%/77%

Type	Appositive	Comparison	ConjunctLink	FrameLink	GroupExclusion	HyphenVerb
Zephyr-Trig	5.4%/49%/9.7%	0.5%/19%/0.9%	30%/62%/41%	33%/35%/34%	1.3%/58%/2.6%	3.0%/37%/5.6%
GPT-3.5T-Trig	6.0%/59%/11%	0.5%/30%/1.1%	35%/77%/48%	28%/61%/38%	1.8%/84%/3.6%	2.8%/52%/5.4%
Zephyr-Roles	25%/24%/25%	0%/0%/0%	19%/15%/17%	18%/12%/15%	6.9%/5.3%/6.0%	69%/78%/73%
GPT-3.5T-Roles	29%/27%/28%	6.1%/3.6%/4.5%	29%/25%/27%	17%/12%/14%	15%/13%/14%	87%/87%/87%

Type	ItCopular	ItemConstraint	ListContain	Membership	MultiSentLink	NounAdjunct
Zephyr-Trig	4.1%/83%/7.8%	2.4%/20%/4.3%	31%/56%/40%	4.1%/45%/7.6%	3.6%/65%/6.8%	14%/32%/19%
GPT-3.5T-Trig	2.4%/96%/4.7%	3.0%/36%/5.5%	20%/77%/32%	4.8%/66%/8.9%	7.8%/74%/14%	13%/51%/21%
Zephyr-Roles	45%/40%/42%	6.9%/6.9%/6.9%	5.0%/3.3%/4.0%	5.6%/4.7%/5.1%	26%/15%/19%	68%/53%/60%
GPT-3.5T-Roles	59%/56%/57%	17%/15%/16%	7.4%/4.5%/5.6%	4.0%/2.7%/3.2%	31%/17%/22%	75%/72%/74%

Type	PartitiveCollect	Possessive	PrepModifier	SuffCondition	Avg Micro	Avg Macro
Zephyr-Trig	0.5%/41%/0.9%	7.5%/31%/12%	1.9%/27%/3.6%	0%/0%/0%	14%/23%/18%	11%/41%/13%
GPT-3.5T-Trig	0.4%/53%/0.8%	6.9%/58%/12%	2.1%/49%/4.0%	0.2%/50%/0.4%	22%/59%/32%	11%/63%/16%
Zephyr-Roles	71%/71%/71%	25%/21%/23%	18%/15%/16%	13%/13%/13%	30%/17%/22%	35%/33%/34%
GPT-3.5T-Roles	70%/68%/69%	42%/39%/41%	21%/18%/19%	14%/13%/13%	27%/21%/24%	41%/38%/39%

a low-occurrence link, such as including a bias score in the prompt, performing an initial filter step, or improving the biasing of the provided examples – may significantly improve precision (and F-measure) moving forward. In general, role identification exhibits a much higher precision compared to trigger identification for infrequent link types. However, the complexity of the SemFrame roles limited the ability of the LLM to reliably identify roles for this type.

7 Future Work

While the structural annotations provided as part of this dataset represent a necessary addition to research in procedure modeling in real-world domains, we intend to publish two additional layers of annotation to further elucidate the complex semantics of procedures in the naval domain. The next layer will be a procedure-specific layer that defines and encodes procedural flow within these documents. This includes distinguishing between procedural steps, conditional steps, external procedural references, and explanatory information orthogonal to the execution of the procedure. Furthermore, this layer will indicate complex relationships between steps including looping behavior, sequence requirements, and alternative satisfaction of procedure requirements.

The final (semantic) layer will include sufficient information to properly model state change among the entities contained in the procedural texts, including (1) entity/event/stative type information linked to Navy-relevant ontology sources, (2) fine-grained sub-categorization of links between frames, and (3) entity and event coreference. As a special focus, we will also include text generated via Large-Language models which seeks to interpret the relationship between noun-adjunct pairings which is otherwise latent in the text. Taken together, we believe these three annotation layers will serve as a significant resource to support future research in procedure modeling, procedure enhancement, and common-sense reasoning over real-world domains.

Acknowledgments. This material is based upon work supported by the Office of Naval Research under Contract No. N00014-20-C-2048. Any opinions, findings and conclusions or recommendations expressed in this material are those of the author(s) and do not necessarily reflect the views of the Office of Naval Research.

Disclosure of Interests. The authors have no competing interests to declare that are relevant to the content of this article.

References

1. Amini, A., Bosselut, A., Mishra, B.D., Choi, Y., Hajishirzi, H.: Procedural reading comprehension with attribute-aware context flow. arXiv preprint arXiv:2003.13878 (2020)
2. Clark, P., Dalvi, B., Tandon, N.: What happened? Leveraging VerbNet to predict the effects of actions in procedural text. arXiv preprint arXiv:1804.05435 (2018)
3. Dalvi, B., Tandon, N., Bosselut, A., Yih, W.t., Clark, P.: Everything happens for a reason: Discovering the purpose of actions in procedural text. In: Proceedings of the 2019 Conference on Empirical Methods in Natural Language Processing and the 9th International Joint Conference on Natural Language Processing (EMNLP-IJCNLP), pp. 4496–4505 (2019)
4. Du, X., et al.: Be consistent! improving procedural text comprehension using label consistency. arXiv preprint arXiv:1906.08942 (2019)
5. Hripcsak, G., Rothschild, A.S.: Agreement, the f-measure, and reliability in information retrieval. JAMIA **12**(3), 296–298 (2005)
6. Jiang, Y., Zaporojets, K., Deleu, J., Demeester, T., Develder, C.: Recipe instruction semantics corpus (RISeC): resolving semantic structure and zero anaphora in recipes. In: Proceedings of 10th IJC-NLP, pp. 821–826 (2020)
7. Johnson, C., Fillmore, C.: The FrameNet tagset for frame-semantic and syntactic coding of predicate-argument structure. In: Proceedings of 1st NAACL, pp. 56–62. Morgan Kaufmann Publishers Inc. San Francisco, CA, USA (2000)
8. Kazeminejad, G., Palmer, M.: Event semantic knowledge in procedural text understanding. In: Proceedings of the The 12th Joint Conference on Lexical and Computational Semantics (* SEM 2023), pp. 388–398 (2023)
9. Kim, E., Huang, K., Saunders, A., McCallum, A., Ceder, G., Olivetti, E.: Materials synthesis insights from scientific literature via text extraction and machine learning. Chem. Mater. **29**(21), 9436–9444 (2017)
10. Langley, P.: Learning hierarchical problem networks for knowledge-based planning. In: Proceedings of 31st International Conference ILP (2022)
11. Li, T.J.J., Mitchell, T., Myers, B.: Interactive task learning from GUI-grounded natural language instructions and demonstrations. In: Proceedings of 58th ACL (SD), pp. 215–223 (2020)

12. Meyers, A., Reeves, R., et al.: The NomBank project: an interim report. In: Proceedings of Workshop Frontiers in Corpus Annotation at HLT-NAACL 2004, pp. 24–31 (2004)
13. Mishra, B.D., Huang, L., Tandon, N., Yih, W.t., Clark, P.: Tracking state changes in procedural text: a challenge dataset and models for process paragraph comprehension. arXiv preprint arXiv:1805.06975 (2018)
14. Mostafazadeh, N., et al.: CaTeRS: causal and temporal relation scheme for semantic annotation of event structures. In: Proceedings 4th Workshop on Events, pp. 51–61 (2016)
15. Mysore, S., Jensen, Z., et al.: The materials science procedural text corpus: annotating materials synthesis procedures with shallow semantic structures. arXiv preprint arXiv:1905.06939 (2019)
16. Orlando, R., Conia, S., Navigli, R.: Exploring non-verbal predicates in semantic role labeling: challenges and opportunities. arXiv preprint arXiv:2307.01870 (2023)
17. Palmer, M., Gildea, D., Kingsbury, P.: The proposition bank: an annotated corpus of semantic roles. Comput. Linguist. **31**(1), 71–106 (2005)
18. Prasad, R., et al.: The Penn discourse treebank 2.0. In: LREC (2008)
19. Rim, K., Tu, J., Ye, B., Verhagen, M., Holderness, E., Pustejovsky, J.: The coreference under transformation labeling dataset: entity tracking in procedural texts using event models. In: Findings of the ACL, pp. 12448–12460 (2023)
20. Shridhar, M., et al.: ALFRED: a benchmark for interpreting grounded instructions for everyday tasks. In: Proceedings IEEE/CVF Conference on CVPR, pp. 10740–10749 (2020)
21. Tandon, N., Mishra, B.D., Grus, J., Yih, W.t., Bosselut, A., Clark, P.: Reasoning about actions and state changes by injecting commonsense knowledge. arXiv preprint arXiv:1808.10012 (2018)
22. Tandon, N., Mishra, B.D., Sakaguchi, K., Bosselut, A., Clark, P.: WIQA: a dataset for "what if..." reasoning over procedural text. arXiv preprint arXiv:1909.04739 (2019)
23. Tandon, N., et al.: A dataset for tracking entities in open domain procedural text. arXiv preprint arXiv:2011.08092 (2020)
24. Tunstall, L., Beeching, E., et al.: Zephyr: direct distillation of LM alignment. arXiv preprint arXiv:2310.16944 (2023)
25. Valmeekam, K., Sreedharan, S., et al.: On the planning abilities of large language models (a critical investigation with a proposed benchmark). arXiv preprint arXiv:2302.06706 (2023)
26. Wei, X., Cui, X., et al.: Zero-shot information extraction via chatting with ChatGPT. arXiv preprint arXiv:2302.10205 (2023)
27. Xu, J., Sun, M., Zhang, Z., Zhou, J.: ChatUIE: exploring chat-based unified information extraction using large language models. arXiv preprint arXiv:2403.05132 (2024)
28. Zhang, H., Liu, X., Pan, H., Song, Y., Leung, C.W.K.: ASER: a large-scale eventuality knowledge graph. In: Proceedings of the Web Conference, pp. 201–211 (2020)
29. Zhang, L., Lyu, Q., Callison-Burch, C.: Reasoning about goals, steps, and temporal ordering with wikiHow. arXiv preprint arXiv:2009.07690 (2020)
30. Zhang, L., Xu, H., et al.: Causal reasoning of entities and events in procedural texts. arXiv preprint arXiv:2301.10896 (2023)
31. Zhou, S., Zhang, L., et al.: Show me more details: discovering hierarchies of procedures from semi-structured web data. arXiv preprint arXiv:2203.07264 (2022)

Models and Strategies for Russian Word Sense Disambiguation: A Comparative Analysis

Anastasiia Aleksandrova[1](✉)[iD] and Joakim Nivre[1,2][iD]

[1] Uppsala University, Uppsala, Sweden
alanev52@hotmail.com
[2] RISE Research Institutes of Sweden, Stockholm, Sweden

Abstract. Word sense disambiguation (WSD) is a core task in computational linguistics that involves interpreting polysemous words in context by identifying senses from a predefined sense inventory. Despite the dominance of BERT and its derivatives in WSD evaluation benchmarks, their effectiveness in encoding and retrieving word senses, especially in languages other than English, remains relatively unexplored. This paper provides a detailed quantitative analysis, comparing various BERT-based models for Russian, and examines two primary WSD strategies: fine-tuning and feature-based nearest-neighbor classification. The best results are obtained with the ruBERT model coupled with the feature-based nearest neighbor strategy. This approach adeptly captures even fine-grained meanings with limited data and diverse sense distributions.

Keywords: word sense disambiguation · BERT · Russian

1 Introduction

Word sense disambiguation (WSD) constitutes a foundational challenge in natural language processing, which involves determining a word's precise meaning in a given context. It serves as a crucial component in various benchmarks evaluating the performance of language models (LMs) like SemEval-2013 Task 12 [26], SuperGLUE [38], FEWS [6], and XL-WSD [29]. These frameworks test the capacity of models to distinguish between different senses of words.

Human performance in discriminating between word senses in English ranges from 67–80% for fine-grained to 82–96% for coarse-grained semantic distinctions [21,24,27], elucidating to which extent WSD is challenging. Nevertheless, human assessment continues to serve as a reference to evaluate the performance of machines.

The most effective contemporary LMs for English WSD achieve results close to human performance, "breaking through the 80% glass ceiling" [4]. Most of the outstanding results are achieved using BERT-based models [15,21,37]. However, the WSD task is still not considered completely solved [6,25].

Despite the substantial achievement of these LMs, there have been only a few in-depth analyses on the ability of LMs to effectively grasp the inherent ambiguity in words, even in English. Widely used evaluation benchmarks often concentrate on sentence-level representations, offering limited insight into how these models capture the intricate semantic properties of individual words. They primarily assess the models' capability to discern whether a target word holds the same meaning in varying contexts, rather than defining a particular sense. Replicating such analyses for languages other than English poses a greater challenge due to the scarcity of available annotated data specific to this task. To the best of our knowledge, there has been a lack of comprehensive research into the capacity of such LMs to capture word sense ambiguity in the Russian language.

In this article, our aim is to explore the influence of fine-tuning and feature extraction strategies on WSD in Russian using language models possessing diverse degrees of knowledge about the language. Our focus lies on exploring the nuances of word representations encoded by three distinct models: base (baseBERT), multilingual BERT (mBERT) [12], and Russian BERT (ruBERT), the latter being part of the DeepPavlov framework [19]. Our examination encompasses two primary LM-based WSD strategies: fine-tuning (FT) and feature-based nearest neighbor (FBNN).

The key contributions of our work can be summarized as follows: the development of a new dataset featuring 21 nouns, specifically compiled for exploring WSD systems and a quantitative analysis testing the limitations of BERT-based models in WSD for Russian using this dataset.[1]

2 Related Work

The extensive development of transformer-based LMs has stimulated research examining the models' capacity to encapsulate various linguistic phenomena. Considerable attention has been devoted to exploring syntax by using probing techniques to showcase the proficiency of transformers in encoding human-like parse trees to a significant extent [14,17].

Initially, the exploration of the semantic competence of LMs prioritized lexical semantics over contextual factors. Lexical semantics refers to individual words, leading numerous research efforts to focus on exploring static representations [9,34,40].

For instance, exploring the effectiveness of word2vec [22] in capturing various coarse-grained senses for sufficiently frequent and evenly distributed senses [40]. However, this scenario involves probing these words in isolation and does not entirely mirror the real-world challenge of WSD in context.

Another direction of research demonstrating impressive results examines contextualized word embeddings and ambiguity using lexical substitution as a key test [41]. Although lexical substitution serves as a useful approximate measure for WSD, it does not offer a comprehensive understanding of the specific challenges that LMs face in handling lexical ambiguity.

---

[1] All the data and code are available at https://github.com/alanev52/WSD.

Several surveys have focused on diverse approaches to the extraction of contextualized word embeddings, using varied sense inventories. These approaches range from knowledge-based methods [1,10,23,36] to purely supervised [3,13] and hybrid supervised/knowledge-based methods [2,4,7,11,18,21].

Supervised LMs treat the WSD task as a classification problem and utilize sense-annotated data for fine-tuning. By contrast, knowledge-based approaches rely on sense inventories like WordNet, drawing on encoded knowledge without the requirement of labeled training data. However, even before the emergence of large LMs, they underperformed supervised models [32,33]. Nowadays BERT-based models are approaching human-level performance [21,29] and hybrid WSD techniques have achieved superior performance [2].

In the context of our research, a survey by Loureiro et al. [21] stands out due to its clear methodology for analyzing the pre-trained BERT models' effectiveness in handling lexical ambiguity in English. Specifically, the authors examine two primary LM-based WSD strategies, namely fine-tuning (FT) and feature-based nearest neighbor classification (FBNN). They exploit a dataset based on Wikipedia, designed specifically for the study.

Going beyond English, there are not many comprehensive studies evaluating LMs on WSD for different languages. A relatively recent initiative like the cross-lingual evaluation suite XL-WSD [29] represents an important step towards creating a large-scale multilingual WSD benchmark, which however does not include Russian.

With respect to WSD in Russian, the research community is dedicated to overcoming the scarcity of semantically annotated data by developing tools for generating and labeling training collections with comprehensive coverage of polysemous words in RuWordNet [8] or by introducing an advanced Russian general evaluation benchmark known as RussianSuperGLUE [37].

In our work, we aim to explore the extent to which BERT models with diverse knowledge of Russian encode lexical meaning for polysemous words in the language. Specifically, we focus on uncovering their practical disambiguation capacities with various WSD strategies.

3 Method

3.1 Language Models

Among the prominent choices for semantic tasks, masked LMs like BERT [12] are in widespread use, especially in WSD [15,21,35]. We use mBERT, pre-trained on 104 languages, including Russian, with data sourced from Wikipedia. To compare mBERT with a language-specific model, we employ ruBERT [19]. This model is based on mBERT with continued pre-training on Russian, with a subword vocabulary derived from 80% Russian Wikipedia and 20% Russian news data, resulting in an augmentation of longer Russian words and subwords compared to the multilingual version. Finally, we utilize baseBERT, trained exclusively on English, as a baseline with no direct knowledge of Russian.

3.2 WSD Strategies

One prevalent approach for leveraging LMs for WSD is FT, i.e., supervised training of a pre-trained BERT model using labeled training data. This procedure adapts the model's parameters to align with the specific classification objectives required for the WSD task.

Given ample training data and computational resources, BERT has the potential to achieve human-comparable performance in identifying noun senses (evidenced by an F-score surpassing 80%) [21]. However, a significant limitation of such supervised models is their need for a substantial number of samples to build separate models for each individual word, rendering this ideal scenario impractical in real-world applications.

An encouraging finding of previous studies highlights the robustness of FBNN approach based on k nearest neighbor classification (FBNN, typically with $k = 1$) [30]. This strategy constructs sense vectors by averaging contextual embeddings from instances in the training set labeled with the same sense.

Both approaches assume that each sense is an independent class, and the architecture cannot exploit any knowledge beyond those inferred from the training corpus. This issue is not merely theoretical but also practical, given the extreme class imbalance where many senses occur sporadically in training datasets, closely mirroring reality [5].

3.3 Lexical Resources and Datasets

WSD as a knowledge-intensive task, relies on two types of data: (1) sense inventories or lexicons, containing possible senses for a given lexeme; and (2) annotated corpora, which consist of extensive texts where words are marked and labeled with meanings from the provided inventory [5].

The most useful sense inventory for the language is RuWordNet [20].[2] This semantic network is structured similarly to WordNet with a total database of 111.5 thousand words and word combinations.

For Russian, there exist only two small semantically annotated corpora. One of them is RUSSE from the RussianSuperGLUE benchmark [37]. The dataset is inherited from RUSSE-2018 [28] and transformed into the WiC task for nouns. Interestingly, the assessment of RUSSE for binary classification task using both Russian and multilingual BERT achieved significantly higher levels of performance than humans, reaching an accuracy of 89% and 84%, respectively, compared to the human baseline of 74% [37].

The second dataset, known as Corpus-1000 [16], is an "all-words" sense-annotated dataset, which encompasses 46k manually annotated lemmas. The corpus served as a test set for the pseudo labeling method, yielding a precision of 74.1%, and for the personalized PageRank method with 75.4% [16].

Although a larger amount of samples is beneficial for LM performance, previous research shows that optimal accuracy can be obtained with only 50

---

[2] https://ruwordnet.ru/ru.

examples per word [21]. Even this modest sample size could only be achieved by consolidating RUSSE and Corpus-1000. To create the new benchmark for WSD we extracted the instances of all available polysemous nouns (excluding "true homographs" or homonyms) from both corpora, ensuring that each word had 2−5 senses and occurred in at least 50 sentences. In total, we obtained 21 nouns: "dejstvie" (*activity*), "delo" (*case*), "den' " (*day*), "disk" (*disk*), "dokument"(*document*), "dolja" (*part*), "dom" (*house*), "doroga" (*road*), "duh" (*spirit*), "mesto" (*place*), "moment" (*instant*), "pravo" (*right*), "set' " (*network*), "sistema" (*system*), "stat'ja" (*article*), "vremja" (*time*), "zadacha" (*exercise*), "zakon" (*law*), "zaschita" (*protection*), "zemlja" (*land*), "zhizn' " (*life*). The labeling schema was harmonized with respect to RuWordNet to guarantee uniformity across the combined dataset. Therefore, we constructed an individual dataset for each word, encompassing files with sentence instances and labels based on a 70/30 train/test balanced split. Table 1 presents a comprehensive overview of the words included in the collection, along with the corresponding number of senses, dataset size, and normalized entropy values for each word, with higher values indicating a more balanced sense distribution (compare "dokument" and "pravo").

4 Experiments

4.1 Experimental Setup

Our experimental study involves comparing the performance of FBNN classification using features from the three BERT models (baseBERT, mBERT, and ruBERT) against FT for WSD with the same three models. Typically, applying FT to a model for each target word results in numerous models, which becomes expensive considering the vocabulary size. Regarding the small size of our dataset, we think the computational costs are acceptable.

In the experiments with BERT, we use the Transformers framework (v2.5.1) [39] and models described in Sect. 3.1. As in previous research on FBNN classification [21], we extract the last four layers of the BERT models as features. For FT we concatenate the average embedding of the target word with the [CLS] token's embedding and feed them into a classifier. Our experiments adopt a uniform default hyper-parameter configuration, and due to varying results in FT, we average the outcomes from three independent runs for consistency. We report micro-F_1 and macro-F_1 scores as the evaluation metrics.

4.2 Experimental Results

The main results of our experiment for all six systems - ruBERT-FBNN, mBERT-FBNN, baseBERT-FBNN, ruBERT-FT, mBERT-FT, and baseBERT-FT are presented in Table 1. Except for baseBERT-FBNN, all models achieve a micro-F_1 in the range 66–73% indicating a moderate WSD accuracy. Moreover, in macro-F_1, the performance is notably lower. The reason for this discrepancy lies in the distinction between the metrics. The F_1-score is computed as a weighted average of precision and recall, with micro-F_1 using weights based on the number of instances for each class, while macro-F_1 assigns equal weights. The macro-F_1 scores are generally about 4% points lower, which reflects the skewed distributions over different senses, where some senses only have 1–3 instances per sense. This may lead to very low F_1 scores for these senses, which decreases the macro-average more than the micro-average. Despite this effect, we retained these limited samples in the dataset to maintain the overall number of words.

As expected, ruBERT-FBNN surpasses by at least 4 points all other models in both evaluation metrics (with a micro-F_1 score of 73% vs. 69% and a macro-F_1 of 69% vs. 65%), which reinforces the potential of language-specific models for WSD in light-weight FBNN settings. On average, the FBNN strategy consistently yields higher results, which is particularly evident for the language-specific models, where the results surpass FT by 7 points in both micro-F_1 and macro-F_1 scores. This trend extends to multilingual configurations, showing a 2-point difference in favor of the FBNN approach.

Table 1. Micro-F_1 and macro-F_1 scores for all models FBNN and FT of ruBERT(ru), mBERT(mult), and baseBERT (base). Transliterated Russian words (word), number of senses (N), normalized entropy (entr.), and total size of the set (size) for each word in the dataset. The color gradient red-white-blue signifies the progression of values, with smaller values represented in red and the highest values in blue.

MERGED DATA				micro-F_1						macro-F_1					
				FBNN			FT			FBNN			FT		
word	N	entr.	size	ru	mult	base	ru	mult	base	ru	mult	base	ru	mult	base
dejstvie	5	0,84	139	84%	70%	24%	53%	51%	51%	82%	72%	26%	44%	47%	47%
delo	4	0,93	173	72%	70%	40%	79%	76%	76%	61%	62%	38%	66%	67%	67%
den'	3	0,84	152	72%	70%	43%	72%	74%	74%	68%	64%	38%	59%	68%	68%
disk	3	0,72	61	89%	89%	28%	83%	83%	83%	59%	60%	15%	54%	55%	55%
dokument	3	1	59	71%	53%	24%	57%	53%	53%	69%	54%	23%	56%	50%	50%
dolja	2	0,83	64	84%	58%	84%	79%	79%	79%	78%	46%	74%	60%	63%	63%
dom	4	0,85	84	69%	77%	62%	74%	76%	76%	52%	57%	47%	47%	49%	49%
doroga	3	0,83	63	74%	74%	42%	79%	67%	67%	64%	64%	38%	71%	55%	55%
duh	4	0,72	68	50%	60%	35%	78%	72%	72%	37%	41%	19%	56%	43%	43%
mesto	3	0,76	107	88%	91%	50%	89%	88%	88%	75%	83%	45%	67%	59%	59%
moment	2	0,35	51	80%	80%	78%	96%	100%	100%	64%	64%	58%	88%	100%	100%
pravo	2	0,25	93	100%	92%	79%	96%	96%	96%	100%	48%	44%	49%	49%	49%
set'	2	0,86	71	95%	95%	76%	84%	81%	81%	94%	94%	69%	76%	69%	69%
sistema	2	1	62	79%	79%	63%	88%	79%	79%	79%	79%	63%	88%	79%	79%
stat'ja	2	0,82	170	92%	90%	63%	92%	93%	93%	91%	88%	62%	91%	91%	91%
vremja	4	0,82	215	61%	55%	45%	80%	80%	80%	61%	58%	40%	72%	76%	76%
zadacha	2	0,9	54	88%	81%	44%	88%	83%	83%	87%	81%	38%	86%	81%	81%
zakon	2	0,61	109	70%	76%	61%	91%	87%	87%	64%	69%	55%	84%	75%	75%
zaschita	4	0,91	93	71%	68%	39%	56%	70%	70%	71%	68%	33%	46%	68%	68%
zemlja	4	0,91	87	65%	62%	19%	62%	72%	72%	64%	59%	14%	51%	60%	60%
zhizn'	3	0,88	119	69%	72%	31%	57%	62%	62%	68%	73%	30%	51%	55%	55%
			average	73%	69%	44%	66%	67%	67%	69%	65%	38%	62%	63%	63%

An intriguing discovery emerges as mBERT slightly outperforms ruBERT with a marginal difference of 63% vs. 62% when employing the FT strategy. This

Table 2. Average sense bias values for the two WSD strategies across all compared models. The minimal value, indicating the most robust system, is highlighted in bold.

	ruBERT	mBERT	bBERT	Average
FT	0.837	0.837	0.837	0.837
FBNN	**0.555**	0.625	0.741	0.640

prompted our investigation into the performance of baseBERT for both strategies, to study the impact of Russian language data in pre-training mBERT and ruBERT. Interestingly, when using the FT strategy, the absence of Russian pre-training data has a negligible impact on the results, as the fine-tuned baseBERT model achieves a micro-F_1 of 67% and a macro-F_1 of 63%. This contrasts sharply with the FBNN results for the same model, where the score drops significantly to a micro-F_1 of 44% and a macro-F_1 of 38%.

4.3 Bias Analysis

The observations so far inspired us to conduct a bias analysis to examine the robustness of the compared models against biases. To quantify biases in a disambiguation system toward a specific sense for a polysemous word we are using the pipeline proposed by Loureiro et al. [21]. The procedure involves calculating the disambiguation bias for each sense employing normalization of the total number of misclassifications toward that sense n_{ij} by the total number of instances with the golden label $\sum_j n_{ij}$. The overall bias toward a specific sense B_j is then computed by summing the biases from all other senses. The sense bias B_w is determined as the maximum value of B_j across all senses for a specific word:

$$
\begin{aligned}
B_j &= \sum_{i=1, i \neq j}^{n} \frac{n_{ij}}{\sum_j n_{ij}} \\
B_w &= \max(B_j) \\
B_\mathrm{m} &= \overline{B}_w
\end{aligned}
\tag{1}
$$

The higher the B_j value, the more biased the disambiguation model is toward a particular sense. The total model bias B_m is calculated as the average of B_w values across all words in the dataset. Fluctuations in the results of FT models are addressed by taking the median of three runs.

The overall results for all the models are presented in Table 2. The closer the bias value is to 1, the stronger the tendency is for a model to be biased toward specific senses. FT BERT models tend to be less robust with respect to sense biases. The average sense bias value for FBNN BERT models is 0.640 in comparison to 0.837 for the FT systems. Furthermore, our findings indicate that FBNN models demonstrate diverse biases depending on the language proficiency of the pre-trained model. The sense bias diminishes with the incorporation of the Russian language into the pre-trained model, with ruBERT-FBNN emerging as the most robust in this respect.

5 Discussion

The overall results of our study are consistent with previous research comparing the effectiveness of FT for WSD with multilingual and language-specific BERT-based models, indicating that the performance of the former is notably lower [29].

5.1 FT vs. FBNN Strategy

The comparison between FT and FBNN has already been explored in the existing literature across various tasks [21,31]. While the conventional belief suggests that FT models excel with adequate training data, in the realm of WSD, the FBNN strategy explored in this article emerges as the more robust choice, given the available data.

Surprisingly, an increased quantity of training instances (*size* in Table 1) does not consistently improve performance, especially in the context of the FBNN WSD strategy. This is indicated by the very low Pearson correlation (0.14 or lower) between *size* and all FBNN disambiguation outcomes achieved using ruBERT and mBERT. By contrast, FT systems exhibit a moderately stronger positive relationship, showcasing a Pearson's coefficient of 0.42. This reaffirms the crucial impact of training data size on FT models.

Furthermore, higher polysemy tends to correspond with lower performance across all systems. Notably, our observations underscore a more pronounced accuracy difference when disambiguating words with 2 and 3 senses compared to those with 3 and 4 senses. The negative Pearson's coefficients (-0.57 for ruBERT-FBNN, -0.50 for mBERT-FBNN, -0.69 for ruBERT-FT, -0.63 for mBERT-FT) indicate a moderate negative linear relationship observed across various models. This suggests that, as the quantity of senses increases, F_1 scores typically decrease. This trend is particularly prominent in all FT models and marginally less so in FBNN.

Moreover, within our experimental setup, FT models are observed to display stronger biases. Regardless of the Russian language proficiency of the employed BERT model, all FT systems exhibit identical biases for all the words in the dataset. By contrast, FBNN models show that models with diverse Russian knowledge might not necessarily have matching biases toward the same senses.

A potential explanation for these discrepancies between the strategies lies in the nature of FT, which entails the optimization of a loss function during training, rendering it more prone to overfitting on the most frequent sense (MFS). This is substantiated by the slightly more pronounced inclination of FT models to favor the MFS. Specifically, in 13 out of 21 words, FT models exhibit a preference for the MFS in the dataset, in contrast to 6 out of 21 words for ruBERT-FBNN. Conversely, the FBNN approach operates by memorizing states, disregarding sense distributions.

5.2 Multilingual vs Language-Specific Models

The outcomes of our experiments partially validate the trend of superior performance of multilingual models over language-specific models, possibly influenced by inherent noise and skewed data distributions [29]. This trend aligns with the FT approach in our research (macro-F_1: ruBERT-FT: 62%, mBERT-FT: 63%, and baseBERT-FT: 63%). However, the results from the FBNN strategy exhibit contrary behavior, revealing a strong correlation between WSD performance and the inclusion of Russian in the LLM's training data (macro-F_1: ruBERT-FBNN: 69%, mBERT-FBNN: 65% and baseBERT-FBNN: 38%). Moreover, the bias analysis demonstrates the enhanced robustness of language-specific FBNN models concerning biases, as compared to their FT counterparts.

5.3 Limitations of the Study

In our study, we do not explore levels of sense interpretability (fine-grained or coarse-grained) due to the absence of objective annotation. Regrettably, the limitations of our research prevent us from conducting a human accuracy test that could serve as a reference.

The dataset is constructed on the basis of two existing collections. Therefore, it inherits the predetermined constraints of the original corpora. This should be taken into account, particularly in terms of the results of the extremely data-dependent FT strategy.

Furthermore, independent factors, such as the absence of a BERT model exclusively trained on Russian, hindered our ability to fully evaluate the impact of non-Russian data on the systems' performance.

6 Conclusion

This article extensively analyzes how various pre-trained BERT-based LMs capture lexical ambiguity in Russian. For noun WSD with diverse sense granularity and limited training data availability, BERT-based models are effective in disambiguating some individual words, although their overall performance is not always solid.

An encouraging finding of our study is the tendency of the FBNN strategy to outperform FT, demonstrating higher results, greater adaptability to diverse sense distributions within the training data, and robustness towards biases even with a limited number of instances per word sense. Hence, our findings underscore the promising potential of the FBNN approach, especially when employing language-specific models like RuBERT.

As future work, it would be interesting to explore the impact of sense granularity on WSD performance, expand the dataset to reveal FT potential, and establish human evaluation baselines.

References

1. Agirre, E., de Lacalle, O.L., Soroa, A.: Random walks for knowledge-based word sense disambiguation. Comput. Linguist. **40**(1), 57–84 (2014)
2. Barba, E., Pasini, T., Navigli, R.: ESC: redesigning WSD with extractive sense comprehension. In: Proceedings of the 2021 Conference of the North American Chapter of the Association for Computational Linguistics: Human Language Technologies, pp. 4661–4672 (2021)
3. Bevilacqua, M., Navigli, R.: Quasi bidirectional encoder representations from transformers for word sense disambiguation. In: Proceedings of the International Conference on Recent Advances in Natural Language Processing (RANLP 2019), pp. 122–131. Varna, Bulgaria (2019)
4. Bevilacqua, M., Navigli, R.: Breaking through the 80% glass ceiling: raising the state of the art in word sense disambiguation by incorporating knowledge graph information. In: Proceedings of the 58th Annual Meeting of the Association for Computational Linguistics, pp. 2854–2864 (2020)
5. Bevilacqua, M., Pasini, T., Raganato, A., Navigli, R.: Recent trends in word sense disambiguation: A survey. In: Proceedings of the Thirtieth International Joint Conference on Artificial Intelligence, IJCAI-21, pp. 4330–4338 (2021)
6. Blevins, T., Joshi, M., Zettlemoyer, L.: FEWS: Large-scale, low-shot word sense disambiguation with the dictionary. In: Proceedings of the 16th Conference of the European Chapter of the Association for Computational Linguistics: Main Volume, pp. 455–465 (2021)
7. Blevins, T., Zettlemoyer, L.: Moving down the long tail of word sense disambiguation with gloss informed Bi-encoders. In: Proceedings of the 58th Annual Meeting of the Association for Computational Linguistics, pp. 1006–1017 (2020)
8. Bolshina, A., Loukachevitch, N.V.: Generating training data for word sense disambiguation in Russian. In: International Conference on Computational Linguistics and Intellectual Technologies Dialogue-2020, pp. 119–132 (2020)
9. Camacho-Collados, J., Pilehvar, M.T.: From word to sense embeddings: a survey on vector representations of meaning. J. Artif. Intell. Res. **63**(1), 743–788 (2018)
10. Chaplot, D.S., Salakhutdinov, R.: Knowledge-based word sense disambiguation using topic models. In: Proceedings of the AAAI Conference on Artificial Intelligence, pp. 5062–5069 (2018)
11. Conia, S., Navigli, R.: Framing word sense disambiguation as a multi-label problem for model-agnostic knowledge integration. In: Proceedings of the 16th Conference of the European Chapter of the Association for Computational Linguistics: Main Volume, pp. 3269–3275 (2021)
12. Devlin, J., Chang, M.W., Lee, K., Toutanova, K.: BERT: Pre-training of deep bidirectional transformers for language understanding. In: Proceedings of the 2019 Conference of the North American Chapter of the Association for Computational Linguistics: Human Language Technologies, Volume 1 (Long and Short Papers), pp. 4171–4186 (2019)
13. Hadiwinoto, C., Ng, H.T., Gan, W.C.: Improved word sense disambiguation using pre-trained contextualized word representations. In: Proceedings of the 2019 Conference on Empirical Methods in Natural Language Processing and the 9th International Joint Conference on Natural Language Processing (EMNLP-IJCNLP), pp. 5297–5306 (2019)

14. Hewitt, J., Manning, C.D.: A structural probe for finding syntax in word representations. In: Proceedings of the 2019 Conference of the North American Chapter of the Association for Computational Linguistics: Human Language Technologies, Volume 1 (Long and Short Papers), pp. 4129–4138 (2019)
15. Huang, L., Sun, C., Qiu, X., Huang, X.: GlossBERT: BERT for word sense disambiguation with gloss knowledge. In: Proceedings of the 2019 Conference on Empirical Methods in Natural Language Processing and the 9th International Joint Conference on Natural Language Processing (EMNLP-IJCNLP), pp. 3509–3514 (2019)
16. Kirillovich, A., Loukachevitch, N., Kulaev, M., Bolshina, A., Ilvovsky, D.: Sense-annotated corpus for Russian. In: Proceedings of the 5th International Conference on Computational Linguistics in Bulgaria (CLIB 2022), pp. 130–136 (2022)
17. Kulmizev, A., Ravishankar, V., Abdou, M., Nivre, J.: Do neural language models show preferences for syntactic formalisms? In: Proceedings of the 58th Annual Meeting of the Association for Computational Linguistics, pp. 4077–4091 (2020)
18. Kumar, S., Jat, S., Saxena, K., Talukdar, P.: Zero-shot word sense disambiguation using sense definition embeddings. In: Proceedings of the 57th Annual Meeting of the Association for Computational Linguistics, pp. 5670–5681 (2019)
19. Kuratov, Y., Arkhipov, M.: Adaptation of deep bidirectional multilingual transformers for Russian language. In: Computational Linguistics and Intellectual Technologies: Proceedings of the International Conference Dialogue-2019, pp. 333–339 (2019)
20. Loukachevitch, N.V., G. Lashevich, A.A.G., Ivanov, V.V., Dobrov, B.V.: Creating Russian wordnet by conversion. In: Proceedings of Conference on Computational Linguistics and Intellectual Technologies Dialog-2016, pp. 405–415 (2016)
21. Loureiro, D., Rezaee, K., Pilehvar, M.T., Camacho-Collados, J.: Analysis and evaluation of language models for word sense disambiguation. Comput. Linguist. **47**(2), 387–443 (2021)
22. Mikolov, T., Chen, K., Corrado, G., Dean, J.: Efficient estimation of word representations in vector space. In: 1st International Conference on Learning Representations (ICLR), Workshop Track Proceedings (2013)
23. Moro, A., Raganato, A., Navigli, R.: Entity linking meets word sense disambiguation: a unified approach. Trans. Assoc. Comput. Linguist. **2**, 231–244 (2014)
24. Navigli, R.: Word sense disambiguation: a survey. ACM Comput. Surv. **41**(2), 1–69 (2009)
25. Navigli, R.: Natural language understanding: Instructions for (present and future) use. In: Proceedings of the Twenty-Seventh International Joint Conference on Artificial Intelligence, IJCAI-18, pp. 5697–5702 (2018)
26. Navigli, R., Jurgens, D., Vannella, D.: SemEval-2013 task 12: multilingual word sense disambiguation. In: Second Joint Conference on Lexical and Computational Semantics (*SEM), Volume 2: Proceedings of the Seventh International Workshop on Semantic Evaluation (SemEval 2013), pp. 222–231 (2013)
27. Palmer, M., Dang, H.T., Fellbaum, C.: Making fine-grained and coarse-grained sense distinctions, both manually and automatically. Nat. Lang. Eng. **13**(2), 137–163 (2007)
28. Panchenko, A., Lopukhina, A., Ustalov, D., Lopukhin, K., Arefyev, N., Leontyev, A., Loukachevitch, N.: RUSSE 2018: A Shared Task on Word Sense Induction for the Russian Language. In: Computational Linguistics and Intellectual Technologies: Papers from the Annual International Conference Dialogue-2018, pp. 547–564 (2018)

29. Pasini, T., Raganato, A., Navigli, R.: XL-WSD: an extra-large and cross-lingual evaluation framework for word sense disambiguation. In: Proceedings of the AAAI Conference on Artificial Intelligence, pp. 13648–13656 (2021)
30. Peters, M.E., et al.: Deep contextualized word representations. In: Proceedings of the 2018 Conference of the North American Chapter of the Association for Computational Linguistics: Human Language Technologies, Volume 1 (Long Papers), pp. 2227–2237 (2018)
31. Peters, M.E., Ruder, S., Smith, N.A.: To tune or not to tune? Adapting pretrained representations to diverse tasks. In: Proceedings of the 4th Workshop on Representation Learning for NLP (RepL4NLP-2019), pp. 7–14 (2019)
32. Pilehvar, M.T., Navigli, R.: A large-scale pseudoword-based evaluation framework for state-of-the-art word sense disambiguation. Comput. Linguist. **40**(4), 837–881 (2014)
33. Raganato, A., Camacho-Collados, J., Navigli, R.: Word sense disambiguation: a unified evaluation framework and empirical comparison. In: Proceedings of the 15th Conference of the European Chapter of the Association for Computational Linguistics: Volume 1, Long Papers, pp. 99–110 (2017)
34. Reisinger, J., Mooney, R.J.: Multi-prototype vector-space models of word meaning. In: Kaplan, R., Burstein, J., Harper, M., Penn, G. (eds.) Human Language Technologies: The 2010 Annual Conference of the North American Chapter of the Association for Computational Linguistics, pp. 109–117 (2010)
35. Scarlini, B., Pasini, T., Navigli, R.: SensEmBERT: context-enhanced sense embeddings for multilingual word sense disambiguation. In: Proceedings of the AAAI Conference on Artificial Intelligence, pp. 8758–8765 (2020)
36. Scozzafava, F., Maru, M., Brignone, F., Torrisi, G., Navigli, R.: Personalized PageRank with syntagmatic information for multilingual word sense disambiguation. In: Proceedings of the 58th Annual Meeting of the Association for Computational Linguistics: System Demonstrations, pp. 37–46 (2020)
37. Shavrina, T., et al.: RussianSuperGLUE: a Russian language understanding evaluation benchmark. In: Proceedings of the 2020 Conference on Empirical Methods in Natural Language Processing (EMNLP), pp. 4717–4726 (2020)
38. Wang, A., et al.: Superglue: a stickier benchmark for general-purpose language understanding systems. In: Proceedings of the 33rd International Conference on Neural Information Processing Systems, pp. 3266–3280 (2019)
39. Wolf, T., et al.: Transformers: state-of-the-art natural language processing. In: Proceedings of the 2020 Conference on Empirical Methods in Natural Language Processing: System Demonstrations, pp. 38–45 (2020)
40. Yaghoobzadeh, Y., Kann, K., Hazen, T.J., Agirre, E., Schütze, H.: Probing for semantic classes: diagnosing the meaning content of word embeddings. In: Proceedings of the 57th Annual Meeting of the Association for Computational Linguistics, pp. 5740–5753 (2019)
41. Zhou, W., Ge, T., Xu, K., Wei, F., Zhou, M.: BERT-based lexical substitution. In: Proceedings of the 57th Annual Meeting of the Association for Computational Linguistics, pp. 3368–3373 (2019)

Open-Source Web Service with Morphological Dictionary–Supplemented Deep Learning for Morphosyntactic Analysis of Czech

Milan Straka[(✉)] and Jana Straková

Institute of Formal and Applied Linguistic, Faculty of Mathematics and Physics, Charles University, Malostranské nám. 25, 11800 Prague, Czech Republic
{straka,strakova}@ufal.mff.cuni.cz

Abstract. We present an open-source web service for Czech morphosyntactic analysis. The system combines a deep learning model with rescoring by a high-precision morphological dictionary at inference time. We show that our hybrid method surpasses two competitive baselines: While the deep learning model ensures generalization for out-of-vocabulary words and better disambiguation, an improvement over an existing morphological analyser MorphoDiTa, at the same time, the deep learning model benefits from inference-time guidance of a manually curated morphological dictionary. We achieve 50% error reduction in lemmatization and 58% error reduction in POS tagging over MorphoDiTa, while also offering dependency parsing. The model is trained on one of the currently largest Czech morphosyntactic corpora, the PDT-C 1.0, with the trained models available at https://hdl.handle.net/11234/1-5293. We provide the tool as a web service deployed at https://lindat.mff.cuni.cz/services/udpipe/. The source code is available at GitHub (https://github.com/ufal/udpipe/tree/udpipe-2), along with a Python client for a simple use. The documentation for the models can be found at https://ufal.mff.cuni.cz/udpipe/2/models#czech_pdtc1.0_model.

Keywords: morphosyntactic analysis · deep learning · morphological dictionary · POS tagging · lemmatization

1 Introduction

Czech landscape of morphosyntactic tools widely available under favorable licensing terms and provided in an off-the-shelf manner is by no means bleak. For morphological analysis, MorphoDiTa (Morphological Dictionary and Tagger) [15] has been widely used in the academic circles in recent years, providing morphological analysis, morphological generation, tagging and tokenization. However, MorphoDiTa's well-known aspect is its reliance on the underlying morphological dictionary, which leads to limited performance for words not included in said dictionary, the out-of-vocabulary (OOV) words. Released in 2014, MorphoDiTa also lacks the incorporation of recent advancements such as deep learning techniques and contextualized embeddings. For both morphological and syntactic analysis, we refer to UDPipe [12], which provides tagging, lemmatization and syntactic analysis for tens of languages, Czech included. This tool, on the other hand, depends solely on deep learning.

In this paper, we present an open-source web service and Python client for morphosyntactic analysis, which combines deep learning architecture of UDPipe 2 [12] with a rescoring by a morphological dictionary MorfFlex [7] (the core of MorphoDiTa [15]) to enhance the effectiveness of the deep learning model. Our evaluation shows that the combined system improves over both a deep learning system and a dictionary-based system by themselves. The deep learning architecture ensures generalization for dictionary OOVs and better disambiguation, while the morphological dictionary promotes consistent outputs by disallowing invalid analyses at inference time. This leads to 50% error reduction in lemmatization accuracy in comparison with MorphoDiTa and 35% error reduction in lemmatization accuracy in comparison with UDPipe 2. For POS tagging accuracy, we achieve 58% and 16% error reduction in comparison with MorphoDiTa and UDPipe 2, respectively.

Moreover, the new model is trained on one of the largest Czech morphosyntactic resources, the PDT-C 1.0 [5].

To sum up, the released tool provides segmentation, tokenization, morphological analysis, lemmatization, POS tagging and syntactic analysis. It does so by combining a deep learning model with a morphological dictionary at inference time.

2 Related Work

MorphoDiTa (Morphological Dictionary and Tagger) [15] is an open-source tool for morphological analysis, which performs morphological analysis, morphological generation, tagging and tokenization of natural texts, and relies on an underlying morphological dictionary (MorfFlex [7] for Czech). MorphoDiTa uses the Czech morphological system by Jan Hajič [6].

UDPipe [12,14] is an open-source tool for segmentation, tokenization, lemmatization, POS tagging, morphological analysis, and dependency parsing of natural texts. UDPipe models are available for 131 datasets of 72 languages of the Universal Dependencies project, using the universal morphosyntactic tagging system of the Universal Dependencies project [11].

Majka [17,18], with its free version Fajka, is a morphological analyser, which assigns a lemma and all possible grammatical tags to each word form on the input. Majka is available for 15 languages. Czech Majka uses a Czech morphological tagset by Jakubíček et al. [9].

In this work, we combine the Czech morphological dictionary MorphoDiTa with UDPipe 2 trained on the Prague Dependency Treebank – Consolidated 1.0 (PDT-C 1.0, [5]). For details on the data and the morphological tagset used, see the following Sect. 3.

3 Data

Our model is trained on The Prague Dependency Treebank - Consolidated 1.0 (PDT-C 1.0, [5]), which has been recently released, and is, to our knowledge, one of the largest manually annotated Czech morphosyntactic resources.[1] The project includes and consolidates several existing Czech corpora, giving rise to the following sections:

- PDT: Prague Dependency Treebank 3.5, written texts,
- PCEDT: Czech part of Prague Czech-English Dependency Treebank 2.0 and Coref 2.0,
- PDTSC: Prague Dependency Treebank of Spoken Czech 2.0, spoken data,
- FAUST: PDT-Faust, user-generated texts.

The morphological layer (m-layer) of the PDT-C 1.0 is manually annotated, containing nearly 4M words (m-forms). PDT-C 1.0 uses the PDT-C tag set [10][2] from MorfFlex [7], which is an evolution of the original PDT tag set devised by Jan Hajič [6].

The surface syntax layer (analytical, a-layer) is manually annotated only partially in the 1.0 version, specifically in a part of the PDT section only, and is planned for full manual annotation in the next released version. The PDT-C 1.0 employs dependency relations from the PDT analytical level [8].[3]

4 Methods

Our architecture is a deep learning model jointly learning morphosyntactic analysis and dependency parsing, with additional rescoring of the morphological outputs by the morphological dictionary MorfFlex [7]. The deep learning architecture is identical to the architecture of UDPipe 2 [12], with RobeCzech [13], a monolingual Czech pretrained language model, as a foundation. We refer to the baseline system without morphological dictionary and trained on PDT-C 1.0 as *UDPipe 2* and to our system with an added morphological dictionary as *Our system*. The overview of our architecture is outlined in Fig. 1.

4.1 Morphological Dictionary–Supplemented Deep Learning

Before we proceed to describe the employment of the morphological dictionary MorfFlex [7] at the model inference time, we first briefly describe the inference process without the presence of a morphological dictionary.

When performing POS tagging, the input to the deep learning model is a surface word form and the output is a POS tag. The model predicts a probability distribution over all POS tags, and the most probable POS tag is selected as the

---

[1] https://ufal.mff.cuni.cz/pdt-c.
[2] https://ufal.mff.cuni.cz/pdt-c/publications/Appendix_M_Tags_2020.pdf.
[3] https://ufal.mff.cuni.cz/pdt-c/publications/Appendix_A_Tags_2020.pdf.

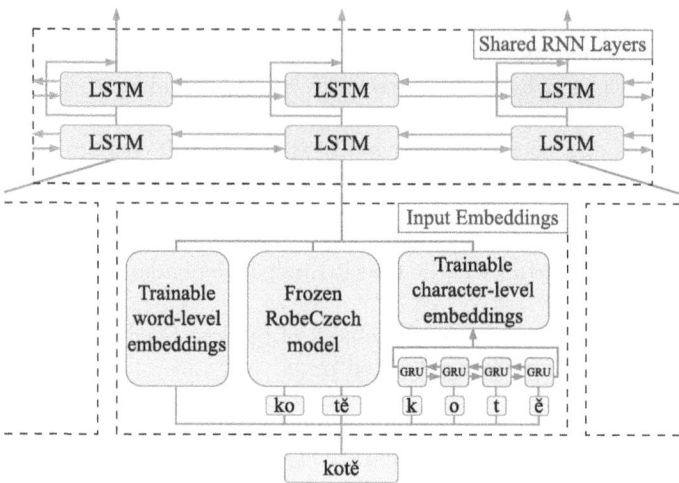

A. Input embeddings and shared RNN layers.

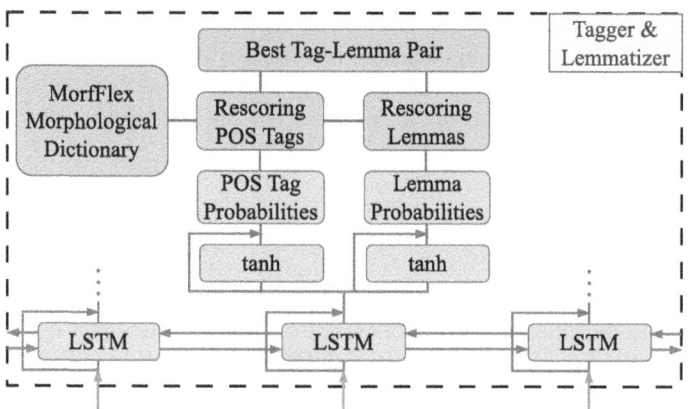

B. Tagger and lemmatizer.

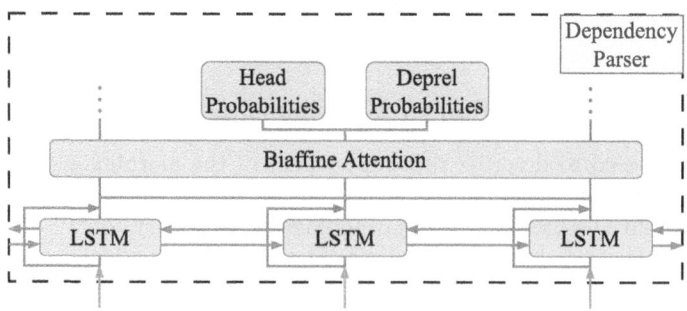

C. Dependency parser.

Fig. 1. An illustration of the proposed model architecture.

output. To perform lemmatization, the very same instance of the model produces also a second output, a character edit rule to convert the word form to the output lemma.[4] Precisely, the model again predicts a probability distribution over all edit rules as its second output, and consequently, the most probable edit rule is applied on the input form to produce an output lemma.

Now with a morphological dictionary, though, we jointly rescore the two probability distributions, both the POS tags and the edit rule probability distributions, in the following way:

1. If the input surface word form is out-of-vocabulary of the dictionary, nothing happens, and the original probability distributions predicted by the model are used without modification.
2. If the dictionary recognizes the word form, we proceed to disambiguation via the dictionary. Most importantly, from now on, we consider only valid form-tag–edit rule entries found in the dictionary, disallowing any invalid form-tag–edit rule pairs suggested by the model. From the valid pairs, the one with the largest product of the tag probability and the edit rule probability, as predicted by the model, is selected.

The succession of the deep learning model and the morphological dictionary ensures that an analysis is generated even for OOVs of the dictionary. Also, the deep learning model ranks the POS tags and lemmas by predicting their probability distributions for a given word form in context, using deep learning techniques and contextualized embeddings. Finally, the important contribution of the morphological dictionary is the pruning of the invalid combinations of forms, tags and lemmas.

4.2 Re-parsing the Automatically Annotated Dependency Trees

Only a part of the PDT section of the PDT-C 1.0 is annotated manually with dependency trees, while the other sections were annotated automatically at the time of their original creation using various automatic dependency parsers available at the time or lack syntactic annotation altogether. However, as the morphological and syntactic analyses are trained jointly in our system [12], the heterogenous quality of various sources of automatic parsing may indirectly influence also the results of morphological analysis.

Therefore, before training the system presented here, we automatically re-parsed these sections in order to improve and level the quality of the automatic annotations. We started our system by training it on the training portion of the syntactically annotated PDT section. We then used the first version of our system to automatically parse the remaining training sections of PDT-C 1.0, and then we re-trained our system on the entire training data of the PDT-C 1.0.

---

[4] On construction of edit rules, consult [12].

Table 1. Lemmatization accuracy [%] on PDT-C 1.0. We consider lemmas with sense disambiguation numbers but without additional dictionary comments.

System	Dictionary	PDT	PCEDT	PDTSC	FAUST	Macro Avg	Error Reduction MorphoDita	UDPipe 2
MorphoDiTa	✓	98.69	98.85	98.18	97.53	98.31		
UDPipe 2	✗	98.76	99.12	98.94	97.98	98.70	23%	
Ours	✓	99.19	99.40	99.23	98.78	99.15	50%	35%

Table 2. POS tagging accuracy [%] on PDT-C 1.0.

System	Dictionary	PDT	PCEDT	PDTSC	FAUST	Macro Avg	Error Reduction MorphoDita	UDPipe 2
MorphoDiTa	✓	96.29	97.00	96.90	94.87	96.27		
UDPipe 2	✗	98.53	98.64	98.55	96.84	98.14	50%	
Ours	✓	98.78	98.80	98.77	97.42	98.44	58%	16%

5 Results

5.1 Lemmatization and POS Tagging Results

Tables 1 and 2 display lemmatization and POS tagging results on PDT-C 1.0, respectively. Our system achieves 50% error reduction for lemmatization accuracy and 58% error reduction for POS tagging accuracy as compared with dictionary-based MorphoDiTa. A comparison with UDPipe 2, a deep learning model without any dictionary, shows that additional employment of a morphological dictionary at inference time reduces lemmatization error by 35% and POS tagging error by 16%. We provide a more detailed error analysis in Sect. 6.

5.2 The Effect of Including Automatically Re-parsed Sections

In Table 3, we show how the addition the automatically re-parsed sections of PDT-C 1.0 to training data on top of the syntactically annotated part of the PDT section improves the lemmatization and POS tagging accuracy. The gain is notable, with 50% error reduction in lemmatization accuracy and 37.16% error reduction in POS tagging accuracy. The improvement is naturally caused by the increased amount of data and domains, but the re-parsing was necessary to allow joint training of our model without deterioration on the lower-quality or missing syntactic annotations.

5.3 Parsing Results

For completeness, we also declare the parsing UAS and LAS score of our system, evaluated on the only section of PDT-C 1.0 with the manually annotated a-layer so far, the PDT section (see also Sect. 3). The system, trained on the whole PDT-C 1.0 with re-parsed treebanks, achieves UAS 94.41% and LAS 91.48% on the PDT section. When trained purely on the PDT section, both scores were lower by a 0.1 percent points.

6 Error Analysis

6.1 Lemmatization Improvement over UDPipe 2

We are now interested in error analysis between UDPipe 2 and our system, in which the added morphological dictionary rescored the model outputs at inference time. On PDT-C 1.0 test data, UDPipe 2 without dictionary made 4 689 lemmatization errors out of all 422 540 lemmas. Our system with a subsequent rescoring by dictionary fixed 1 692 errors, while introducing only 143 new errors. This lead to 35% error reduction of macro lemmatization accuracy, as shown in Table 1 in Sect. 5.

Table 3. The effect of including automatically re-parsed sections on lemmatization and POS tagging accuracy [%]. The first system is trained on the syntactically annotated part of the PDT section. The second system is trained on the entire PDT-C 1.0 with re-parsed sections. Both systems are UDPipe 2 without the morphological dictionary.

Train Section	PDT		PCEDT		PDTSC		FAUST		Macro Avg	
	Lemmas	POS	Lemmas	POS	Lemmas	POS	Lemmas	POS	Lemmas	POS
PDT	98.64	98.41	98.25	97.86	97.04	96.73	96.28	95.16	97.55	97.04
PDT-C	98.76	98.53	99.12	98.64	98.94	98.55	97.98	96.84	98.70	98.14

Apparently, UDPipe 2 is willing to generate and prefer a non-existent lemma, as 49% of its lemma errors are hallucinated non-lemmas (not in dictionary). Fortunately, it is also able to generate the correct lemma among other options, and its score is often close to the leading positions, so after the morphological dictionary prunes the most likely but invalid options, the correct lemma emerges as the winner. Consequently, the majority of dictionary-based corrections (1147, 68%) were corrections of such fictitious lemmas.

This is particularly well observable in cases when only one valid analysis is possible according to the dictionary: See Table 4, where UDPipe 2 does not achieve 100% lemmatization and POS tagging accuracy, despite only one possible valid analysis. After correction with the dictionary, our system reached 100% accuracy in the single analysis setting.

In more detail, of all the dictionary-fixed errors, be it lemmas or non-lemmas, 759 lemmatization errors (45%) were sense corrections—the lemma was correctly generated, but the lemma sense needed to be disambiguated by the morphological dictionary ("ještě-1" → "ještě-2", "Lincoln-2" → "Lincoln-3", "jak-3" → "jak-2", etc.). In a few other cases (169 errors, which equals to 10% errors), the lemma was also almost correctly generated, but the casing had to be corrected to fit the actual dictionary entry ("lovochemie" → "Lovochemie", "Fytoplankton" → "fytoplankton", "Kozoroh" → "kozoroh"). The remaining dictionary-fixed 764 errors (45%) were completely incorrectly generated lemmas. We print the most frequent dictionary corrections of the completely incorrect lemmas (without lemma senses and capitalization errors) in Table 5.

6.2 Lemmatization Improvement over MorphoDiTa

The error reduction between MorphoDiTa and our system is 50% in lemmatization macro accuracy and 58% in POS tagging macro accuracy, as shown in Table 1 and Table 2, respectively. MorphoDiTa makes 5 380 lemmatization errors on the PDT-C 1.0 test data, and our system made only 3 140 lemmatization errors. Our system was correct in 3 274 (61%) lemmatization errors made by MorphoDiTa. But, on the other hand, our system made other, new 1 034 errors where MorphoDiTa had the correct lemma.

A detailed analysis of errors made at varying levels of word ambiguity is shown in Table 6. Besides an expected improvement of error rate in the OOV condition (0 analyses), we interestingly see an increasing trend of improvement in the more ambiguous situations, where more lemmas or lemma senses need to be disambiguated (2 and more analyses, with maximum gain at 9+ analyses). The former can be explained by the neural network generating better lemmas than the MorphoDiTa guesser in the OOV condition, as our system corrected 70% of these errors from MorphoDita. The latter shows better disambiguation on our part.

Indeed, the qualitative analysis of both outputs confirms that MorphoDiTa does assign some lemma from the dictionary (unlike the deep learning model

Table 4. Micro average accuracies [%] for varying level of ambiguity, reflected in number of analyses; plus absolute differences between UDPipe 2 and our system, on PDT-C 1.0 test data.

Analyses	Weight	UDPipe 2		Ours		Abs. Delta	
		POS	Lemma	POS	Lemma	POS	Lemma
0	0.85%	91.01	91.71	91.01	91.71	0.00	0.00
1	41.14%	99.75	99.75	100.00	100.00	0.10	0.10
2	13.89%	97.82	98.00	98.16	98.59	0.05	0.08
3	11.99%	98.61	98.86	98.79	99.32	0.02	0.06
4	9.68%	98.26	98.55	98.45	99.07	0.02	0.05
5	4.99%	96.55	96.90	96.69	97.19	0.01	0.01
6	3.08%	97.01	97.49	97.22	98.07	0.01	0.02
7	1.93%	97.10	97.70	97.33	98.22	0.00	0.01
8	2.52%	97.80	98.81	97.96	99.26	0.00	0.01
9+	9.92%	97.41	99.23	97.53	99.47	0.01	0.02

Table 5. Most frequent dictionary corrections of completely incorrect lemmas by UDPipe 2 on PDT-C 1.0 test data. Asterisk marks generated lemmas not in dictionary.

Forms by frequency	UDPipe 2 lemma	Dictionary-corrected lemma
úhlům, úhlech	úhlo*	úhel
Angl	Ang-2*	Anglie
Kan	kan*	Kanada
kateg	kateg*	kategorie
zataženo	zataženo-2	zatáhnout
el	el-88*	elektrický
Nig	nig*	Nigérie
dožínky, Dožínky, dožínkách	dožínka*	dožínky
nedaleko	nedaleko	daleko-1
Kristem	Krist*	Kristus-3
mg	mgetr*	miligram
proklel	proklet*	proklít
nenesli, nenesly, Nenesla	nenést*	nést
jehož	jenž	jehož
Pierce, Piercem	Pierc*	Pierce
nindžové, nindžů	nindž*	nindža
g	gok*	gram
prostřednictvím	prostřednictvím	prostřednictví
dešťů	dešť*	déšť
studiích, studií	studie	studium
MW	MW*	megawatt
přímek	přímek*	přímka
Maď	maď*	Maďarsko
kpt	kpt*	kapitán
So	so-1*	sobota
Út	Út*	úterý

of UDPipe 2, which is perfectly content with generating a hallucinated lemma), but struggles with selecting the correct lemma and/or disambiguation of lemma senses. In this, our system based on UDPipe 2 has several fundamental advantages: (i) vastly larger capacity of the neural network, (ii) pre-training of the language model on large data, and (iii) better contextualization, because it uses the contextualized word embeddings produced by a BERT-like [2], Transformer-based [16] Czech model RobeCzech [13], and these embeddings are further contextualized with a bidirectional RNN [4]. MorphoDiTa, on the other hand, is an older tagger implemented as supervised, rich feature averaged perceptron [1] with the Viterbi algorithm [3].

Table 6. Micro average accuracies [%] for varying level of ambiguity, reflected in number of analyses; plus absolute differences between MorphoDiTa and our system, on PDT-C 1.0 test data.

Analyses	Weight	MorphoDiTa		Ours		Abs. Delta	
		POS	Lemma	POS	Lemma	POS	Lemma
0	0.85%	81.38	84.93	91.01	91.71	0.08	0.06
1	41.14%	100.00	100.00	100.00	100.00	0.00	0.00
2	13.89%	95.45	98.00	98.16	98.59	0.38	0.08
3	11.99%	96.98	98.89	98.79	99.32	0.22	0.05
4	9.68%	94.98	98.06	98.45	99.07	0.34	0.10
5	4.99%	94.24	96.00	96.69	97.19	0.12	0.06
6	3.08%	91.84	95.22	97.22	98.07	0.17	0.09
7	1.93%	92.88	96.23	97.33	98.22	0.09	0.04
8	2.52%	95.45	98.58	97.96	99.26	0.06	0.02
9+	9.92%	90.17	99.07	97.53	99.47	0.73	0.04

7 Limitations

The increased capacity and contextualization, and consequently, improved disambiguation and accuracy, come at a price in terms of computational demand and efficiency: MorphoDiTa's throughput is 10-200K words per second,[5] while the throughput of UDPipe 2 is 60 words per second using 1 CPU thread, or 300 words per second using 8 CPU threads, or 2k words per second on a GPU.[6] In conclusion, the selection of a tradeoff between efficiency and effectiveness is a consideration essential for the given task.

8 Conclusions

We presented an open-source web service and tool for morphosyntactic analysis. It combines a deep learning model with additional rescoring by a morphological dictionary at inference time.

In comparison with dictionary-based MorphoDiTa, we achieved 50% error reduction in lemmatization and 58% error reduction in POS tagging. By employing a morphological dictionary at inference time, we observed lemmatization error reduction by 35% and POS tagging error reduction by 16%, as compared with the deep-learning-only model UDPipe 2.

The model is deployed at https://lindat.mff.cuni.cz/services/udpipe/, the source code along with a Python client at https://github.com/ufal/udpipe/tree/udpipe-2, and the trained models at https://hdl.handle.net/11234/1-5293, under MPL 2.0 license for the source code and CC BY-NC-SA 4.0 license for the

---

[5] Source: https://ufal.mff.cuni.cz/morphodita.
[6] Source: https://ufal.mff.cuni.cz/udpipe.

models. The documentation for the models can be found at https://ufal.mff.cuni.cz/udpipe/2/models#czech_pdtc1.0_model.

Acknowledgments. This work has been supported by the Grant Agency of the Czech Republic under the EXPRO program as project "LUSyD" (project No. GX20-16819X). The work described herein has also been using data provided by the LINDAT/CLARIAH-CZ Research Infrastructure (https://lindat.cz), supported by the Ministry of Education, Youth and Sports of the Czech Republic (Project No. LM2023062).

We are grateful to our three knowledgeable and competent reviewers for their helpful comments and writing suggestions.

References

1. Collins, M.: Discriminative training methods for hidden markov models: theory and experiments with perceptron algorithms. In: Proceedings of the 2002 Conference on Empirical Methods in Natural Language Processing, pp. 1–8. Association for Computational Linguistics (2002). https://doi.org/10.3115/1118693.1118694, http://www.aclweb.org/anthology/W02-1001
2. Devlin, J., Chang, M.W., Lee, K., Toutanova, K.: BERT: pre-training of deep bidirectional transformers for language understanding. In: Burstein, J., Doran, C., Solorio, T. (eds.) Proceedings of the 2019 Conference of the North American Chapter of the Association for Computational Linguistics: Human Language Technologies, Volume 1 (Long and Short Papers), pp. 4171–4186. Association for Computational Linguistics, Minneapolis, Minnesota (2019). https://doi.org/10.18653/v1/N19-1423, https://aclanthology.org/N19-1423
3. Forney, G.D.: The Viterbi algorithm. Proc. IEEE **61**(3), 268–278 (1973). https://doi.org/10.1109/PROC.1973.9030
4. Graves, A., Schmidhuber, J.: Framewise phoneme classification with bidirectional LSTM networks. In: Proceedings. 2005 IEEE International Joint Conference on Neural Networks, vol. 4, pp. 2047–2052 (2005). https://doi.org/10.1109/IJCNN.2005.1556215
5. Hajič, J., et al.: Prague dependency treebank - consolidated 1.0. In: Calzolari, N., et al. (eds.) Proceedings of the Twelfth Language Resources and Evaluation Conference, pp. 5208–5218. European Language Resources Association, Marseille, France (2020). https://aclanthology.org/2020.lrec-1.641
6. Hajič, J.: Disambiguation of Rich Inflection (Computational Morphology of Czech). Linguistic Data Consortium, University of Pennsylvania (2004)
7. Hajič, J., Hlaváčová, J., Mikulová, M., Straka, M., Štěpánková, B.: MorfFlex CZ 2.0 (2020). http://hdl.handle.net/11234/1-3186, LINDAT/CLARIN digital library at the Institute of Formal and Applied Linguistics (ÚFAL), Faculty of Mathematics and Physics, Charles University
8. Hajič, J., et al.: Annotations At Analytical Level. Charles University, Instructions for annotators. Institute of Formal and Applied Linguistics (1999)
9. Jakubíček, M., Kovář, V., Šmerk, P.: Czech morphological Tagset revisited. In: Horák, A., Rychlý, P. (eds.) Proceedings of Recent Advances in Slavonic Natural Language Processing, RASLAN 2011. pp. 29–42 (2011)
10. Mikulová, M., et al.: Manual for Morphological Annotation, Revision for the Prague Dependency Treebank - Consolidated 2020 release. Charles University, Institute of Formal and Applied Linguistics (2020)

11. Nivre, J., et al.: Universal Dependencies v2: an evergrowing multilingual treebank collection. In: Calzolari, N., et al. (eds.) Proceedings of the Twelfth Language Resources and Evaluation Conference, pp. 4034–4043. European Language Resources Association, Marseille, France (2020). https://aclanthology.org/2020.lrec-1.497
12. Straka, M.: UDPipe 2.0 prototype at CoNLL 2018 UD shared task. In: Zeman, D., Hajič, J. (eds.) Proceedings of the CoNLL 2018 Shared Task: Multilingual Parsing from Raw Text to Universal Dependencies, pp. 197–207. Association for Computational Linguistics, Brussels, Belgium (2018). https://doi.org/10.18653/v1/K18-2020, https://aclanthology.org/K18-2020
13. Straka, M., Náplava, J., Straková, J., Samuel, D.: RobeCzech: Czech RoBERTa, a monolingual contextualized language representation model. In: Ekštein, K., Pártl, F., Konopík, M. (eds.) Text, Speech, and Dialogue: 24th International Conference, TSD 2021, Olomouc, Czech Republic, September 6–9, 2021, Proceedings, pp. 197–209. Springer International Publishing, Cham (2021). https://doi.org/10.1007/978-3-030-83527-9_17
14. Straka, M., Straková, J.: Tokenizing, POS tagging, lemmatizing and parsing UD 2.0 with UDPipe. In: Hajič, J., Zeman, D. (eds.) Proceedings of the CoNLL 2017 Shared Task: Multilingual Parsing from Raw Text to Universal Dependencies, pp. 88–99. Association for Computational Linguistics, Vancouver, Canada (2017). https://doi.org/10.18653/v1/K17-3009, https://aclanthology.org/K17-3009
15. Straková, J., Straka, M., Hajič, J.: Open-source tools for morphology, lemmatization, POS tagging and named entity recognition. In: Bontcheva, K., Zhu, J. (eds.) Proceedings of 52nd Annual Meeting of the Association for Computational Linguistics: System Demonstrations, pp. 13–18. Association for Computational Linguistics, Baltimore, Maryland (2014). https://doi.org/10.3115/v1/P14-5003, https://aclanthology.org/P14-5003
16. Vaswani, A., et al.: Attention is all you need. In: Proceedings of the 31st International Conference on Neural Information Processing Systems, pp. 6000—6010. NIPS'17, Curran Associates Inc., Red Hook, NY, USA (2017). https://proceedings.neurips.cc/paper_files/paper/2017/file/3f5ee243547dee91fbd053c1c4a845aa-Paper.pdf
17. Šmerk, P.: Fast morphological analysis of Czech. In: Proceedings of Third Workshop on Recent Advances in Slavonic Natural Language Processing, RASLAN 2009, pp. 13–16 (2009)
18. Šmerk, P., Rychlý, P.: Majka - rychlý morfologický analyzátor. Tech. rep., Masarykova univerzita (2009). http://nlp.fi.muni.cz/ma/

Mistrík's Readability Metric – an Online Library

Mária Pappová(✉) [ID] and Matúš Valko [ID]

Constantine the Philosopher University in Nitra, Tr. A. Hlinku 1,
949 74 Nitra, Slovakia
{maria.pappova2,matus.valko}@ukf.sk

Abstract. The term "readability" describes how simple it is for a reader to understand a written text. This can be measured with a variety of readability metrics. While some tools exist for assessing the readability of Slovak texts, no free or open-source tools currently offer this functionality. This article presents an online Python library that uses Mistrík's readability metric for the Slovak language. We developed an open-source library for measuring the readability score of Slovak texts and evaluated the findings from Mistrik's initial investigation approach.

Keywords: Mistrík's readability metric · text comprehension · Slovak language

1 Introduction

Reading is a fundamental skill that unlocks a world of knowledge and information. However, mastering it requires us to develop a complex set of abilities. Reading comprehension involves constructing a cognitive representation of the text within the reader's mind. As the reader engages with the text, the process continues. The outcome of this procedure is a cohesive cognitive representation of the text. The goal of reading comprehension is to create a precise cognitive representation of the text's intended meaning. [1,2]

Text readability is the comprehensibility of a written text. It depends on the content, as well as a number of linguistic and typographical features. A readable text is one that the intended reader can understand well without excessive effort [3]. Most often, we measure linguistic complexity based on length and frequency, but we also use readability indices. There are typically two key factors that determine readability: Typography and Content and Readability score [4].

Typography covers the presentation of text; attributes such as font size, line height, etc. are taken into account. The content employs a statistical and lexical approach to the words within the text. The readability scores are numbers that tell you how easy it will be for someone to read a particular piece of text. One of their shortcomings is that they do not consider grammar, spelling, intonation, or any other qualitative aspects of the text. A text full of grammatical problems can get the same readability score as a text with perfect grammar [5].

Quantitative techniques, known as readability metrics, estimate the comprehensibility of a written text, indicating its ease or difficulty of understanding. These metrics analyse various linguistic properties, such as word length, sentence structure, and syntactic complexity, and generate numerical scores or levels that indicate the readability of the text [5].

Quantitative readability assessment helps maintain the ideal degree of readability for the target audience. The Flesch Reading Ease formula was the first widely used measure of readability [6]. Since then, researchers have created multiple English readability measures. Previous formulations, designed to be applicable to any language, relied on basic quantitative measurements. Current metrics, such as the one developed by McNamara et al. [7], use linguistic knowledge-based features, making them clearly dependent on language.

The concept of "difficulty of a text" refers to a compilation of attributes that are inherent to the text and influence the learning subject's perception, comprehension, and processing of its content [4,8,9].

In learning systems, when we are talking about text difficulty, there are basically two concepts: complexity and difficulty. They are sometimes used as synonyms, but they are actually more like pseudo-synonyms.

Complexity refers to the inherent structure and intricacy of a task, regardless of student performance. Difficulty, on the other hand, is relative and takes into account how challenging a task is for a specific student group based on their skills, knowledge, and environment. Understanding this distinction is critical to designing effective learning systems that cater to a variety of student needs [10].

To assess text difficulty objectively, researchers rely on inherent properties like sentence length, comprehensibility, and concept count. These factors, along with lexical elements, determine the inherent difficulty a text presents to readers. Formulas like Flesch, Challa, Mikko, Pisarek, and Mistrík leverage these characteristics to calculate a text's difficulty score. While formulas vary in emphasis on specific parameters, they all aim to gauge a text's readability based on its objective characteristics [3,10].

This article aims to create an online library for the Python programming environment that includes Mistrík's readability formula. Currently, there is no freely available utility that does this. The first section of this article will analyse and discuss readability metrics for the Slovak language. Next, we will automate the metric for Slovak texts and review the library's development to ensure proper implementation of Mistrík's level of legibility in Python. We will evaluate the results by comparing them to Jozef Mistrík's original research.

2 Related Work

In contrast to international research, there has been little interest in studying Czech or Slovak text comprehension. The Czech language does not have a consistent tradition of readability research or a plain-language planning system, even though there is a lot of research on styles, including administrative styles (Mistrík [11], Exner [12], or Těšitelová [13–15]).

Bendová and Cinková [16] focused on the adaptation of well-established readability metrics for the Czech language. The authors propose methods to adapt four classic readability formulas to the Czech context. They analyse the sensitivity of these adapted metrics to textual paraphrases. Furthermore, the study investigates the relationship between Czech readers' readability scores and reading comprehension. The study specifically focuses on adapting the Flesch Reading Ease formula from Russian to Czech. The study emphasises the continued relevance of traditional readability metrics in professional writing, language instruction, and assessment, even in the age of neural network-based formulas.

The most popular Polish formula for computing readability was proposed in the 1960 s by Walery Pisarek [17] based on the research of Rudolf Flesch [6] and Jozef Mistrík [18]. The formula presented below calculates the text difficulty level T [17, 19]:

$$T = \frac{\sqrt{T_s^2 + T_w^2}}{2} \qquad (1)$$

where:

- T_s^2 - percentage of 4-or-more-syllable words (in lemmatized form),
- T_w^2 - average sentence length (in words),

In other words, the difficulty formula takes into account only two features of the text: the mean length of a sentence and the percentage of "potentially difficult" words (longer than three syllables). Existing methods, such as the Pisarek method, were established during a time when neither frequency lists nor large text corpora were available. So, it was not possible to apply the methods on a large scale.

Flesch's readability formula, from which Pisarek's formula is also derived, originated from Flesch's Reading Ease test of text complexity. They are commonly designed to measure the level of difficulty and they are widely used for English text. It is one of the most widely used metrics for English texts. The Flesch Reading Ease score, which reflects how readable the text is, assigns different weighting factors to the same basic metrics, namely word length and sentence length. This metric is described using the following formula [6]:

$$FRES = 2206,835 - 1,015(\frac{total words}{total sentences}) - 84,6(\frac{total syllabes}{total words}) \qquad (2)$$

The Flesch Reading Ease score ranges from 1 to 100, with higher readability scores indicating easier text reading, in contrast to other metrics that favor lower scores. Regarding the limitations of the metric, since it was developed for school teaching aids, it shows weaknesses compared to direct usability testing with typical readers. It ignores the differences between readability and content, text layout, and information retrieval aids. For example, the pangram "Cwm 9ordbank glyphs vext quiz" has a readability score of 100 and a level score of 0.52 despite its obscure words [20]. Another problem with the Flesch Reading-Ease is, for example, that it was developed for the English language and its results are influenced by the linguistic characteristics of this language [20]. That means, for example, that it may not evaluate highly inflected languages in which higher specificities are present, such as declensions or conjugations, which can affect the text's readability.

Despite the low interest in research in the field of text readability, there are currently two readability metrics for the Slovak language as well. The first of them is Mistrík's measure of comprehensibility [18], and the second is the measure T, which was created by modifying Nestler's readability metric [5, 21].

In connection with metric T, a diploma thesis was created by Drahošová [22], where the relationship was investigated between the difficulty of the text of textbooks and the readability of the text of textbooks by students. The modified Nestler's method determined the difficulty, while the Cloze test and Fog index determined the students' comprehension and readability of the textbook text. The results of the study showed that textbook difficulty increases depending on the year. The level of comprehensibility and readability in the evaluated textbooks was higher than the highest recommended values.

The previous diploma thesis dealt with the T metric precisely because of the study by Průcha [5]. There were several reasons why Průcha [21] searched for a different metric of text difficulty. He aimed to develop a metric that was not only easy to apply but also capable of accurately identifying two or three factors that contribute to the text's difficulty, such as the length of words, the length of sentences, and the number of words in a sentence. The text difficulty metric should meet demanding criteria, namely [5, 22]:

- Complexity: to incorporate both quantitative and qualitative (semantic) aspects of textbook text,
- Validity: in relation to the real difficulty of the text in learning,
- Operativity: is the ability to manage tasks, even for practical purposes, without using complex computing techniques.

Průcha found that these requirements were met by the degree of difficulty and complexity of the text of the German researcher Käte Nestler, which was found in 1982 [1, 5]. Nestler's formula is considered to be more complex than, for example, Místrík's measure of complexity, Pluskal's metric or Pisarek's metric, and it was also modified for Czech and Slovak texts as well. This formula has been extensively used in Czech and Slovak educational research to measure the "difficulty" of reading textbooks [1, 5].

In these countries, it has been considered a standard instrument for measuring the readability of expository texts. As a result, it was used to evaluate many primary and secondary school textbooks as part of quality analyses (e.g., Průcha, 1989; Hrabí, 2004; Janoušková, 2009). These measurements helped to improve the expository parts of primary and secondary school textbooks.

Průcha [5, 23] chose this method to study Czech didactic texts because it was good at showing how difficult they really are. For instance, experiments with German students of various grades demonstrated that the objectively measured difficulty levels corresponded to the difficulty of the texts. Later, it was adapted to be used in the Czech and Slovak educational environments by Průcha [5], and even later, it was slightly modified by Pluskala [24]. Currently, the updated

Nestler - Průcha - Pluskala metric is used in practice, which is abbreviated as the T measure (the symbol T indicates the degree of difficulty of the text) [5]. The T measure is calculated from the syntactic and semantic difficulty of the text, as it follows in the formula below [5,22]:

$$T = T_s + T_p, \tag{3}$$

where:

- Syntactic difficulty (T_s) - is determined by the syntactic structure of the text, which affects the reception and understanding of the text and is influenced by the average length of sentences and sentence segments. The following formula describes how T_s is calculated [22]:

$$T_s = 0.1 * V * U, \tag{4}$$

where:
  - V - the average sentence length is calculated according to the following formula:
  $$V = \frac{totalsentences}{totalwords} \tag{5}$$
  - U - the average length of sentence segments, which is calculated according to the following formula:
  $$U = \frac{totalwords}{totalverbs} \tag{6}$$

In each sample, the number of verbs is first determined, but only in a certain form (indefinite verbs are not counted). Subsequently, the total number of verbs in the word is determined and the parameter U is calculated.

- Semantic (conceptual) difficulty (T_p) - is caused by the semantic structure of the text. Five quantitative characteristics reflect the high value of the semantic factor. The types of concepts and quantities represented in the text determine its level of difficulty. The text distinguishes between technical, factual, numerical, and repeated terms. The formula below explains how T_p is calculated, but Drahošová [22] and Průcha [5,23] provide detailed descriptions of each parameter.

$$T_p = 100 * \frac{\sum N}{\sum P} * \frac{\sum P_1 + 3\sum P_2 + 2\sum P_3 + 2\sum P_4 + \sum P_5}{\sum N} \tag{7}$$

Gavora [1] investigated the link between text comprehension and readability among lower secondary school students. Gavora emphasises the importance of text comprehension in various disciplines, such as didactics, linguistics, and text-based learning theory. The study assessed comprehension, not knowledge, using older geography textbooks and an upper secondary school textbook. Texts were evaluated using two readability measures: Nestler's adjusted formula and Mistrík's formula. The study compared students from different localities, with the expectation that students from larger cities would perform better than those from smaller towns. However, the study found no statistical difference between the two groups. The paper emphasises the importance of text comprehension and readability in educational settings.

Rochovská and Huľová [25] analysed three educational resources (two textbooks and a workbook) for technical education in elementary schools. The analysis examined various aspects, including content scope, structure, learning activities, illustrations, and the use of didactic materials. Additionally, the Mistrík's measure was employed to assess text comprehensibility. The results reveal significant variability across these aspects of the analysed resources. Based on these findings, the authors proposed developing a new textbook that integrates the strengths of all three analysed materials.

Pappová [26] analysed the readability of economic texts in Slovak and English. The study compared various readability metrics with regard to the reference, which was Mistrík's readability measure, and examined whether there was agreement in the assessment of text complexity based on the language. Mistrík's readability measure is not appropriate for the English language, according to the research. Nevertheless, other metrics assessed both texts as highly professional and appropriate for readers with a university education. Certain metrics, including the SMOG Index and the Gunning Fog Index, responded to the text's complexity without regard to language, with little variation in ratings across languages. The Flesch Reading Ease metric found the Slovak text difficult to comprehend, attributed to its inflexibility.

3 Mistrík's readability formula

In our article, we use the terms comprehensibility and readability interchangeably and synonymously. It should be pointed out, as Mistrík himself pointed out, that in American works he often only uses the term readability. However, Mistrík preferred the term comprehensibility when describing his metric because it can be used equally for oral expression as well as for written expression, while readability can only refer to written text [18, 27].

The Mistrík's readability formula is calculated using the Phrase Repetition Index [18]. This implies that a text becomes easier to read the more words it repeats. Mistrík defines extratextual fact as information that is stored in the subconscious, is not realised verbally or by visual means, but only based on circumstances that are evident in a specific situation, self-evident. He distinguishes two types of texts: objective texts and literary texts. The factual texts are austere; their function is to inform. They are divided into three groups: journalistic, professional-educational, and administrative-legal texts [18, 27].

Literary texts mainly have an aesthetic function; their task is to evoke unusual feelings, moods, and aesthetic experiences in the reader. According to the processing method, we distinguish three types of literature, namely prose, poetry, and drama [18]. The following formula calculates the formula R, where R stands for readability [18, 27, 28]:

$$R = 50 - \frac{V * S}{I}, \tag{8}$$

where:

- V - average length of sentences in number of words,
- S - average length of words in number of syllables,
- I - word repetition index, which is calculated as:

$$I = \frac{N}{L}, \tag{9}$$

where:

- N - number of all lexical units (words) of the text,
- L - number of different lexical units (unique words).

The word repetition index has an effect on the difficulty of the text, which means that repeating words more often reduces its difficulty. This fact is especially important for explanatory texts. The measure of readability (R) is compared to a scale where texts with the lowest difficulty have values from 40 to 50 points and texts with the highest difficulty (at the limit of comprehensibility) have 0–10 points [18].

Mistrík's formula highlights the text's lexical diversity. Examining school textbooks reveals the presence of diverse language, which correlates with the language proficiency expected of students in a specific age group. However, this formula does have a limitation. It excludes qualitative lexicon measurements and text syntax. For the reader attempting to comprehend the read text, not only the quantitative range of the text's vocabulary is important, but also its qualitative parameters - whether or not technical terms are used in the text and how frequently they are repeated. Similarly, the syntactic factor considers not only the length of the sentences but also the specific sentence constructions, such as whether the sentence has a larger or smaller number of propositions [1,5,22].

3.1 Automatisation of the Metric

As we mentioned earlier, the Mistrík's readability metric is used in some contexts, but no public tool or library can be found on the internet that can use this metric. What we want to achieve is to create a public open-source library using the Python programming language. The library will have an algorithm that can count the syllables of words and apply the Mistrík's formula to any given text. The algorithm will produce the readability rate (R) and all the variables used in the formula. This could help people preparing Slovak school textbooks to quickly estimate the text's readability and ensure that the content is also readable for a specific group of readers.

We chose two fairy tales for children and two scientific papers for testing the algorithm. We chose these texts because the author of the readability formula, Jozef Mistrík, also tested very similar texts by himself. We also wanted to approach approximately the same number of words as Mistrík did in his work, which is about 300 words per sample [18]. It should be noted that although Mistrík described the use of excerpts of texts, but we do not know which words or sentences he specifically used.

We thoroughly examined the selected texts and then prepared them for manipulation by converting them to a text file and removing all unnecessary characters. The next step was to develop an algorithm for Mistrík's readability metric. We chose to write the algorithm in Python because it is well-known for its usability in data science, and our library can be easily installed later. The algorithm was based on an article by Mistrík [18], which is the original author's description of the metric. The description of our algorithm is found below:

1. Text preprocessing
   (a) Remove all quotation marks, parentheses and other symbols
   (b) Split all compound words
2. Syllabification algorithm
   (a) Segment the text into words
   (b) Calculate each word's syllable count
   (c) Remove punctuation marks from the words
   (d) Get list of all the words, cleaned from unwanted characters
3. Calculate the variables in the Mistrík's readability measure

Accurate syllable determination was critical during the metric's development. The Mistrík's formula uses a variable (S) to represent the average sentence length in syllable count. The Slovak language employs two fundamental methods for dividing a word.

1. Syllable division: In this division, we focus on the rhythmic division of the word into syllables. A syllable consists of [29]:

   – A vowel by itself (e.g., "a", "e", "i", "o", "u").
   – Combining a vowel with a consonant (e.g., "ma", "le", "ni", "to").
   – Syllable consonants "ŕ", "ĺ", "ř", "ľ" (e.g., "vŕ-ba", "mĺk-vy").

   We divide the word:

1. In places where two vowels belonging to different syllables meet (e.g., "ide-ál").
2. Before a consonant between two vowels (e.g., "že-na", "pra-co-vať").
3. At the interface of two consonants between two vowels (e.g., "lás-ka", "mas-lo").
4. In compound words on the border of composition (e.g., "Bra-ti-sla-va").

2. Morphematic division: In this division, we focus on dividing the word into morphemes (the smallest meaningful parts of the word). We separate:

   – Monosyllabic and polysyllabic prefix morphemes (e.g., "vy-brať", "pre-brať").
   – Derivative morphemes (word-forming suffixes) beginning with a consonant (e.g., "voľ-ba", "sud-ca").

- Grammatical morphemes (case and personal suffixes) starting with a consonant (e.g., "chlap-mi", "pri-stúp-me").
- Words in compound words on the border of composition (e.g., "Bra-ti-sla-va").

These methods are useful for dividing a word into syllables, but we do not need the position of the syllables, where the syllables are starting and ending. The number of syllables would be sufficient for calculating the formula. We investigated two methods for determining the count of syllables in the words. The first approach involves creating a dictionary with pre-calculated syllable counts. We could not find this type of dictionary anywhere online, and making one for the inflectional Slovak language would be too difficult and time-consuming.

We chose the second approach, which is based on counting vowels, diphthongs, and consonants. This approach is similar to that of the English language, but there are some subtle differences. From a pragmatic perspective, we can simplify the rules for counting syllables as follows:

- Count the number of vowels in the word (a,á,ä,e,é,i,í,y,ý,o,ó,u,ú,),
- Add 1 for each syllable that also defines the syllable in the word (ŕ,ĺ,ô),
- Add 1 for each letter (r and l) that define the syllable in the word,
- Subtract 1 for each diphthong and other vowel combinations in the word (ia,ie,iu,au,ou,oo).

The number at the end indicates how many syllables there are in the word. If the number is 0, it is a one-syllable word, just like if the word has less than three letters. If we counted only vowels and diphthongs, we would incorrectly calculate six words from the given text. These words are: "netrpezlivo", "zamrmlala", "chlpatá", "hlbokom", "najprv", and "zhrmotalo". These words have one common thing.

If the syllable in the word does not have a vowel or diphthong, its function is taken over by a consonant. In this case, we say that they are syllabic. It is about 'l', 'ĺ', 'r' and 'ŕ' [29]. We already defined the letters 'ĺ' and 'ŕ' as always counting to make a syllable in the word, but not the 'l' and 'r'. Thus, they become the bearers of syllabicity when they are between consonants in the same word. It does not matter whether the consonant that follows the syllable-forming 'l' or 'r' belongs to the same or the following syllable.

After the implementation of the rules for counting the syllables, the created algorithm is accurate on all of the unique words from the text sample by M. Ďuríčková. We then tested the algorithm on the four selected texts.

- Danka a Janka v rozprávke - Mária Ďuríčková,
- Ahoj, prvá trieda - Eleonóra Gašparová,
- Morfológia slovenského jazyka - text from Milan Urbančok,
- Morfológia slovenského jazyka - text from Jozef Mistrík.

The two fairy tale books have readability numbers of 42.2 and 41.4, respectively, indicating that they are very easy to read. This is a similar number that Mistrík (44.5 and 44.1) found when he was calculating the readability of the children's books. Also, the scientific texts have a similar readability number (32.2

Table 1. The results of algorithm for analysed texts

Texts (Authors)	Sentences	V	S	N	L	I	R
M. Ďuríčková	44	7 (6.909)	1.9 (1.875)	304	182	1.67	**42** (42.243)
E. Gašparová	35	9 (8.657)	1.7 (1.749)	303	172	1.76	**41** (41.406)
M. Urbančok	29	11 (10.643)	2.3 (2.332)	298	213	1.4	**32** (32.258)
J. Mistrík	23	13 (12.913)	2.3 (2.296)	297	195	1.52	**31** (30.53)

and 30.5) that was already found by author Mistrík (31.5 and 31.6). The readability score R between 30 and 40 denotes average, easy-to-understand texts. We rounded some values as recommended by the formula's author, despite not rounding the readability measure R during verification. We achieved satisfactory accuracy with the metric, but we expect that we will have to improve the algorithm for different texts in the future.

Table 2. Comparison of different metrics for analysed texts

Texts (Authors)	Mistrik	Flesch	Flesch grade level	Gunning	Gunning grade level
M. Ďuríčková	42.2	68.3	8/9	7.7	8
E. Gašparová	41.4	67.0	8/9	7.5	8
M. Urbančok	32.2	38.8	college	13.2	college
J. Mistrík	30.5	47.2	college	12.1	12

We applied the Flesch Reading Ease and Gunning Fog Index readability metrics, designed for English, to Slovak texts. Both metrics indicate that our chosen fairy tale books require eight years of formal education to understand the text on the first reading. This is a relatively high number because the texts are simple to read. The two scientific papers received higher ratings. Understanding them necessitates at least, or nearly, a college degree. These metrics are not particularly effective at estimating Slovak texts, demonstrating the utility of Mistrik's readability metric.

4 Conclusion

Mistrík's measure of comprehensibility is a straightforward readability assessment method. In Slovakia, readability measures are somewhat common, but not as widespread as they are abroad. Our goal was to support the use of readability measures, especially Mistrík's, by creating an open-source Python library, since there is still no library or tool that could apply a given metric to selected texts. At the same time, we wanted to make this metric more accessible because improving reading comprehension skills not only improves comprehension, but also supports lifelong learning by enabling individuals to effectively absorb information in a variety of areas.

In our research, based on the goal of making this metric more accessible, we first focused on analysing the current state, where we tried to find as much available information as possible about readability metrics used in Slovakia and also about metrics used in the world.

Our work led to the creation of an online Python library, which will be available for both research and general use. We compared the validity of the metric with similar text samples that the author also used.

The paper highlights the importance of text comprehension and readability in educational contexts, emphasising the need for tools and metrics such as Mistrík's metric for evaluating and improving the quality of educational materials. The source code and documentation for Mistrík's readability measure are available at https://github.com/MatusValko/Mistrik.

Disclosure of Interests. This work was supported by the Slovak Research and Development Agency under grant No. APVV-18-0473, University Grant Agency under grant No. VII/11/2024 and University Grant Agency under grant No. VII/14/2024.

References

1. Gavora, P.: Text comprehension and text readability: findings on lower secondary school pupils in Slovakia. Wydawnictwo Uniwersytetu Kazimierza Wielkiego w Bydgoszczy (2012)
2. Tillman, R., Hagberg, L.: Readability algorithms compability on multiple languages (2014)
3. LiFR Linguistic Factors of Readability in Czech Administrative and Educational Texts. https://ufal.mff.cuni.cz/grants/lifr. Accessed 18 Apr 2024
4. Gavora, P., et al.: Ako rozvíjať porozumenie textu u žiaka. Enigma, Nitra (2008). 978-80-89132-57-7
5. Průcha, J.: Hodnocení obtížnosti učebnic: struktury a parametry učiva. Výzkumný ústav odborného školství, Praha (1984)
6. Flesch, R.: A new readability yardstick. J. Appl. Psychol. **32**(3), 221–233 (1948). https://doi.org/10.1037/h0057532
7. McNamara, D. S., et al.: Automated Evaluation of Text and Discourse with Coh-Metrix. Cambridge University Press (2014)
8. Průcha, J.: Učebnice: Edukace zprostředkovaná médiem. 2. aktualizované vydání. In: Moderní pedagogika. Praha: Portál (2002)
9. Gavora, P.: Úvod do pedagogického výskumu. 2. vyd., Bratislava: Univerzita Komenského, (1999b). ISBN 80-223-1342-4
10. Pelánek, R., Effenberger, T., Čechák, J.: Complexity and difficulty of items in learning systems. Int. J. Artif. Intell. Educ. **32**(1), 196–232 (2022). https://doi.org/10.1007/s40593-021-00252-4
11. Mistrík, J.: Štylistika. Bratislava: Slovenské pedagogické nakladateľstvo, ISBN: 80-08-02529-8 (1968)
12. Exner, O.: Některé charakteristické rysy úředního jazyka komunistické éry. In: In: Naše řeč, v.75, n. 2, pp. 91-98 (1992)

13. Těšitelová, M.: Využití statistických metod v gramatice. Praha (1980)
14. Těšitelová, M. et al.: Psaná a mluvená odborná čeština z kvantitativního hlediska (v rámci věcného stylu). Praha: Linguistica, 4. ÚJČ ČSAV (1983)
15. Těšitelová, M.: O frekvenci slov v dialogu a v monologu. In: SaS, v.52, pp. 109–122 (1991)
16. Bendová, K., Cinková, S.: Adaptation of classic readability metrics to Czech. In: Ekštein, K., Pártl, F., Konopík, M. (eds.) Text, Speech, and Dialogue: 24th International Conference, TSD 2021, Olomouc, Czech Republic, September 6–9, 2021, Proceedings, pp. 159–171. Springer International Publishing, Cham (2021). https://doi.org/10.1007/978-3-030-83527-9_14
17. Pisarek, W.: Jak mierzyć zrozumiałość tekstu. Zeszyty Prasoznawcze, No. **4**, 35–48 (1969)
18. Mistrík, J.: Meranie zrozumiteľnosti prehovoru [Measurement of speech clarity]. In: Slovenská řeč, 33, 3 (1968)
19. Gruszczyński, W., Broda, B., Charzyńska, E., et al.: Measuring readability of Polish texts (2015). https://nlp.ipipan.waw.pl/Bib/gru:etal:15:ltc.pdf. Accessed 10 Jun 2024
20. Redish, J.: Readability formulas have even more limitations than Klare discusses. In: ACM J. Comput. Docum. **24**(3), 132–137 (2000). Brno: Paido,. 148 p. ISBN 80-85931-49-4. https://doi.org/10.1145/344599.344637
21. Průcha, J.: Učebnice: teorie a analýzy edukačního média. Brno: Paido,. 148 p. ISBN 80-85931-49-4 (1998)
22. Drahošová, R.: Hodnotenie zrozumiteľnosti textu učebnice. Master's thesis. Brno: Masaryk University, Faculty of Education (2014). https://is.muni.cz/th/bsmyn. Accessed 17 Apr 2024
23. Průcha, J.: Učení z textu a didaktická informace. Academia, Praha (1987)
24. Pluskal, M.: Zdokonalení metody pro měření obtížnosti didaktických textů. In: Pedagogika, vol. 66, no. 1, pp. 62–76 (1996)
25. Rochovská, I. and Huľová, Z.: An Analysis if the technical education textbooks for 3rd grade of primary school in Slovakia. J. Technol. Inf. **13**(1), 23–40 (2021). https://doi.org/10.5507/jtie.2021.006
26. Pappová, M.: Vplyv zložitosti jazyka na čitateľnosť ekonomických textov. In: Študentská Vedecká Konferencia 2024 - Zborník recenzovaných príspevkov, vol. 10, p. 482 (2024). https://doi.org/10.24040/2024.9788055721477
27. Mistrík, J.: Efektívne čítanie. Bratislava: Veda, p.117. ISBN 80-224-0454-3 (1996)
28. Mistrík, J.: Rýchle čítanie. SPN, Bratislava (1982)
29. SAV - Jazykovedný ústav Ľudovíta Štúra.: Pravidlá slovenského pravopisu. 3., upravené a doplnené vyd. Bratislava: Veda (2000)

Author Index

A

Abed, Mohammed Hamzah II-58
Aleksandrova, Anastasiia I-267
Alex Raj, S. M. II-161
Aller, Sven II-13
Altinok, Duygu I-196, II-236
Álvarez, Aitor II-105
Anetta, Krištof I-110
Arias-Vergara, Tomas II-149
Aziz, Dosti II-24
Azrou, Lilia I-134

B

Bañeras-Roux, Thibault II-174
Benko, Vladimír I-55
Berberich, Klaus II-288
Bergler, Christian II-252
Blache, Philippe I-134
Boman, Magnus I-71
Bordea, Georgeta I-97
Božič, Martin I-85
Bracewell, David I-252
Brunson, Mary I-252
Brychcín, Tomáš I-227

C

Carson-Berndsen, Julie II-81
Cazenave, Tristan I-3
Chenebaux, Maixent I-3
Chng, Eng Siong II-70
Chung, Fu-lai II-210
Chuy, Chang Nian I-121

D

David Rios-Urrego, Cristian II-252
De Viron, Olivier I-97
Delaunay, Julien I-17, I-97
Denisová, Michaela I-30
Ding, Cherie I-121
Doucet, Antoine I-17, I-97
Druart, Lucas II-199

Ducos, Mathilde I-97
Dufour, Richard II-174

E

Escobar-Grisales, Daniel II-313
Estève, Yannick II-199

F

Feld, Michael II-288, II-300
Fishel, Mark II-13

G

Gallo-Aristizábal, Jeferson David II-313
Georges, Munir I-214, II-3, II-275
Gogoulou, Evangelia I-71
González-Gallardo, Carlos-Emiliano I-17, I-97
Gröttrup, Sören I-214
Guenoune, Hani I-43

H

Hamdine, Israa I-134
Hanzlíček, Zdeněk II-36, II-46, II-118
Harbusch, Karin I-171
Hartung, Kai I-214
Hernandez, Abner II-149
Holdt, Špela Arhar I-85
Horák, Aleš I-110, II-139
Hu, Qinmin Vivian I-121

J

Jakubec, Maroš II-184
Jarina, Roman II-184
John, Vojtěch I-239

K

Kasák, Peter II-184
Kleer, Niko II-288, II-300
Klemen, Matej I-85

Kumar, Ankit II-275
Kwok, Chin Yuen II-70

L
Lafourcade, Mathieu I-43
Lee, Sandra I-252
Lehečka, Jan II-46
Lesort, Timothée I-71
Lieskovská, Eva II-184
Liyanage, Chamila I-159
Lopez-Santander, Diego Alexander II-252

M
Maier, Andreas II-149
Mala, J. B. II-161
Mallick, Sambit I-214
Manfredi, Ilaria II-130
Martín-Doñas, Juan M. II-105
Matoušek, Jindřich II-36, II-46, II-94
Memmesheimer, Denis I-171
Mištera, Adam I-227
Mohler, Michael I-252
Moot, Richard I-43
Moreno-Acevedo, Santiago A. II-105

N
Nevěřilová, Zuzana I-147
Nivre, Joakim I-71, I-267
Nöth, Elmar II-252, II-313

O
Orozco-Arroyave, Juan Rafael II-149, II-252, II-313
Ortega, Daniel II-222
Oufaida, Houda I-134

P
Pappová, Mária I-291
Perez-Toro, Paula Andrea II-149
Plhák, Jaromír II-263
Pollak, Senja I-17, I-97
Porteš, David II-139
Pramodya, Ashmari I-159
Pushpananda, Randil I-159

R
Rajan, Rajeev II-161
Ríos-Urrego, Cristian David II-313
Robnik-Šikonja, Marko I-85
Rouvier, Mickael II-174
Roux, Jérémie I-43
Rychlý, Pavel I-30, I-184

S
Schmidt, Marisa I-171
Shams, Erfan A. II-81
Sidere, Nicolas I-97
Signoroni, Edoardo I-184
Šmahel, David II-263
Söhnel, Steven II-222
Sotolář, Ondřej II-263
Spišiak, Michal II-184
Srinivasagan, Gokul II-3
Stephen, Abishek I-239
Straka, Milan I-279
Straková, Jana I-279
Sztahó, Dávid II-24, II-58

T
Tihelka, Daniel II-36, II-46
Tran, Hanh Thi Hong I-17, I-97

V
Valko, Matúš I-291
Vasquez-Correa, Juan Camilo II-105, II-149
Vielzeuf, Valentin II-199
Vladař, Lukáš II-36, II-94
Vu, Ngoc Thang II-222

W
Weerasinghe, Ruvan I-159
Weyand, Leon II-288
Wolter, Julian II-300
Wong, Kwan-yeung II-210
Wottawa, Jane II-174

Y
Yang, Seung Hee II-149
Yip, Jia Qi II-70

Z
Žabokrtský, Zdeněk I-239
Žižková, Hana I-147

SPRINGER NATURE

GPSR Compliance

The European Union's (EU) General Product Safety Regulation (GPSR) is a set of rules that requires consumer products to be safe and our obligations to ensure this.

If you have any concerns about our products, you can contact us on ProductSafety@springernature.com

In case Publisher is established outside the EU, the EU authorized representative is:

Springer Nature Customer Service Center GmbH
Europaplatz 3
69115 Heidelberg, Germany

The manufacturer's authorised representative in the EU is Springer Nature Customer Service Centre GmbH, Europaplatz 3, 69115 Heidelberg, Germany. If you have any concerns regarding our products, please contact ProductSafety@springernature.com

Printed and bound by CPI Group (UK) Ltd, Croydon, CR0 4YY

25/03/2026

02078191-0014